AF792514

SUSTAINABLE CONSTRUCTION AND BUILDING MATERIALS

Edited by **Sayed Hemeda**

Sustainable Construction and Building Materials
http://dx.doi.org/10.5772/intechopen.78713
Edited by Sayed Hemeda

Contributors

Xiaoniu Yu, Yan Gao, Abel Olorunnisola, Alper Bideci, Sabit Oymael, Özlem Sallı Bideci, Zeli Que, Liuliu Zhang, Feibin Wang, Yifan Gao, Weizhen Cai, Xinmeng Wang, Haitao Zhang, Mingyang Gong, Gonzalo Ruiz, Lucía Garijo, Xiaoxin Zhang, José Joaquín Ortega, Rena C. Yu, Miloš Drdácký, Martin Šperl, Ning Zhang, Mohammad Akhtar, Paulo Brito, Pedro Romano, Alireza Moghayedi, Eleni Tsangouri, Mehrdad Honarmand, Javad Tanzadeh

© The Editor(s) and the Author(s) 2019
The rights of the editor(s) and the author(s) have been asserted in accordance with the Copyright, Designs and Patents Act 1988. All rights to the book as a whole are reserved by INTECHOPEN LIMITED. The book as a whole (compilation) cannot be reproduced, distributed or used for commercial or non-commercial purposes without INTECHOPEN LIMITED's written permission. Enquiries concerning the use of the book should be directed to INTECHOPEN LIMITED rights and permissions department (permissions@intechopen.com).
Violations are liable to prosecution under the governing Copyright Law.

(cc) BY

Individual chapters of this publication are distributed under the terms of the Creative Commons Attribution 3.0 Unported License which permits commercial use, distribution and reproduction of the individual chapters, provided the original author(s) and source publication are appropriately acknowledged. If so indicated, certain images may not be included under the Creative Commons license. In such cases users will need to obtain permission from the license holder to reproduce the material. More details and guidelines concerning content reuse and adaptation can be found at http://www.intechopen.com/copyright-policy.html.

Notice

Statements and opinions expressed in the chapters are these of the individual contributors and not necessarily those of the editors or publisher. No responsibility is accepted for the accuracy of information contained in the published chapters. The publisher assumes no responsibility for any damage or injury to persons or property arising out of the use of any materials, instructions, methods or ideas contained in the book.

First published in London, United Kingdom, 2019 by IntechOpen
IntechOpen is the global imprint of INTECHOPEN LIMITED, registered in England and Wales, registration number: 11086078, The Shard, 25th floor, 32 London Bridge Street
London, SE19SG – United Kingdom
Printed in Croatia

British Library Cataloguing-in-Publication Data
A catalogue record for this book is available from the British Library

Additional hard copies can be obtained from orders@intechopen.com

Sustainable Construction and Building Materials, Edited by Sayed Hemeda
p. cm.
Print ISBN 978-1-78985-749-8
Online ISBN 978-1-78985-750-4

We are IntechOpen,
the world's leading publisher of
Open Access books
Built by scientists, for scientists

4,000+
Open access books available

116,000+
International authors and editors

120M+
Downloads

151
Countries delivered to

Our authors are among the
Top 1%
most cited scientists

12.2%
Contributors from top 500 universities

WEB OF SCIENCE™

Selection of our books indexed in the Book Citation Index
in Web of Science™ Core Collection (BKCI)

Interested in publishing with us?
Contact book.department@intechopen.com

Numbers displayed above are based on latest data collected.
For more information visit www.intechopen.com

For EU product safety concerns:
IN TECH d.o.o., Prolaz Marije Krucifikse Kozulić 3, 51000 Rijeka, Croatia,
info@intechopen.com or visit our website at intechopen.com.

Meet the editor

Dr. Sayed Hemeda is a doctor of the Civil Engineering Department, Aristotle University of Thessaloniki, Greece. Currently, he occupies the position of Professor of Geotechnical Engineering and Structural Restoration of Architectural Heritage, Conservation Department, Faculty of Archaeology, Cairo University, Egypt. He is also the Vice Manager of the Historic Buildings Conservation Center in Cairo University. He was awarded Cairo University's prize for scientific excellence in 2017, Cairo University's prize for courage in 2014, and Cairo University's prize for the best PhD theses, 2009–2010. He has published some 60 articles, including 14 books, and has been cited around 80 times. He has given over 18 invited lectures in 12 countries. His interests are primarily in geotechnical engineering for architectural heritage preservation and engineering data analysis, including pattern recognition as applied primarily to analytical data from various sources, comprising objects of cultural significance. He is the Editor in Chief of the *Journal of Geological Research*. He is an Editorial Board Member of the Sustainable Civil Infrastructures book series, published by SpringerNature, IntechOpen England, book series publisher, Progress of Electrical and Electronic Engineering in Singapore, the *Geoscience Journal* in Singapore, and the *Alexandria Engineering Journal*.

Contents

Preface

Sustainable Construction and Building Materials provides an overview of the complete range of modified and modern sustainable construction and building materials available to civil engineers and all those involved in the building and construction industries. The book sheds light on recent advances in sustainable construction and building materials with special emphasis on the characterization of natural and composite hydraulic mortars, advanced concrete technology, green building materials, and the application of nanotechnology to the improvement of the design of building materials. The book covers in detail the characterization of natural hydraulic lime mortars, a decade of research on self-healing concrete, biocomposite cement binding process and performance, development of sustainable building materials from agro-industrial wastes, applications of sugarcane biomass ash for developing sustainable construction materials, oil-contaminated sand: sources, properties, remediation, and engineering applications, oil shale ash addition effect in concrete to freezing/thawing, connection node design and performance optimization of girders, functionally graded concrete structures, cumulative tensile damage and consolidation effects on fracture properties of sandstone, key performance criteria influencing the selection of construction methods used for the fabrication of building components in the Middle East, fly ash as a resource material for the construction industry, degradation monitoring systems for a building information modeling maintenance approach, durability of composite-modified asphalt mixtures based on inherent and improved performance, and bitumen and its modifiers.

With its distinguished editor and eminent editorial committee, *Sustainable Construction and Building Materials* is a standard introductory reference book on the complete range of construction and building materials. It is aimed at students of civil engineering, construction engineering, and allied courses, including water supply and drainage engineering. It also serves as a source of essential background information for engineers and professionals in the civil engineering and construction sector.

Sayed Hemeda
Doctor of Civil Engineering
Aristotle University of Thessaloniki, Greece

Professor of Geotechnical Engineering
and Structural Restoration of Historic Buildings
Faculty of Archaeology
Cairo University, Egypt

Chapter 1

Mechanical Behavior of Natural Hydraulic Lime Mortars

Lucía Garijo, XiaoXin Zhang, Gonzalo Ruiz,
José J. Ortega and Rena C. Yu

Additional information is available at the end of the chapter

http://dx.doi.org/10.5772/intechopen.80852

Abstract

Natural hydraulic lime (NHL) mortars are widely used for restoration works due to their good compatibility with the substrate material in terms of physical, chemical, and mechanical properties. Regarding their mechanical characterization, there is still a need for further understanding of their fracture behavior and the influence of their dosage methodology on the mechanical properties. Thus, this chapter focuses on the mechanical characterization of NHL mortars, such as flexural, compressive, and splitting tensile strengths, elastic modulus, and fracture energy. Moreover, the influence of the composition and production process on such properties was studied as well. Furthermore, the loading rate effect on the fracture behavior was also presented. The results show that NHL mortars have shape and size effect on the compressive strength. In addition, NHL mortar is rate sensitive, mainly due to the viscous effects caused by the presence of free water in the porous structure.

Keywords: mechanical characterization, dosage, empirical equations, size effect, loading rate effect

1. Introduction

Natural hydraulic lime (NHL) is a binding material formed by burning of argillaceous or siliceous limestones with reduction to powder by slaking with or without grinding [1]. From a mineralogical point of view, it is mainly composed of portlandite ($Ca(OH)_2$), dicalcium silicates (C_2S), gehlenite (C_2AS), and small amounts of tricalcium silicates (C_3S). Tricalcium aluminate (C_3A) and tetracalcium aluminoferrite (C_4AF) can be present as well [2]. Moreover,

IntechOpen

© 2018 The Author(s). Licensee IntechOpen. This chapter is distributed under the terms of the Creative Commons Attribution License (http://creativecommons.org/licenses/by/3.0), which permits unrestricted use, distribution, and reproduction in any medium, provided the original work is properly cited.

calcite ($CaCO_3$) can also appear in NHL as a result of a slight carbonation (reaction with carbon dioxide from the air) of portlandite during storage [3]. When mixed appropriately with water and aggregates, NHL produces mortars which are able to harden and gain strength with time [1]. As a hydraulic material, NHL mortars have the property of setting and hardening under water and reacting with carbon dioxide from the air (carbonation) [1, 4]. According to the standard EN 459-1 [1], NHL is classified as NHL 2, NHL 3.5, and NHL 5; the corresponding compressive strength at 28 days in MPa is as follows: 2 ≤ NHL 2 ≤ 7, 3.5 ≤ NHL 3.5 ≤ 10, and 5 ≤ NHL 5 ≤ 15.

The use of NHL mortars prevails presently for restoration works due to their good compatibility with the substrate material in terms of physical, chemical, and mechanical properties [5, 6]. That is to say, NHL mortars interact quite well with stones and blocks of masonry walls; the interventions with the materials are durable in time and do not originate spalling of the stones [7]. Moreover, NHL mortars are more appropriate than air-hardening or lime pozzolana ones when the early strength gain is essential [6]. Furthermore, NHL mortars are eco-efficient as they require low amount of energy during their production process and absorb CO_2 from the air while carbonating [6].

Knowledge on the mechanical properties of mortar is crucial to ensure a good performance of masonry structures [8]. In general, NHL mortars are well investigated in terms of compressive and flexural strengths [5, 9–11]. Recently, Garijo, Zhang, and Ruiz *et al.* [12, 13] have measured fracture energy, splitting tensile strength and characteristic length of such materials. At present time, the dynamic mechanical behavior of lime mortars is getting more attention due to the fact that most masonry structures are situated in zones of seismic activity, such as Lisbon (Portugal), L'Aquila (Italy), Lorca (Spain), Kathmandu (Nepal), Nairobi (Kenya), etc. Pereira and Lourenço [14] studied the dynamic compressive behavior of masonry specimens, clay brick, and mortar prisms by using a drop-weight tower. The corresponding strain rate range was from 2 to 200 s^{-1}. A commercial ready-mix mortar (Mapei Mape-Antique MC) was used for the fabrication of masonry and mortar specimens. The dynamic increase factor (DIF, a ratio of the dynamic response over the corresponding quasi-static one) 2.73 was obtained for the compressive fracture energy of the mortar specimens. Asprone *et al.* [15] analyzed the tensile behavior of a basalt fiber-reinforced natural hydraulic lime mortar at medium and high strain rates, using a hydropneumatic machine and a modified Hopkinson bar apparatus, respectively. The DIF for tensile strength was 5.11 at the strain rate 90 s^{-1}. For bending behavior, Chan and Bindiganavile [16] studied the strain rate sensitivity of plain and fiber-reinforced hydraulic lime mortar by adopting a universal testing machine and a drop-weight impact machine at strain rate range from 10^{-6} to 10 s^{-1}. Impact tests on notched beams with dimensions 100 mm × 100 mm × 350 mm were conducted; the notch depth was 12.5 mm and 2 mm in width, and the span was 300 mm. For plain hydraulic lime mortar, the results show that the DIF for modulus of rupture was 12 at the drop height 500 mm (corresponding to strain rate 10 s^{-1}), while it was 53 for fracture toughness. Moreover, the flexural behavior of hydraulic lime mortar is more sensitive to strain rate than fiber-reinforced hydraulic lime mortar. Later, the bond behavior between the stone masonry block and the plain and fiber-reinforced hydraulic lime mortar was also investigated by Bindiganavile *et al.* [17, 18]. The results show that there was an improvement in the bond strength due to polypropylene microfibers (20 mm in length) but a difference in the fracture performance between the Portland cement-lime and hydraulic lime mortars. Whereas with the former, the fibers promoted failure through fracture in the stone block especially under dynamic loading conditions; with the latter, the fibers moved the failure plane from the interface to within the bulk mortar.

Nevertheless, there is still a need for the further understanding of the mechanical behavior of NHL mortars under quasi-static and dynamic loading conditions, especially for the latter. In this chapter, under quasi-static loading condition, an advanced mechanical characterization of NHL 3.5 mortars has been performed, such as elastic modulus, fracture energy, splitting tensile strength, and compressive and flexural strengths. Particularly, compressive strengths from prisms and cylinders were compared to study the size and shape influence. Moreover, the effects of dosage and production process on these properties were analyzed as well. Under dynamic loading condition, NHL mortar beams were tested at various loading rates (loading-point displacement rates) by using a servo-hydraulic testing machine, from the quasi-static one, 5.0×10^{-4} mm/s, to rate sensitive ones, 5.0×10^{-1} mm/s (intermediate loading rate) and 1.6×10^{1} mm/s (fast loading rate). The results show that the peak load and fracture energy are rate sensitive, which is mainly due to the viscous effect caused by the presence of free water in the porous structure.

2. Experimental procedure

2.1. Raw materials

The binder used for all seven types of mortars was a NHL 3.5, according to EN 459-1 [1]. It was supplied by "Socli, Italcementi group (France)" with an apparent particle density of 2.58 g/cm^3 and apparent density of 0.85 g/cm^3. Three types of aggregates were adopted in the fabrication process. The usual one was a commercial crushed limestone with maximum size of 4 mm; the rest were the crushed limestones with a maximum size of 2 mm and river sand with a maximum size of 4 mm. The particle-size distribution curves of the aggregates were determined in accordance with EN 1015-1 [19] and are shown in **Figure 1**. The apparent density and apparent particle density of each aggregate listed in **Table 1** were measured according to EN 1097-3 [20] and EN 1097-6 [21], respectively.

2.2. Mortar preparation

In total, seven types of mortars were prepared and tested (see **Table 2**). All of them presented a lime/aggregate volume ratio 1:3 as it is the one commonly used for the historical mortars [22, 23]. Volume proportions of compounds were converted to weight to obtain a convenient measurement for the mixing process. First of all, a benchmark mortar, named as NHL09C04M as a simplification, was fabricated with a water/lime volume ratio of 0.9, which provided a plastic consistency (between 140 and 200 mm), determined by the flow table test according to EN 1015-3 [24] and EN 1015-6 [25]. In addition, a crushed limestone aggregate with a maximum grain size of 4 mm and a metallic mold were used. The rest of the mortars were prepared by varying one aspect of the benchmark mortar, such as the water/lime ratio, the

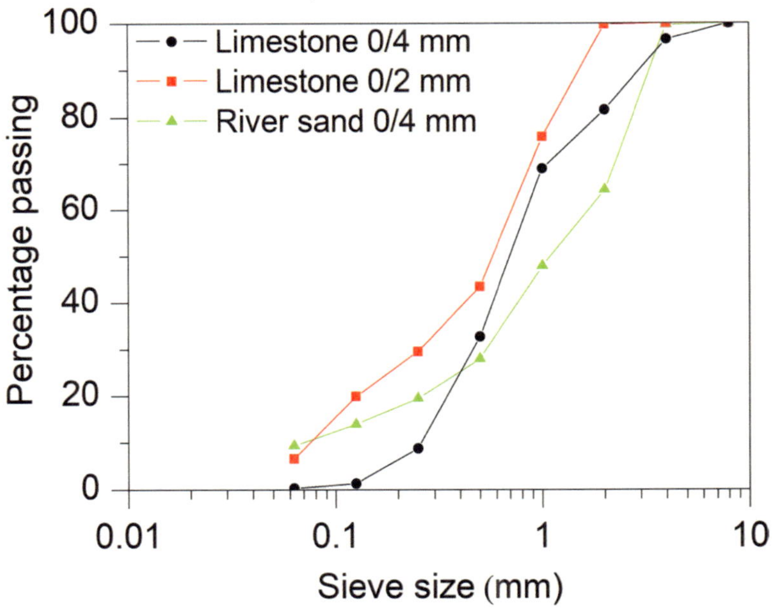

Figure 1. Aggregates grading curves.

Density	Standards	Crushed limestone 0/4 mm	Crushed limestone 0/2 mm	River sand 0/4 mm
Apparent density (g/cm³)	EN 1097-3 [20]	1.82	1.81	1.46
Apparent particle density (g/cm³)	EN 1097-6 [21]	2.68	2.74	2.59

Table 1. Apparent density and apparent particle density of the aggregates used.

maximum aggregate size, the aggregate type, the material of the mold, and the curing conditions (see **Table 2**). In order to isolate and quantify the influence of the type and maximum size of aggregate, materials of mold, and curing conditions on the mechanical properties of mortars, the water/lime ratio was kept constant as 0.9 instead of following the conventional procedure by adapting the water/lime ratio to obtain an approximately constant consistency as presented in the Refs. [5, 6, 9].

The mixing process was performed according to EN 1015-2 [26]. Both prismatic (40 × 40 × 160 mm) and cylindrical (75 mm in diameter and 150 mm in height) specimens were prepared. The mortar was poured in two layers when using the prismatic molds and in three layers instead when the cylindrical molds were used; each layer was compacted with 25 strokes of the tamper. All specimens were removed from the molds in 2 days after the fabrication according to EN 1015-11 [27].

2.3. Mechanical tests

The flexural, compressive, and splitting tensile strengths, elastic modulus, and fracture energy were measured by various types of tests as shown in **Figure 2**, at an age of 56 days. Moreover,

Mortar composition	Volumetric water/lime ratio	Consistency (mm)	Type of aggregate	Maximum grain size (mm)	Material of the mold	Curing conditions
NHL09C04M	0.9	150–155	Crushed limestone	4	Metallic	56 days in CC
NHL08C04M	0.8	130–135	Crushed limestone	4	Metallic	56 days in CC
NHL11C04M	1.1	238–240	Crushed limestone	4	Metallic	56 days in CC
NHL09C02M	0.9	120–125	Crushed limestone	2	Metallic	56 days in CC
NHL09R04M	0.9	180–187	River sand	4	Metallic	56 days in CC
NHL09C04W	0.9	150–155	Crushed limestone	4	Wooden	56 days in CC
NHL09C04MA	0.9	150–155	Crushed limestone	4	Metallic	7 days in CC and 49 in AC

CC, climatic chamber (RH: 97 ± 0.5%, 20 ± 0.5°C); AC, ambient laboratory conditions (RH: 50 ± 10%, 23 ± 3°C).

Table 2. Characteristics of the seven mortar compositions prepared.

the loading rate effect on fracture energy was studied by performing three-point bending tests on beams at various loading rates (loading-point displacement rates), 5.0×10^{-4} mm/s (quasi-static loading rate), 5.0×10^{-1} mm/s (intermediate loading rate), and 1.6×10^{1} mm/s (high loading rate), respectively.

2.3.1. *Flexural and compressive strength*

The flexural and compressive strengths were measured according to EN 1015-11 [27] by using the Instron 1011 testing machine. The flexural strength was obtained by a three-point bending test on three beams at a loading rate of 10 N/s and a span of 100 mm (see **Figure 2(a)**). It is worth noting that the anti-torsion supports were used for the test, which is specially important for quasi-brittle materials. That is to say, the beam rests on two rigid-steel cylinders placed on two supports which permit rotation out of the plane of the beam and rolling along the longitudinal axis of the beam with negligible friction.

The compressive tests were performed at a loading rate of 50 N/s on the six half prisms remaining from the bending tests, as presented in **Figure 2(b)**. The load was centered in the middle of the longest side by using a steel plate (40 mm × 40 mm × 10 mm). Moreover, an individualized ball-and-socket joint over the steel plate was used to reduce the eccentricity during the loading process.

2.3.2. *Elastic modulus and compressive strength from cylinders*

For studying the size and shape effects on the compressive strength, compressive tests were also performed on four cylinders (75 mm in diameter and 150 mm in height) at a loading rate of 10 N/s by using the Instron 8805 testing machine. Moreover, the elastic modulus was determined following the principles of EN 12390-13 [28] (see **Figure 2(c)**). Two clips (strain gauge

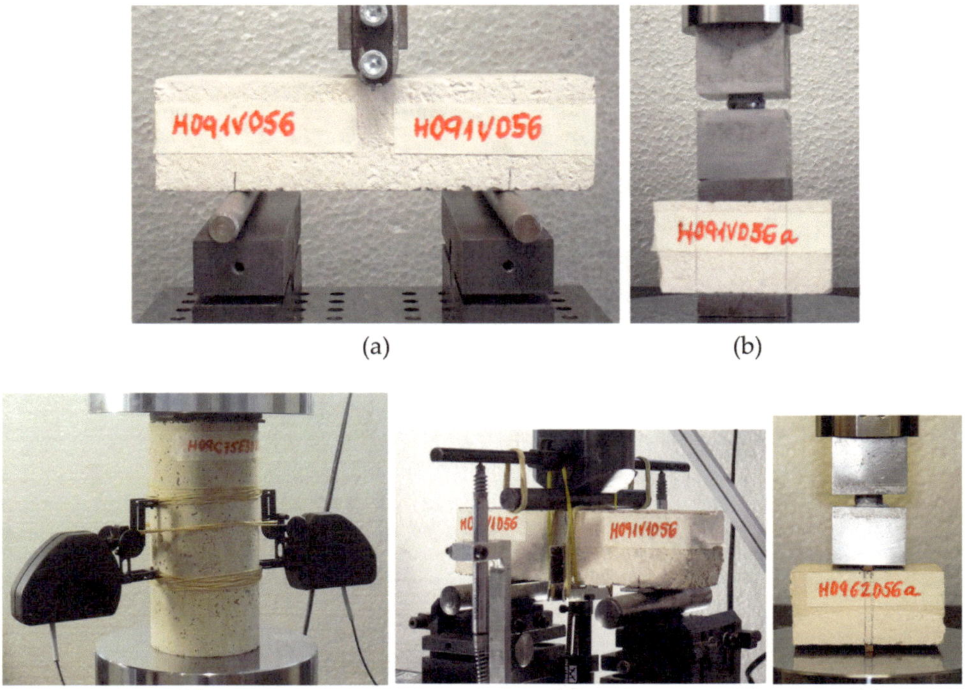

(a) (b) (c) (d) (e)

Figure 2. Tests for (a) flexural strength, (b) compressive strength, (c) elastic modulus and compressive strength from cylinders, (d) fracture energy and elastic modulus from prisms, and (e) splitting tensile strength.

extensometers Instron 2630) were used to measure the axial deformation. They were placed centrally in opposite generatrices of the specimen covering a span of 50 mm in order to avoid local constrictions caused by the friction of the specimen surface and the steel platens of the machine. Two rubber layers with 2 mm thickness each were used between the upper surface of the sample and the steel platen to keep away from contact problems due to the irregular roughness of the sample. In order to obtain a stable value, five repeat monotonic tests were conducted up to 30% of the compressive strength at a loading rate 10 N/s for each cylinder; the mean of the last three values measured was taken as the elastic modulus.

2.3.3. Fracture energy and loading rate effect

The fracture energy, defined as the energy required to open a unit area of crack surface, was measured through a three-point bending test by using the Instron 8805 testing machine, following the procedure recommended by the RILEM [29] and the improvements proposed by Planas, Guinea, and Elices [30–32] (see **Figure 2(d)**). The beams were the same size as those used for the flexural tests. A precast notch in the middle of the specimens was introduced by using a cardboard piece (2 mm in width and 20 mm in depth) during the fabrication. More detailed information on how to determine the fracture energy can be found in the Refs. [12, 13].

The tests were performed in the displacement control. In order to study the rate effect on the fracture energy, three various loading rates (loading-point displacement rates) were applied during the test from a quasi-static level (5.0×10^{-4} mm/s) to rate dependent (dynamic) levels (5.0×10^{-1} mm/s and 1.6×10^{1} mm/s). In such way, it took around 30 minutes and 0.3 seconds to finish the tests for the lowest and fastest loading rates, respectively. Three beams were tested at each loading rate. The corresponding strain rate, $\dot{\varepsilon}$, can be calculated approximately according to the Eq. (1):

$$\dot{\varepsilon} = \frac{6(D-a)\,\dot{\delta}}{S^2} \tag{1}$$

where $\dot{\delta}$ stands for the loading rate; D and S are the beam depth and span, respectively; and a is the notch depth.

2.3.4. Elastic modulus from prisms

When performing the three-point bending tests at quasi-static loading rate, an extensometer (strain gauge extensometer Instron 2620) was attached to the lower surface of the specimen to obtain the crack-mouth opening displacement (CMOD). In this way, the elastic modulus (E_{pr}) could be obtained from prisms by applying general Eqs. (2) and (3) for span/depth (S/D) ratios (β) between 2.5 and 16 [33]:

$$E_{pr} = 6\frac{Sa}{C_i B D^2}\,\upsilon_\beta(\alpha) \tag{2}$$

$$\upsilon_\beta(\alpha) = \upsilon_\beta(a/D) = 0.8 - 1.7\alpha + 2.4\alpha^2 + \frac{0.66}{(1-\alpha)^2} + \frac{4}{\beta}(-0.04 - 0.58\alpha + 1.47\alpha^2 - 2.04\alpha^3) \tag{3}$$

where C_i is the initial compliance determined from load-CMOD curve, $\upsilon_\beta(\alpha)$ is a dimensionless shape function depending on β, and the relative notch/depth ratio α. The other parameters of the beam have been previously defined.

2.3.5. Splitting tensile strength (indirect tensile strength)

The splitting tensile strength was measured through quasi-static splitting tensile tests (Brazilian tests) on four prismatic halves resulting from the preceding bending test for measuring fracture energy, following the procedure recommended by EN 12390-6 [34]. To conduct the test, the Instron 1011 testing machine was used, and the loading rate was set at 50 N/s. The proportion between the load-bearing width and the height of the specimens was maintained as low as 1/10 following the recommendations in [35–37]. The bearing strips were made of plywood, and they were placed in the middle of the longest side of the halves (see **Figure 2(e)**). The splitting tensile strength is determined as

$$f_t = 2F/\pi BD \tag{4}$$

where f_t is the splitting tensile strength, F is the maximum load, and B and D are the specimen width and depth, respectively, as mentioned previously.

2.3.6. Characteristic length

Once the splitting tensile strength (f_t), elastic modulus from cylinders (E_{cy}), and quasi-static fracture energy (G_F) are obtained, the characteristic length, l_{ch}, is calculated in accordance with Eq. (5). It is a parameter proposed by Hillerborg *et al.* [38] for fracture behavior, which is related to the length of the fracture process zone. It could be used to predict the brittleness of a material. As it decreases, brittle nature dominates and vice versa:

$$l_{ch} = E_{cy} G_F / f_t^2 \tag{5}$$

3. Results and discussion

In this section, the results of the experimental campaign described in Section 2 are presented. It is organized as follows: influence of dosage and production process on the mechanical properties of seven types of NHL mortars, size effect on the compressive strength, and the loading rate effect on the fracture behavior.

3.1. Influence of dosage and production process on the mechanical properties

In total, five influence factors on the mechanical properties of seven types of NHL mortars were studied, i.e., water/lime ratio, material of the mold, type and maximum size of the aggregates, and curing conditions. As explained in Section 2, a benchmark mortar was prepared, named as NHL09C04M (see **Table 2**). The influence of each factor was analyzed by varying an aspect of the mortar. That is to say, the influence of the water/lime ratio was studied among NHL08C04M, NHL09C04M, and NHL11C04M; the effect of the material of the mold was investigated by comparing NHL09C04W and the benchmark mortar, the influence of the type and maximum size of the aggregates through mortars NHL09R04M and NHL09C02M with the benchmark mortar, respectively. Finally, the impact of the curing conditions was obtained by comparing NHL09C04MA with the benchmark one. The detailed experimental results on the mechanical properties are listed in **Table 3**, f_{flex}, f_{cpr}, f_{ccy}, G_F, f_t, E_{cy}, E_{pr}, and l_{ch}, respectively, the flexural strength, compressive strength from prisms, compressive strength from cylinders, fracture energy at quasi-static loading rate, elastic modulus from cylinders, elastic modulus from prisms, and characteristic length. It is worth noting that for NHL09C04W, only the flexural and splitting tensile strengths (**Figure 3(a)** and **(c)**) were measured as the cylindrical wooden molds were not fabricated. Moreover, the notched beams were damaged when demolding the specimens from the prismatic wooden molds [12].

It is observed from **Table 3**, in general, the mechanical properties of NHL mortars are improved for low water/lime ratio when comparing the experimental results among NHL08C04M, NHL09C04M, and NHL11C04M, due to the fact that the open porosity increases with an

Mortar composition		f_{flex} (MPa)	f_{cpr} (MPa)	f_{ccy} (MPa)	G_F (N/m)	f_t (MPa)	E_{cy} (GPa)	E_{pr} (GPa)	l_{ch} (mm)
NHL09C04M	Mean	1.3	3.2	2.0	12	0.39	5.0	5.2	390
	Std. dev.	0.1	0.1	0.2	3	0.02	0.2	0.5	
NHL08C04M	Mean	1.3	4.2	2.7	13	0.51	5.4	6.0	260
	Std. dev.	0.1	0.3	0.3	1	0.01	0.6	0.2	
NHL11C04M	Mean	0.89	1.7	1.4	4.9	0.24	2.8	3.8	240
	Std. dev.	0.04	0.1	0.1	0.8	0.03	0.7	1.0	
NHL09C04W	Mean	1.7	3.5	—	—	0.57	—	—	—
	Std. dev.	0.1	0.1			0.05			
NHL09C02M	Mean	1.1	3.2	2.0	12	0.49	4.6	5.1	220
	Std. dev.	0.1	0.2	0.1	1	0.05	0.2	0.6	
NHL09R04M	Mean	0.96	2.3	1.5	10	0.38	4.2	4.4	280
	Std. dev.	0.06	0.1	0.1	2	0.03	0.2	0.4	
NHL09C04MA	Mean	0.91	2.4	1.5	8	0.34	2.8	3.2	190
	Std. dev.	0.02	0.1	0.1	1	0.03	0.4	0.6	

Table 3. Mechanical properties of NHL mortars at an age of 56 days.

increase in water/lime ratio which results in a weakening of the mortar structure and its mechanical properties [12]. Moreover, using wooden molds also benefits the mechanical properties as they absorb the excess of free water. The influence of the aggregate size is coupled with the effect of the water/lime ratio. According to the Ref. [6], larger coarse aggregates improve the resistance in a comparison among mortars with similar consistencies. However, this is achieved by adding more water to mortars with smaller aggregates, as they have a higher water demand, which modifies the water/lime ratio. In our case, NHL09C04M and NHL09C02M have the same water/lime ratio as 0.9. As both have similar mechanical properties, it could be due to the positive effect of a lower water/lime ratio offsetting the possible lower capacity of smaller aggregates.

NHL09C04M and NHL09R04M mortars were fabricated with crushed limestone and river sand, respectively. In a fresh state, NHL09R04M has a higher consistency (180–187 mm) than NHL09C04M (150–155 mm) for the same water/lime ratio. It is obvious that the crushed limestone aggregates exhibit better interlocking behavior than the river sands with round particles [39]. It should be noted if less water were added to NHL09R04M to get similar consistency as NHL09C04M, the mortar would be stronger.

The effect of the curing conditions were studied between NHL09C04MA, cured in the climatic chamber (RH: 97 ± 0.5%, 20 ± 0.5°C) for the first 7 days and at ambient laboratory conditions (RH: 50 ± 10%, 23 ± 3°C) for the rest 49 days, and NHL09C04M, cured in the climatic chamber for 56 days. The former presents lower mechanical properties than the latter. For example,

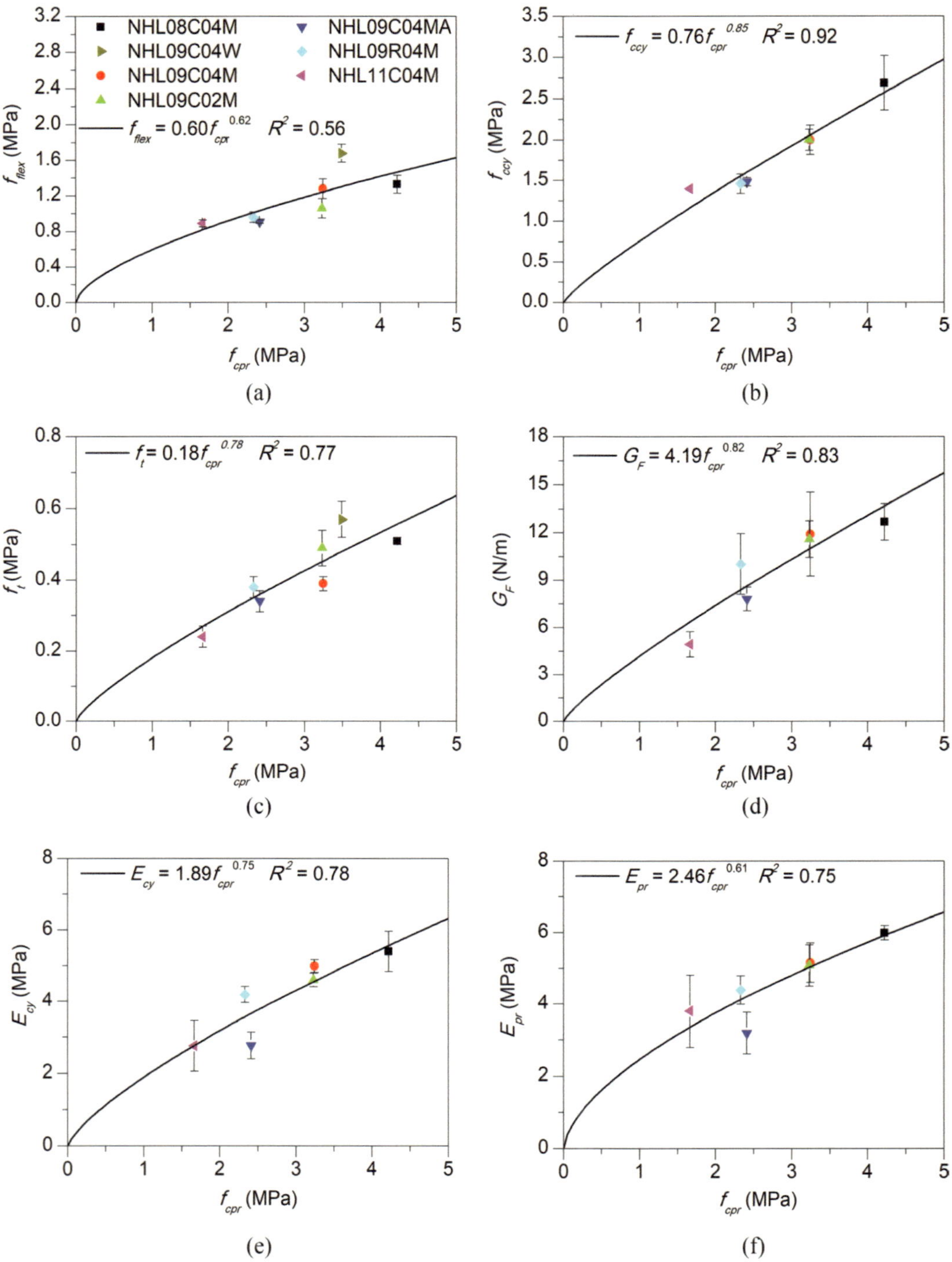

Figure 3. Relationship between compressive strength from prisms and other mechanical properties: (a) flexural strength, (b) compressive strength from cylinders, (c) splitting tensile strength, (d) fracture energy, (e) elastic modulus from cylinders, (f) elastic modulus from prisms.

their flexural strengths are 0.91 and 1.3 MPa, respectively. This is mainly due to the fact that high humid conditions favor the hydration of NHL mortars [9, 10, 12].

As mentioned in Section 2.3.6, characteristic length is an indicator of brittleness of quasi-brittle materials. The shorter a material is, the more brittle it is. It is clear that using river sand in the fabrication and curing at ambient laboratory make the mortars more brittle. Moreover, smaller aggregate size of the mortar favors the brittle behavior.

In **Figure 3**, various empirical equations relating some mechanical properties with the compressive strength from prisms were obtained (Eq. (6); **Table 4**) by fitting the experimental results. This could be useful for numerical simulations and structural design when only the compressive strength is measured owing to the convenience of performing the test. It is similar with the application of equations provided by the FIB Model Code [40] and ACI Building Code [41] for concrete. In most cases, R^2, the determination coefficient, is over 75%. However, for the flexural strength, it is only 56%, due to the fact that the result of the specimens fabricated in the wooden mold does not follow the trend. The equation used to do the empirical correlations is:

$$Y = m f_{cpr}^{n} \tag{6}$$

3.2. Size effect on compressive strength

From **Table 3** and **Figure 3(b)**, it is observed that there is a big difference between the compressive strength obtained from prisms and cylinders. For example, they are 4.2 and 2.7 MPa, respectively, for NHL08C04M, while 1.7 and 1.4 MPa for NHL11C04M, respectively. This could be due to improper curing throughout the specimen in the case of the cylinders, which are more massive than prisms [42]. However, this would not be the main cause as both specimens present similar densities and open porosities [12]. Instead, the difference is mainly because of geometry and size effects, which are confirmed by the following numerical simulation on the compressive tests of prisms.

The numerical model was created by using a commercial software, ATENA [43]. This program uses the finite element method and has nonlinear constitutive material models implemented,

Y	m	n
f_{flex}	0.60	0.62
f_{ccy}	0.76	0.85
f_t	0.18	0.78
G_F	4.19	0.82
E_{cy}	1.89	0.75
E_{pr}	2.46	0.61

Table 4. Values of adjusting parameters according to Eq. (6).

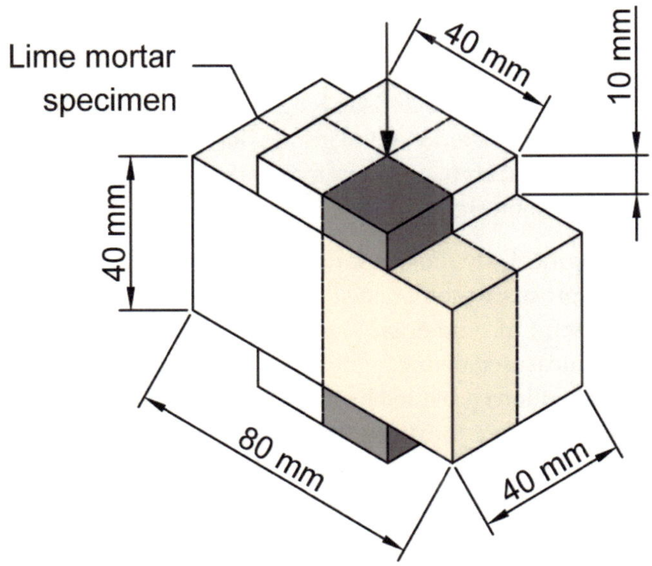

Figure 4. Geometry of the numerical model.

especially aimed for cohesive materials [13, 44]. The geometry of the model is formed by the mortar specimen itself and two metallic plates as used in the experimental test (see **Figure 4**). In order to reduce the number of finite elements and computing cost, symmetry was used in the model, defined by two vertical axes of symmetry. The material properties (E_{cy}, G_F, f_t, and f_c) were taken from the experimental results of mortars NHL08C04M (dry consistency) and NHL11C04M (fluid consistency), respectively (**Table 3**). Under the hypothesis that the compressive strength of cylinder is closer to the intrinsic compressive strength of the material, we adopted their corresponding values as f_c, instead of the corresponding to prisms. The metallic plates were defined as perfect linear elastic as their only function was to transmit the load and their strength was much higher than the one of the mortar [13].

In order to complete the comprehension of the possible size effect, two additional models of both mortars were performed by doubling the size of the original one once (80 mm × 80 mm × 160 mm) and twice (160 mm × 160 mm × 320 mm), respectively, and it followed the procedure described by del Viso *et al.* in [45]. Then, it was possible to obtain the nominal strength of each model, σ_N, as the ratio of the corresponding maximum loads and their respective bearing surfaces.

In **Figure 5**, the square of the nominal strength of the three models for each mortar, σ_N^2, is plotted against the inverse of Hillerborg's brittleness number, $1/\beta_H$ or l_{ch}/D. By adjusting a linear fitting of these three points, it is possible to obtain the asymptotic value of the compressive strength when the specimen depth, D, tends to infinite, that is to say, the intrinsic compressive strength of the material, σ_{inf} [13, 45]. This fitted line is defined in Eq. (7):

$$\sigma_N^2 = \sigma_{inf}^2 + \sigma_{inf}^2 A \frac{1}{\beta_H} \tag{7}$$

http://dx.doi.org/10.5772/intechopen.80852

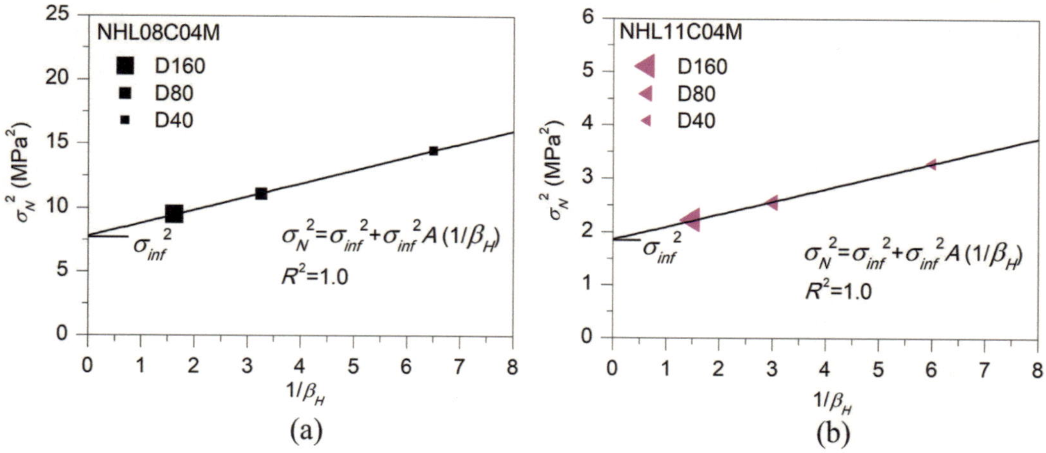

Figure 5. Regression for setting σ_{inf} and A for (a) NHL08C04M and (b) NHL11C04M.

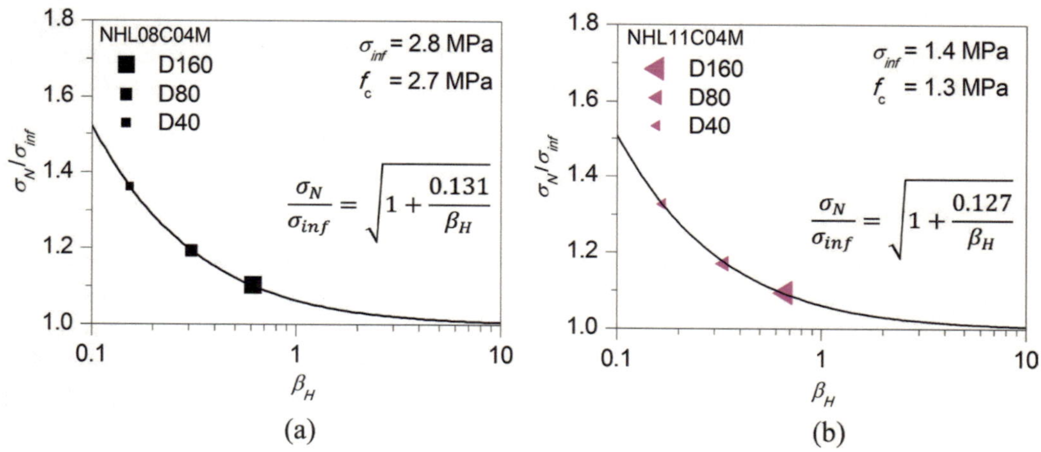

Figure 6. Size effect law for (a) NHL08C04M and (b) NHL11C04M.

where A is obtained from the slope of the fitted line of **Figure 5** for each mortar. With σ_{inf}, it is possible to obtain the non-dimensional values of σ_N (the ratio σ_N/σ_{inf}) and plot them related to β_H with a logarithmic scale (**Figure 6**). Then, it is possible to adjust one curve per mortar mix fitting these points. These two curves represent their size effect laws. As shown by them, for large specimen sizes, the size effect tends to disappear. In both cases, σ_{inf} is close to the compressive strength of cylinders, meaning that this value can roughly be considered as the intrinsic compressive strength of the material, instead of the compressive strength of the prisms.

3.3. Loading rate effect on the fracture behavior

In order to study the rate effect on the fracture behavior of NHL mortars, the benchmark mortar, NHL09C04M, was tested in three-point bending under three different loading rates,

Type of NHL specimens	Loading rate, $\dot{\delta}$ (mm/s)	Strain rate, $\dot{\varepsilon}$ (s^{-1})	Maximum load, P_{max} (N)	DIF for P_{max}	Fracture energy, G_F (N/m)	DIF for G_F	Water content (%)
Normal	5.0×10^{-4}	6×10^{-6}	99 (10)	1.0	12.2 (1)	1.0	2.52 (0.24)
	5.0×10^{-1}	6×10^{-3}	120 (10)	1.2	17.3 (1)	1.4	
	1.6×10^{1}	2×10^{-1}	140 (15)	1.4	22.8 (10)	1.9	
Dry	5.0×10^{-4}	6×10^{-6}	110 (20)	1.0	17.9 (2)	1.0	
	1.6×10^{1}	2×10^{-1}	110 (10)	1.0	22.5 (7)	1.3	

Table 5. Experimental results at various loading rates for NHL09C04M.

5.0×10^{-4} mm/s (quasi-static loading rate), 5.0×10^{-1} mm/s, and 1.6×10^{1} mm/s. **Table 5** presents the experimental results, where the values in brackets refer to the corresponding standard deviation. Moreover, an estimation of the strain rate, $\dot{\varepsilon}$, is included as well for comparison with results of the literature under other loading methods. The DIF is defined by the ratios of peak load (P_{max}) and fracture energy (G_F) to their corresponding quasi-static values. **Figure 7** shows the peak load and fracture energy obtained under three different loading rates.

From **Table 5** and **Figure 7**, it is observed that the maximum load and fracture energy of the NHL mortar increase with the loading rate. For example, their DIFs at the highest rate are 1.4 and 1.9, respectively.

The tendency of the rate effect on the peak load and fracture energy is also presented in **Figure 8**. Moreover, following Eqs. (8) and (9), curve fittings of the DIFs are derived from the experimental results in regard to the peak load and the fracture energy, respectively, by using the least squares method. These adjustments provide a correlation coefficient higher than 95%. Though the format of such equations is original for plain and steel fiber-reinforced concretes [46–49], they are still valid for NHL mortars:

$$DIF_{P_{max}} = 1 + 0.19\left(\frac{\dot{\delta}}{\dot{\delta}_0}\right)^{0.29} \tag{8}$$

$$DIF_{G_F} = 1 + 0.44\left(\frac{\dot{\delta}}{\dot{\delta}_0}\right)^{0.25} \tag{9}$$

where $\dot{\delta}_0$ is set as 1 mm/s. Such equations may be useful to predict the rate effect on the peak load and the fracture energy and when performing numerical simulations involving fracture on NHL mortars.

For understanding the cause of the loading rate dependency, six specimens (three per loading rate) were dried for 48 hours at 105°C and tested at the lowest and highest loading rates. The experimental results of the dry specimens are also presented in **Table 5**. Moreover, the comparison of the maximum DIF between the normal and dry lime mortars is shown in **Figure 9**. It is obvious that the rate effect of NHL mortar is mainly related with the movement of free water in pores. For example, for the dry NHL mortar, the peak load is almost

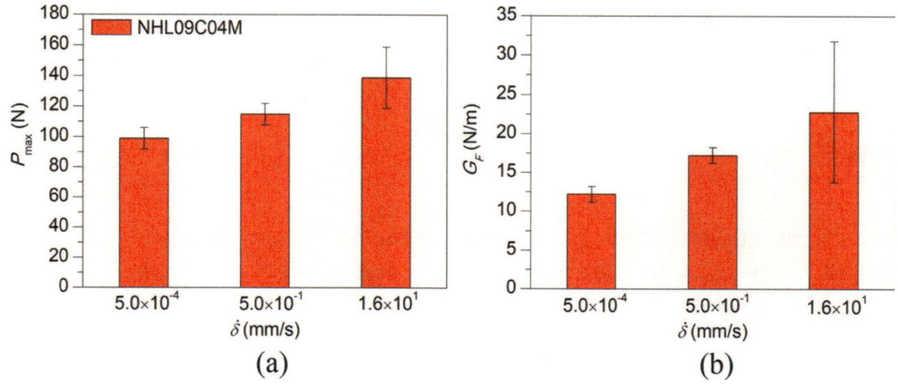

Figure 7. Peak load and fracture energy at various loading rates. (a) Peak load. (b) Fracture energy.

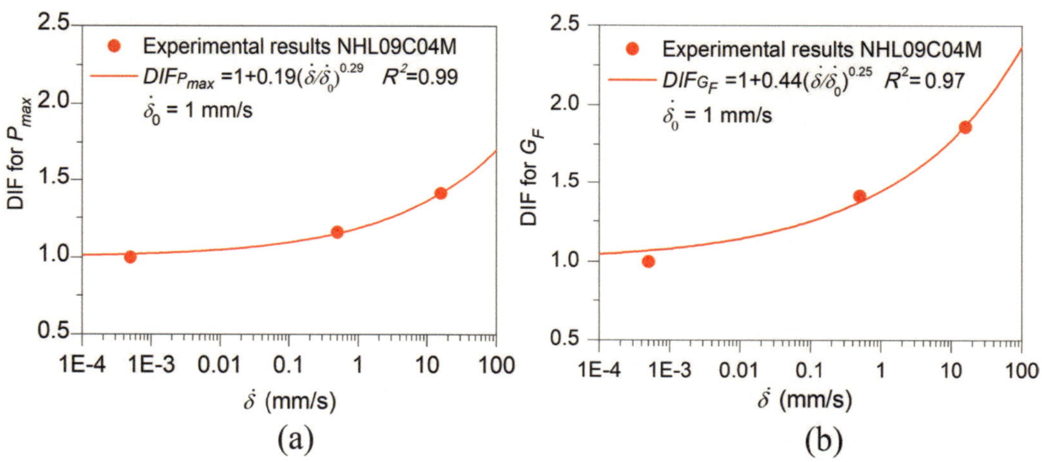

Figure 8. Loading rate effect on (a) peak load and (b) fracture energy.

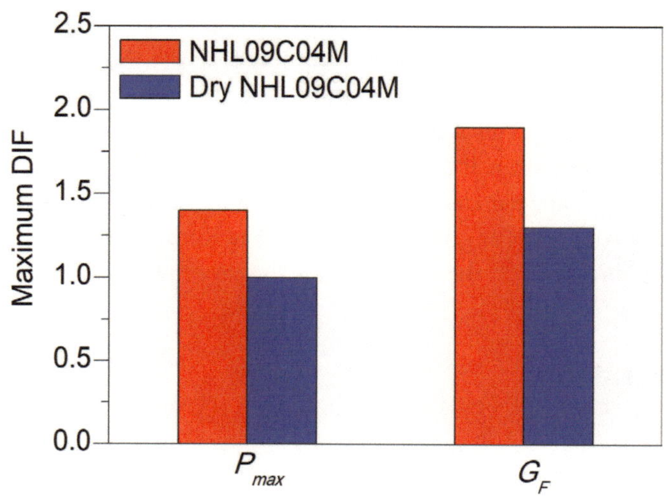

Figure 9. Maximum DIF of the normal and dry specimens of NHL09C04M.

constant at different loading rates, a 30% increase in fracture energy. However, for the normal NHL mortar, the DIFs of the peak load and fracture energy are 1.4 and 1.9, respectively.

4. Conclusions

In the chapter, the influence of five factors affecting the dosage and fabrication process of seven types of NHL 3.5 mortars on their mechanical properties was studied. Moreover, the shape and size effect on the compressive strength was disclosed by numerical simulation and experiments. Furthermore, the loading rate effect on the fracture energy was investigated as well.

Regarding the five influence factors on the mechanical properties, the results show that high water/lime ratios produce structural weakening and reduce mechanical properties. High relative humidity (97 ± 0.5%) is more suitable than ambient laboratory conditions for the hydration of the compounds of NHL mortars and for the increase of its ductility. Moreover, using wooden molds also improves these properties as they absorb the excess of free water. When maintaining constant water/lime ratios, using aggregates with higher grain size increases the mechanical properties. Mortars with river sand have lower mechanical properties. Certainly, if the water/lime ratio also varies at the same time, the tendency would be different.

There exists a big variation on the compressive strengths from prisms and cylinders; the ratio could reach 1.6. A numerical simulation analysis of the compressive test on prisms of two NHL mortars confirms that such variation is mainly due to the geometry and size effects.

With respect to the loading rate effect, from 5.0×10^{-4} to 1.6×10^{1} mm/s, on the fracture behavior of NHL mortar, it is obvious that the material is rate sensitive. The DIFs of the peak load and fracture energy are 1.6 and 1.9, respectively. Such phenomenon is mainly because of the viscous effects caused by the presence of free water in the porous structure.

Acknowledgements

The authors acknowledge the funding from the INCRECYT, the *Ministerio de Economía y Competitividad*, Spain, under grant BIA2015-68678-C2-1-R and from the *Junta de Comunidades de Castilla-La Mancha* (JCCM), *Fondo Europeo de Desarrollo Regional*, Spain, PEII-2014-016-P. L. Garijo thanks the funding from the scholarship FPU014/05186 given by the *Ministerio de Educación, Cultura y Deporte*, Spain, and J. J. Ortega from the scholarship 2017/759 awarded by JCCM, Spain. We also express our gratitude to Prof. Pere Roca, from *Universidad Politécnica de Cataluña*, Spain, for his advice on the dosage of the lime mortars.

Conflict of interest

None.

Author details

Lucía Garijo, XiaoXin Zhang, Gonzalo Ruiz*, José J. Ortega and Rena C. Yu

*Address all correspondence to: Gonzalo.Ruiz@uclm.es

School of Civil Engineering, University of Castilla-La Mancha, Ciudad Real, Spain

References

[1] BS EN 459-1. Building Lime—Part 1: Definitions, Specifications and Conformity Criteria. Brussels, Belgium: BSI; 2015. p. 52

[2] Silva BA, Pinto APF, Gomes A. Influence of natural hydraulic lime content on the properties of aerial lime-based mortars. Construction and Building Materials. 2014;**72**:208-218

[3] Arizzi A, Martinez-Huerga G, Sebastian-Pardo E, Cultrone G. Mineralogical, textural and physical-mechanical study of hydraulic lime mortars cured under different moisture conditions. Materiales de Construcción. 2015;**65**(318):1-11

[4] Lanas J, Arandigoyen M, Alvarez JI, Pérez Bernal JL, Bello MA. Mechanical behavior of masonry repair mortars: Aerial and hydraulic lime-based mixtures. In: 10th International Congress on Deterioration and Conservation of Stones Stockholm. 2004

[5] Lanas J, Bernal JLP, Bello MA, Galindo JIA. Mechanical properties of natural hydraulic lime-based mortars. Cement and Concrete Research. 2004;**34**(12):2191-2201

[6] Kalagri A, Karatasios I, Kilikoglou V. The effect of aggregate size and type of binder on microstructure and mechanical properties of NHL mortars. Construction and Building Materials. 2014;**53**:467-474

[7] Chan R, Bindiganavile V. Toughness of fibre reinforced hydraulic lime mortar. Part-1: Quasi-static response. Materials and Structures. 2010;**43**(10):1435-1444

[8] Haach VG, Vasconcelos G, Lourenço PB. Influence of aggregates grading and water/cement ratio in workability and hardened properties of mortars. Construction and Building Materials. 2011;**25**(6):2980-2987

[9] Lanas J, Sirera R, Alvarez JI. Study of the mechanical behavior of masonry repair lime-based mortars cured and exposed under different conditions. Cement and Concrete Research. 2006;**36**(5):961-970

[10] Grilo J, Silva AS, Faria P, Gameiro A, Veiga R, Velosa A. Mechanical and mineralogical properties of natural hydraulic lime-metakaolin mortars in different curing conditions. Construction and Building Materials. 2014;**51**:287-294

[11] Papayianni I, Stefanidou M. Strength-porosity relationships in lime-pozzolan mortars. Construction and Building Materials. 2006;**20**(9):700-705

[12] Garijo L, Zhang XX, Ruiz G, Ortega JJ, Wu ZM. The effects of dosage and production process on the mechanical and physical properties of natural hydraulic lime mortars. Construction and Building Materials. 2018;**169**:325-334

[13] Garijo L, Zhang XX, Ruiz G, Ortega JJ, Yu RC. Advanced mechanical characterization of NHL mortars and cohesive simulation of their failure behavior. Construction and Building Materials. 2017;**153**:569-577

[14] Pereira JM, Lourenço PB. Experimental characterization of masonry and masonry components at high strain rates. Journal of Materials of Civil Engineering. 2017;**29**(2):1-10

[15] Asprone D, Cadoni E, Iucolano F, Prota A. Analysis of the strain-rate behavior of a basalt fiber reinforced natural hydraulic mortar. Cement and Concrete Composites. 2014;**53**:52-58

[16] Chan R, Bindiganavile V. Toughness of fibre reinforced hydraulic lime mortar. Part-2: Dynamic response. Materials and Structures. 2010;**43**(10):1445-1455

[17] Islam MT, Chan R, Bindiganavile V. Stress rate sensitivity of stone masonry units bound with fibre reinforced hydraulic lime mortar. Materials and Structures. 2012;**45**(5):765-776

[18] Islam MT, Bindiganavile V. Dynamic fracture toughness of sandstone masonry beams bound with fiber-reinforced mortars. Journal of Materials in Civil Engineering. 2014; **26**(1):125-133

[19] BS EN 1015-1. Methods of Test for Mortar for Masonry—Part 1: Determination of Particle Size Distribution (By Sieve Analysis). BSI. 1998/A1; 2006. p. 8

[20] BS EN 1097-3. Test for Mechanical and Physical Properties of Aggregates Part 3: Determination of Loose Bulk Density and Voids. BSI; 1998. p. 10

[21] BS EN 1097-6. Test for Mechanical and Physical Properties of Aggregates Part 6: Determination of Particle Density and Water Absorption. BSI; 2013. p. 54

[22] Moropoulou A, Cakmak AS, Biscontin G, Bakolas A, Zendri E. Advanced byzantine cement based composites resisting earthquake stresses: The crushed brick/lime mortars of Justinian's Hagia Sophia. Construction and Building Materials. 2002;**16**(8):543-552

[23] Marastoni D, Benedetti A, Pela L, Pignagnoli G. Torque Penetrometric test for the in-situ characterisation of historical mortars: Fracture mechanics interpretation and experimental validation. Construction and Building Materials. 2017;**157**:509-520

[24] BS EN 1015-3. Methods of Test for Mortar for Masonry—Part 3: Determination of Consistence of Fresh Mortar (By Flow Table). BSI. 1999/A2; 2006. p. 10

[25] BS EN 1015-6. Merthods of Test for Mortar for Masonry—Part 6: Determination of Bulk Density of Fresh Mortar. BSI. 1998/A1; 2006. p. 8

[26] BS EN 1015-2. Methods of Test for Mortar for Masonry—Part 2: Bulk Sampling of Mortars and Preparation of Test Mortars. BSI. 1998/A1; 2006. p. 8

[27] BS EN 1015-11. Methods of Test for Mortar for Masonry—Part 11: Determination of Flexural and Compressive Strength of Hardened Mortar. BSI. 1999/A1; 2006. p. 12

[28] BS EN 12390-13. Testing Hardened Concrete Part 13: Determination of Secant Modulus of Elasticity in Compression. BSI; 2012. p. 10

[29] RILEM TC 50-FMC. Determination of the fracture energy of mortar and concrete by means of the three-point bend tests on notched beams. Materials and Structures. 1985; **18**:285-290

[30] Elices M, Guinea GV, Planas J. Measurement of the fracture energy using three point bend tests. 3. Influence of cutting the P-δ tail. Materials and Structures. 1992;**25**:327-334

[31] Elices M, Guinea GV, Planas J. Measurement of the fracture energy using three point bend tests. 1. Influence of experimental procedures. Materials and Structures. 1992;**25**:121-218

[32] Planas J, Elices M, Guinea GV. Measurement of the fracture energy using three point bend tests. 2. Influence of bulk energy dissipation. Materials and Structures. 1992;**25**:305-312

[33] Guinea GV, Pastor JY, Planas J, Elices M. Stress intensity factor, compliance and CMOD for a general three-point-bend beam. International Journal of Fracture. 1998;**89**(2):103-116

[34] BS EN 12390-6. Testing Hardened Concrete—Part 6: Tensile Splitting Strength of Test Specimens. BSI; 2009. p. 14

[35] Rocco C, Guinea GV, Planas J, Elices M. Size effect and boundary conditions in the Brazilian test: Theoretical analysis. Materials and Structures. 1999;**32**(220):437-444

[36] Iglesias I, Acosta B, Yu RC, Ruiz G, Aineto M, Acosta A. Study of mechanical characterization of ceramic specimens from a Brazilian test adaptation. Materiales de Construcción. 2011;**61**(303):417-429

[37] Rocco C, Guinea GV, Planas J, Elices M. Size effect and boundary conditions in the Brazilian test: Experimental verification. Materials and Structures. 1999;**32**(217):210-217

[38] Hillerborg A, Modéer M, Peterson PE. Analysis of crack formation and crack growth in concrete by means of fracture mechanics and finite elements. Cement and Concrete Research. 1976;**6**:773-782

[39] Lanas J, Alvarez JI. Masonry repair lime-based mortars: Factors affecting the mechanical behavior. Cement and Concrete Research. 2003;**33**(11):1867-1876

[40] FIB. Model Code 2010, Final Draft. Vol. 2012. Lausanne, Switzerland: FIB-Fédération Internationale du Béton. p. 311

[41] ACI 318-14. Building Code Requirements for Structural Concrete and Commentary. Farmington Hills: American Concrete Institute; 2014. p. 520

[42] Drougkas A, Roca P, Molins C. Compressive strength and elasticity of pure lime mortar masonry. Materials and Structures. 2016;**49**(3):983-999

[43] Cervenka V, Jendele L, Cervenka J. Computer Program of Nonlinear Finite Element Analysis of Reinforced Concrete Structures. Prague: Cervenka Consulting; 2009. pp. 1-142

[44] Poveda E, Ortega JJ, Ruiz G, Porras R, Carmona JR. Normal and tangential extraction of embedded anchor plates from precast facade concrete panels. Engineering Structures. 2016;**110**:21-35

[45] del Viso JR, Carmona JR, Ruiz G. Shape and size effects on the compressive strength of high-strength concrete. Cement and Concrete Research. 2008;**38**(3):386-395

[46] Zhang XX, Abd Elazim AM, Ruiz G, Yu RC. Fracture behavior of steel fibre-reinforced concrete at a wide range of loading rates. International Journal of Impact Engineering. 2014;**71**:89-96

[47] Zhang XX, Ruiz G, Tarifa M, Cendón D, Gálvez F, Alhazmi WH. Dynamic fracture behavior of Steel Fiber Reinforced Self-Compacting Concretes (SFRSCCs). Materials. 2017;**10**(11):1-16

[48] Zhang XX, Ruiz G, Yu RC, Poveda E, Porras R. Rate effect on the mechanical properties of eight types of high-strength concrete and comparison with FIB MC2010. Construction and Building Materials. 2012;**30**:301-308

[49] Zhang XX, Ruiz G, Yu RC, Tarifa M. Fracture behaviour of high-strength concrete at a wide range of loading rates. International Journal of Impact Engineering. 2009;**36**(10-11):1204-1209

Chapter 2

A Decade of Research on Self-Healing Concrete

Eleni Tsangouri

Additional information is available at the end of the chapter

http://dx.doi.org/10.5772/intechopen.82525

Abstract

The main findings of a decade of research on the design and development of the first self-healing concrete are summarized in this chapter. The autonomous healing concept is introduced, and plethora of design campaigns is enlisted. Healing agent encapsulation and agent tubes vascular networks are reported as the most efficient healing configurations for laboratory-scale and real-size applications, respectively. Crack formation, closure after healing and further damage are phenomena tracked by using advanced experimental monitoring methods and their performance is critically revised. The effect of self-healing technology on concrete mechanical response, durability and long-term response to damage are critically discussed. The study contributes to the open discussion in the scientific research community regarding self-healing concrete upscaling feasibility and finally it aims to contribute as a base for the future studies dealing with concrete design optimization.

Keywords: concrete, cracking, repair, autonomous self-healing, damage monitoring

1. Introduction

1.1. A millennium of experience on concrete healing

The invention of construction materials equally strong and efficient to natural ones (i.e. stone, wood and fibers) is an inherent need of humans. Archeological findings prove that the concept of 'concrete' composite material is dating back to Roman Empire period. A variety of pozzolana compositions were established and numerous still standing today monuments are made of it, indicatively the Colosseum in Rome and the Mediterranean Caesarea harbor. The latter Roman concrete infrastructure is built undersea and its revolutionary composition is recently decoded: mineral aluminum tobermorite is exposed to sea water during the service life leading to progressive and post-curing lime crystallization [1]. The composition is reported as

IntechOpen

© 2018 The Author(s). Licensee IntechOpen. This chapter is distributed under the terms of the Creative Commons Attribution License (http://creativecommons.org/licenses/by/3.0), which permits unrestricted use, distribution, and reproduction in any medium, provided the original work is properly cited.

the first historical evidence of concrete design that concerns self-healing mechanisms in an attempt to enhance its durability.

Durability issues arise in concrete when under early age shrinkage or service loads, micro-cracks form and propagate on the cementitious media. Due to material quasi-brittle nature and heterogeneous multi-scale composition, micro-cracks build dense networking, making macro-crack formation inevitable. This complex material response to damage is acquainted since antiquity, therefore smart lime crystallization mechanisms as in Roman concrete were introduced.

Extended literature exists on the autogenous repair mechanisms intrinsically built in concrete dependent on the material composition. The micro-organisms calcite precipitation and the continuous hydration of un-hydrated cement grains are the most promising crack repair mechanisms [2, 3]. Nowadays, the intrinsic autogenous hydration ability of cement is enhanced by embedding superabsorbent polymers at casting [4] and use of high volume fractions of micro-fibers (i.e. engineered cementitious composite [5]). Eventually, autogenous healing delays cracking, but it cannot effectively eliminate the degradation in the presence of macro-cracks. Therefore, the treatment of cracks with size of several hundred micrometers appears the crucial repair mechanisms and plays a key role on the material sustainable design.

1.2. The pioneering autonomous healing concept

Reaching the end of the previous century, technological progress has established easy and cost-effective methods to repair open macro-cracks on concrete. European Standard EN1504 provides guidelines on repair based on adhesive polyurethane or other epoxy agent injection into cracks. Today one finds plethora of conventional repair kits in the market that promise sealing of cracks ranging from capillary to few millimeters opening size. In real cases, the effective damage repair is limited since the agent cannot penetrate and effectively polymerize into narrow crack paths. Furthermore, only the cracks that reach the concrete surface, are visually detected and are easily accessible, can be treated using external agent injection. As a result, repair kits provide partial sealing and limited short-term repair, being therefore not a sustainable healing solution.

The pioneering research at the University of Illinois introduced the idea of autonomous healing design both on concrete [6, 7] and polymer composites (White protocol [8]). Dry et al. designed brittle fibrous tubes that carried sealant agent and embedded them into large concete plates. The cracks developed due to surface shrinkage, propagated through the tubes and broke them releasing the sealant into the crack void. The cracks were filled and repair joints were formed at the concrete surface proving successful sealing. The healing prototype design was poorly investigated from the following years until 2009, when a boost in research studies on autonomous healing appears (see statistical graph in **Figure 1**).

1.3. Boost on research interest: design and composition optimization

Stepping on pioneering work of Dry, Van Tittelboom et al. designed laboratory-scale concrete beams carrying methyl methacrylate and polyurethane agent embedded into short (typical

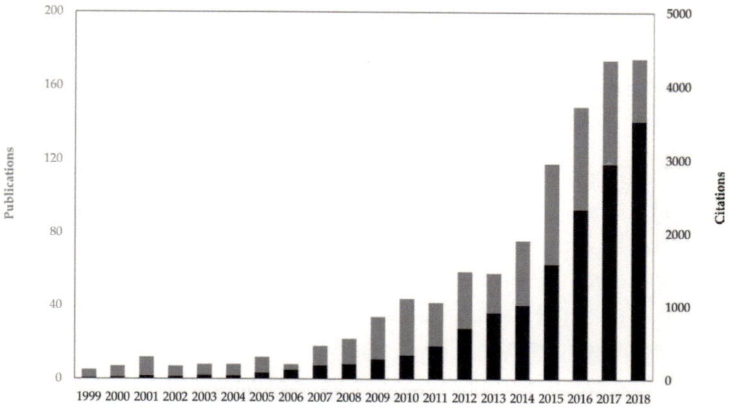

Figure 1. Total publications (right vertical axis) and summary of times cited (left vertical axis) per year in the topic of 'healing concrete'. Source: Web of Science database.

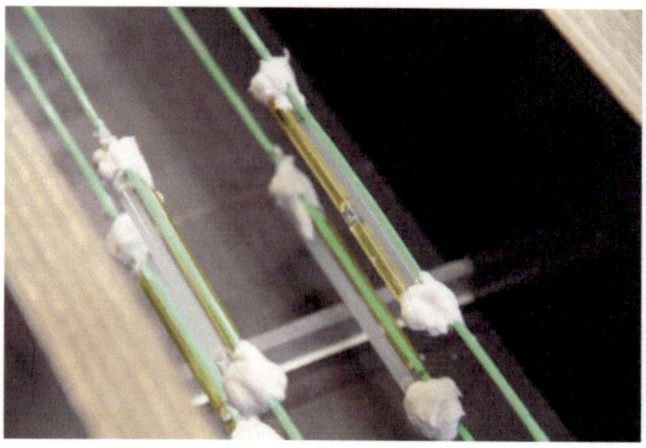

Figure 2. Four pairs of borosilicate glass capsules carrying the healing agent (polyurethane in yellow) and its actuator (white component) are attached using thin plastic wires to the concrete mold. Picture taken before concrete casting.

length at 50 mm), thin (diameter up to 5 mm) brittle glass borosilicate capsules [9–12]. Pelletier et al. manufactured polyurethane short capsules carrying sodium silicate solution with repair ability [13].

The master concept is graphically illustrated in **Figure 2**. Based on it, numerous studies followed in an attempt to optimize the healing efficiency of concrete systems, in all cases following a series of principal requirements:

- Autonomous healing mechanism should be activated only in the presence of damage. The healing agent should be stored within concrete, protected by a pre-cursor that is strong enough to survive concrete mixing process and in parallel brittle enough to instantly break in tension as crack propagates through it.
- The amount of healing agent should be enough to fill up the cracks formed in concrete under service loads, but should remain limited in volume in order to eliminate agent leaching on exterior surfaces. The agent rheological properties should be adjusted to fulfill this request.

- The healing cycles should be repeatable in order to protect concrete throughout the structure's service life. Concerning the time and spatial spread of damage phenomena on concrete, the self-healing concept should be accompanied by a self-sensing mechanism that detects cracks at any moment in order to control the healing activation procedure.

Section 2 presents the optimized and most promising design strategies: healing agent encapsulation (Section 2.1) and agent delivery vascular networks (Section 2.2). The loading configuration and monitoring techniques used on test campaigns to assess damage and consequent healing are enlisted in Section 3 and a critical review of their effectiveness on damage and healing assessment is given. The combination of the optimal self-healing and self-sensing technologies permits us today to test the feasibility of building constructions with autonomous repair ability. The experimental tests on real-size concrete structures loaded under service loads are reported. This chapter closes with a critical and extensive discussion on healing definition, the transition from laboratory testing to industry and the potential contributions of cutting-edge technologies on future concrete healing concepts.

2. Design strategies

2.1. Encapsulated healing agent

2.1.1. Healing agent

Polymer-based agents similar to the ones traditionally used for external manual injection are used. The preliminary studies discussed the self-healing effectiveness of polymers existed in the market (polyurethane, superglue and epoxy) [10]. It was evident that the polymer chemistry had to be modified in order to optimally perform as encapsulated healing agent:

- The agent should react only with specific catalyst, indicatively air, water or other chemical accelerators.
- The agent should be effectively polymerized enclosed into the aggressive alkaline concrete environment.
- The agent curing should be rapid enough to prevent crack propagation and exposure of reinforcement to ambient environment.
- The agent rheology should permit fast penetration through thin cracks (by means of capillary and gravitational forces), but agent leaching due to low viscosity should be eliminated.
- Optimally, the agent should expand in volume in order to fully seal the crack void.

2.1.2. Agent carrier

Borosilicate glass [12], ceramic [14], cementitious [14], polyurethane [13] and other polymeric [15] capsules were used. Glass effectively breaks in the presence of macro-cracks, however, cannot survive mixing and long-term exposure to the aggressive alkaline concrete environment. Polymer-based capsules can survive mixing and embedment into concrete, but often do not rupture in the presence of cracks. Finally, ceramic and cementitious capsules appear the most promising. Minnebo and Tsangouri developed the latter capsules by material extrusion [14]. The capsules can survive the concrete mixing and provide the optimal interfacial bonding with the concrete. Enhanced concrete-capsule interfacial bonding ensures instant capsule rupture as cracks form and eliminate debonding effects.

The capsules were initially designed in spherical shape [16], but later the tubular shape [10] was adopted as the optimal shape since it increases the chances of rupture in cracking stage. The size of capsules varies: Yang introduced micro-capsules [16], Van Tittelboom established the short 25–75 mm long tubular capsules and recently longer capsules up to 1 m long were manufactured [14, 17].

2.2. Agent delivery vascular networks

In the recent decade, an alternative effective healing system is under investigation inspired by the blood circulation in vascular networks. Concrete sections are made porous and healing agent circulates through the material providing repeatable repair at different locations [18]. Kuang designed a porous core in concrete and as a proof-of-concept adhesive epoxy was manually injected in the presence of crack [19]. Shape memory alloys and shrinkable polymer tendons were implemented to enhance the effective agent penetration and curing [19, 20]. Upon damage, externally introduced heating activates the tendons which recover and shrink to their original shape, closing in this manner the crack-void and facilitating the bonding at the fractured surfaces. The aforementioned systems provide passive manual healing, but their contribution should not be underestimated since these prototypes introduced the idea of agent delivery vascular networks.

The research team at the Department of Mechanics of Materials and Constructions (MeMC) at Vrije Universiteit Brussel (VUB) worked the past 5 years on the manufacturing of healing agent tubes network that imitates the vascular network flow. A comparative study on optimal tubes composition that considers the material compatibility and brittleness, the bonding conditions at the concrete tubes interface and durability performance reached the conclusion that ceramic and cementitious long (up to 1 m), extruded capsules with inner diameter up to 10 mm optimally perform [14, 17]. The tubes are interconnected by 3D printed connection nodes that often stand below reservoir deposits. The agent is stored in the deposit and is poured into the tubes only in the presence of cracks, procedure manually triggered by an air pressure sensor. Current research investigates how sensor configurations can be implemented into the healing system introducing autonomous sensing of damage and healing process trigger.

In parallel, the on-going research project materials for life focuses on the development of thin tunnels network embedded into concrete carrying organic, but also inorganic healing agent [21]. The tunnels network is built into thin wall elements and its performance under service loads is currently under investigation.

3. Assessment of healing efficiency

The healing efficiency of encapsulation systems is evaluated in the laboratory by performing tests on small and bigger concrete samples that carry the healing system. At first, the intact concrete sample is loaded until a macro-crack is formed. As the crack propagates, it breaks the capsules that pass through, activating eventually the healing process: due to gravitational and capillary forces, the agent is released in the crack void. A loading pause of few hours (often less than 24 hours) follows during which agent polymerization takes place. After the healing pause, loading is applied once again and at this time the response of the healed zone to further cracking is assessed. In the case that numerous capsules are embedded and new cracks form after healing, more than one reloading cycles are applied to detect potential repeatable healing.

In **Figure 3**, different loading configurations and specimen geometries are illustrated giving an overview of the most popular test setups used up to date for healing assessment. Cases 1–3 stand for short concrete beams (840 × 100 × 100 mm) loaded under 3- and 4-point bending. Only in Case 1, a notch is prepared to guide the crack at the middle section of the beam [22] where limited number of short glass capsules are embedded. Case 2 considers numerous short capsules scattered into concrete and their performance is assessed as multiple cracks form under 4-point bending [23]. Similarly, in Case 3, 4-point bending is applied on a beam that carries a prototype agent delivery tubes network [14]. Cases 4–6 illustrate the tests done on real-size structural elements (beam, slab and wall, respectively). Short glass capsules are embedded along the length of a beam 3 m long and potential multiple cracks healing is assessed under 4-point bending [24, 25]. The prototype tubes vascular network is extended in 2D plane and is embedded into a slab 4 m long and 1 m wide in an attempt to upscale this promising healing system (Case 5, [17]). Finally, the healing mechanisms built in the course of materials for life research project are evaluated by loading and monitoring the cracks formed on 1 m-wide walls as illustrated in Case 6 [21].

3.1. Detection of healing activation

The first step towards autonomous healing is the in time and effective activation of the healing process. The capsules should break in the presence of a macro-crack leading to healing agent

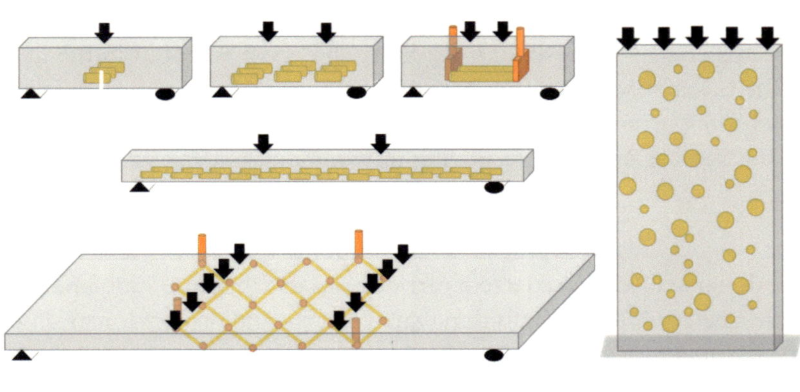

Figure 3. Design and configuration of tests performed to assess the healing efficiency on concrete. Loading is commonly applied in two or more cycles.

release. Continuous monitoring using acoustic emission (AE) is proven to accurately sense the capsules rupture. In **Figure 4a**, the AE hits energy emitted from short concrete beams (design Case 1) is plotted. The beams carry numerous short glass capsules that break as crack forms and propagates. The majority of AE hits are originated from the concrete cracking, however there are series of AE hits instantly emitted carrying energy up to a scale higher than the concrete cracking hits (typically a series consists eight hits captured by the eight sensors attached to the beam). The high energy hits, marked in black in **Figure 4a** are released as the capsules rupture.

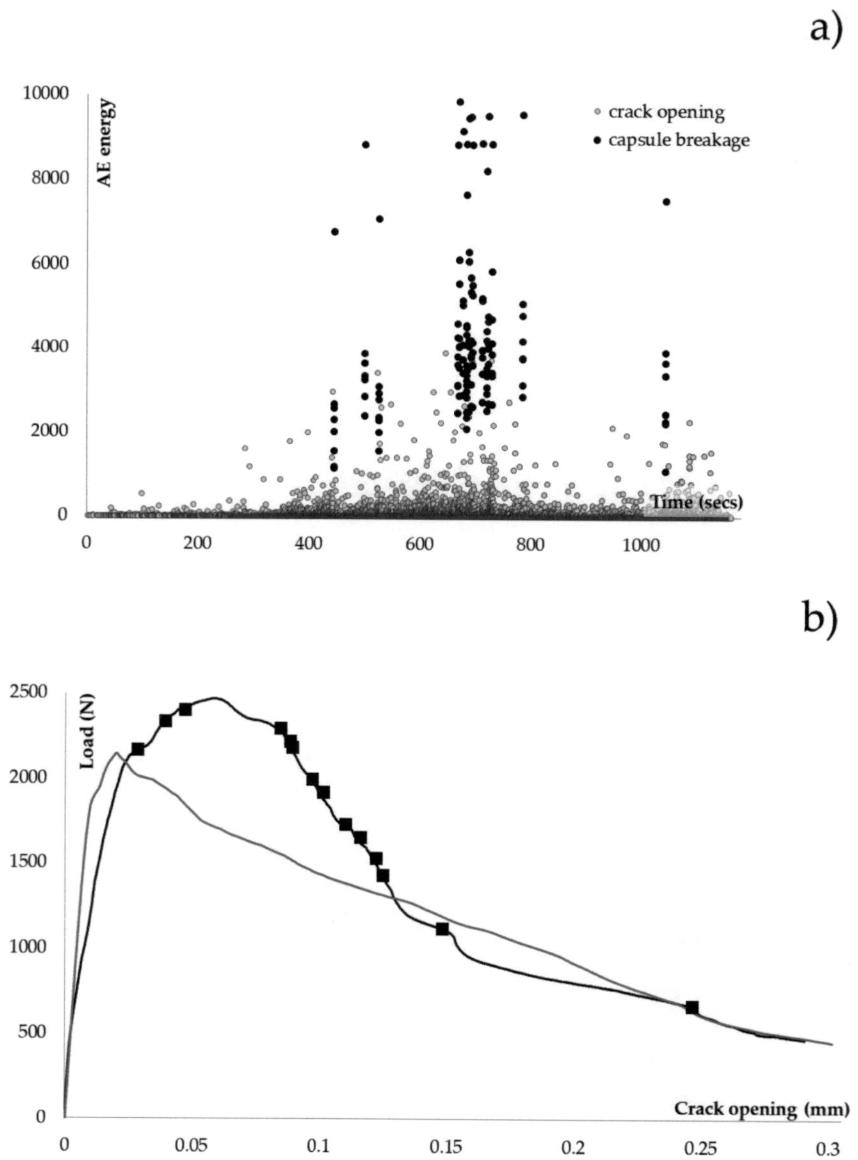

Figure 4. (a) Scatter of AE hits energy where hits emitted due to capsules rupture have values at least a scale greater than the hits originated from concrete cracking; (b) the capsules rupture detected by AE energy analysis are tracked on the load-crack opening curve. The loading curve is compared to reference case at which healing capsules are not used.

AE hits energy-based protocol was used to distinguish the rupture of different types of capsules: glass (**Figure 4a**, [22]), ceramic [14], cementitious [14] and glass with cementitious reinforcement [23]. In all cases, capsule breakages were distinguishable from concrete cracking, however the energy levels drop in the cases that compatible to concrete capsules material is used (i.e. cementitious mortar). In a step further, the capsules rupture can be localized relevant to the fracture stage: in **Figure 4b**, the instants of capsule rupture (detected by AE energy protocol) are plotted along the load-crack opening curve. The AE analysis proves that capsule break in the presence of a macro-crack with size ranging from 40 to 150 μm. Furthermore, the load-crack opening curve of a reference concrete beam (carrying no capsules) is given highlighting the fracture toughness increase in the healing series. Apparently, the capsules enhance the concrete response to fracture and provide an indirect local reinforcement to the structure. The study proves the multi-beneficial contribution of encapsulated agent systems on concrete damage response.

3.2. Sealing efficiency

The healing performance of the proposed technology is variously interpreted. Often in literature, the terms sealing, repair and mechanical restoration are enclosed in the term 'healing'. Effective sealing can be assessed by performing water permeability tests [23], neutron tomography [12] and optical scanning microscopy [26]. The water flow through multiple cracks formed on short and small-scale concrete beams was measured and it was found that only partial sealing can be achieved since the expansive agent only partially covers the crack void [12]. The water permeability configuration loses its accuracy in the presence of several cracks [23]. Scanning microscopy has effectively illustrated the agent release into the crack void. In **Figure 5**, Feiteira et al. visualized the effective sealing of a crack by an elastic polymer agent [26]. In bending after healing, the agent elastically deforms and progressively debonds from the crack faces. The method can provide sealing evidence only if samples of few hundred millimeters are considered, but it loses in accuracy as larger damage zones are inspected [23, 24].

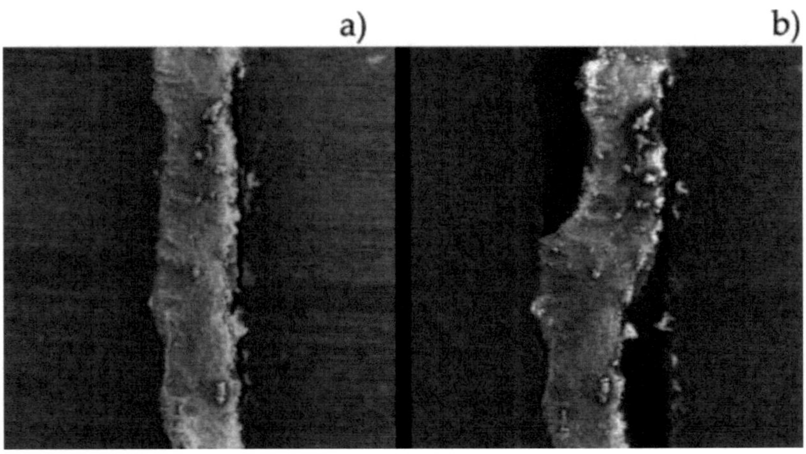

Figure 5. Scanning microscopy: (a) a crack formed on small-scale concrete beam is sealed with an elastic polymer healing agent and (b) at reloading stage, the agent elastically deforms and debonds from the crack faces [26].

3.3. Mechanical reset and repair

Mechanical restoration can be quantified by measuring the stiffness, strength and fracture toughness after healing and compare these values to the ones of intact stage. Extended research was previously performed on short concrete beams that carry a crack control pre-notch and a series of short agent capsules were embedded into them. The analyses have shown that in the ideal case that the unique crack is fully sealed and the agent is well polymerized, stiffness and strength recovery up to 100% can be achieved. In practice and considering an effective healing, typical reset levels are ranging from 40 to 70%.

The mechanical restoration can be associated to ultrasound pulse velocity recovery after healing. In a recent study, a pair of transmitter-receiver sensors is set facing the zone where controlled unique crack will form under load. As shown in **Figure 6a**, the piezoelectric transducers are embedded into concrete beams (design Case 1) during casting, this way eliminating the surface coupling and wave attenuation effects (further setup description can be found in [27]). A damage index that by definition considers both the shift of the arrival time and the amplitude variations is calculated and the outcome is presented in **Figure 6b** for both reference (carrying no healing material) and healing series [27]. It is shown that only in the case of healed samples, the damage index resets at the reloading cycle onset.

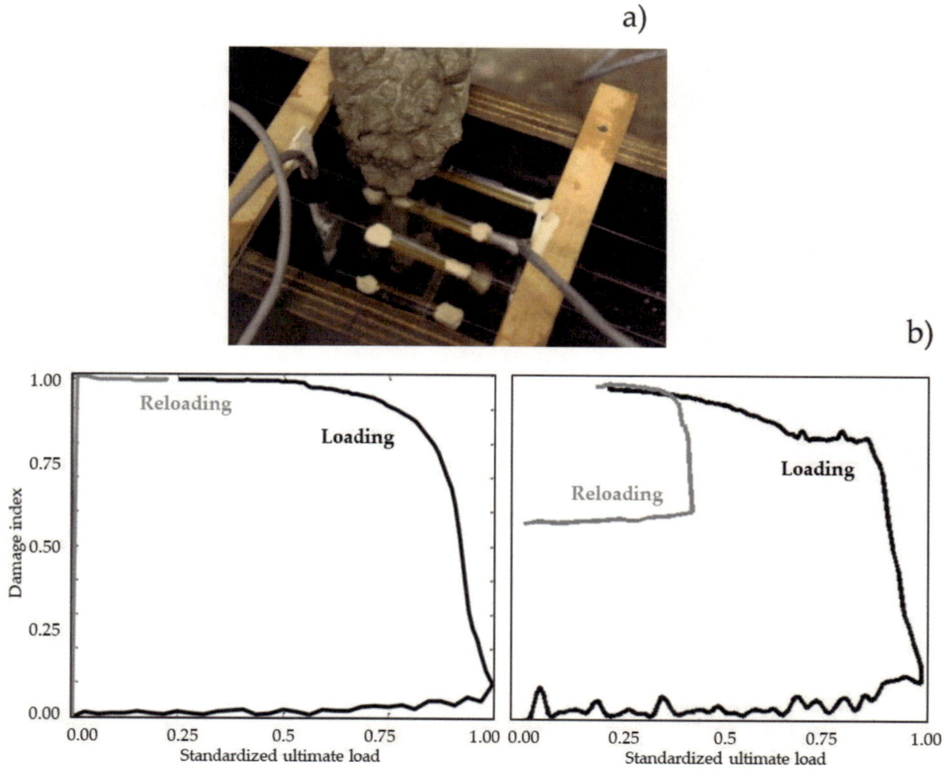

Figure 6. (a) A pair of piezoelectric transducers is detached to the concrete mold. The concrete is cast and the sensors remain embedded into it; (b) damage index evolution at loading/reloading cycles for reference and healing series.

The use of embedded sensors to track damage appears effective monitoring method only when small-scale controlled damage test configurations are considered and significantly loses in accuracy in larger concrete samples (i.e. design Case 4). Monitoring of larger concrete samples requires the development of a well-spread sensors network.

Tracking of repair efficiency appears more challenging as multiple cracks form and interact in concrete (design Case 2). Digital image correlation (DIC) strain concentration mapping can indicate the cracked zones on concrete surface, therefore highlights potential crack closure after healing. In **Figure 7**, a representative cracks pattern is presented. There are five macro-cracks with crack opening up to 300 μm formed on short concrete beams tested under

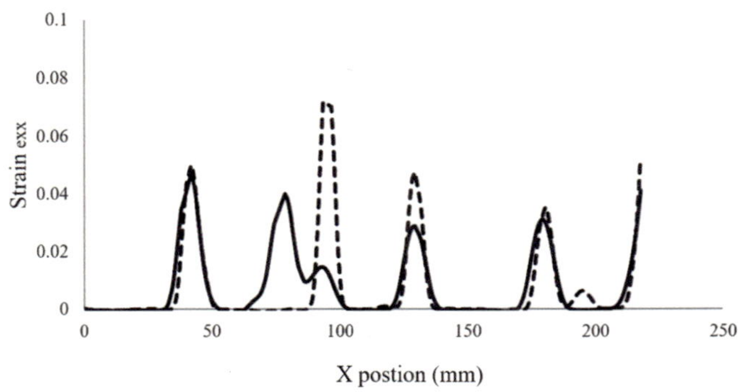

Figure 7. Horizontal strain (parallel to crack opening direction) maps captured by DIC on concrete surface of small-scale concrete beams. One of the cracks formed at the first loading cycle is healed, therefore at reloading stage another crack forms at its vicinity [28].

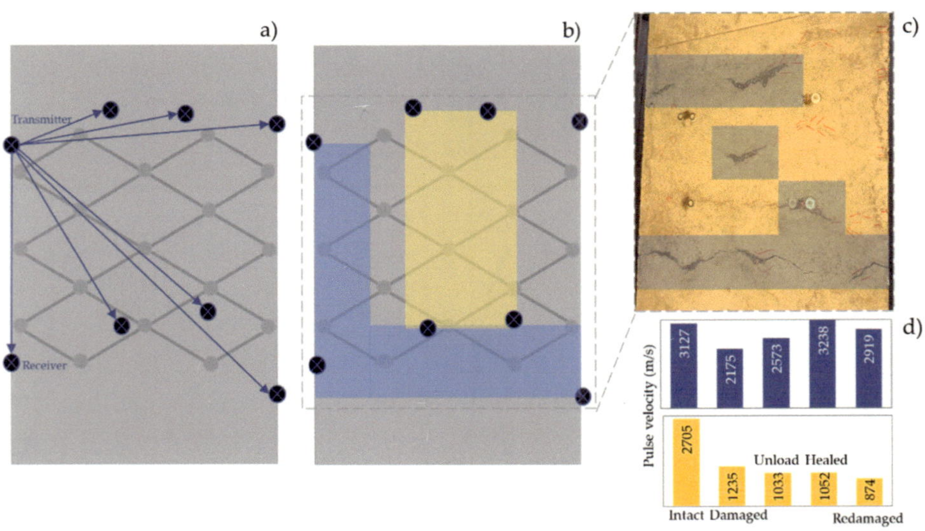

Figure 8. Mapping of ultrasound pulse velocity distribution, indicative of effective healing: (a) transmitter-receivers configuration; (b) maps of healed (blue) and non-healed (yellow) zones; (c) visual inspection of the plate bottom indicates areas where agent is leaching; (d) indicative distribution of velocities for a healed and non-healed zone, respectively.

four-point bending. The majority of cracks reopen at reloading stage after healing, however, one of them is effectively repaired and closed, therefore another crack forms at its surrounding at this second loading stage. The crack closure and opening of new cracks after healing is an evidence of effective material mechanical restoration.

In a recent study, ultrasound pulse velocity is applied to map the structural health condition of a plate (design Case 5) after healing. An array of piezoelectric transducers is attached to concrete surface as illustrated in **Figure 8**. Periodically, each one of the sensors emit a burst pulse (emitter) that travels through the concrete plate and reaches the other seven sensors (receivers). Considering known the spatial distance of the sensors and measuring the pulse arrival time, one can calculate the wave propagation velocity along the transmitter-receiver line. The study has shown that there are zones (marked in blue, **Figure 8b**) where the velocity drops under bending due to cracks formation, but after healing the velocity resets. On the other hand, yellow-marked zones on the plate indicate areas the velocity has not recovered after healing, therefore local healing was not effective. Indicative velocities distribution is illustrated in **Figure 8d**. The healed zones detected by ultrasound pulse velocity measurement are correlated to the zones where agent leaching is detected at the bottom side of the plate (**Figure 8c**), the latter being an indirect evidence of healing activation mechanism.

4. Closing remarks/future perspectives

4.1. Testing protocol establishment

The studies performed to assess the efficiency of newly developed healing technologies are empirically designed. The loading/unloading/reloading testing protocol is well-established since strength, stiffness, toughness or other mechanical feature reset index can be calculated:

$$\text{Repair index (\%)} = \frac{\text{Mechanical feature at intact stage}}{\text{Mechaincal feature after damage and healing}} \times 100\% \qquad (1)$$

However, there is no standardized test protocol based on which the healing index is obtained. The author proposes the design of concrete elements according to Rilem TC 50-FMC recommendation: small-scale unreinforced concrete beams carrying a pre-notch in the middle section that guides and controls the crack propagation, loaded under three-point bending in a quasi-static mode. Rilem TC 221-SHC should extend its recommendation on self-healing concrete testing protocol.

In this direction, the terms healing, repair, sealing and mechanical feature reset should be studied using different measuring and monitoring methodologies. Indicatively, the water permeability testing quantifies the sealing efficiency, digital image correlation can be used to track cracks repair and load-crack opening curve should be plotted to obtain potential stiffness and strength reset after healing.

4.2. Hyper-sensing in the service of healing

The analytical presentation of the monitoring methods outcome highlights the utmost importance of their application for healing assessment on concrete. An integrated methodology that combines optical, acoustic or other acoustic techniques should be designed providing essential structural health assessment.

4.3. Long-term healing efficiency

The progress on self-healing concrete design is impressive the last decade, still the technology has not reached the market since there is an open discussion on the material sustainability. Up to date, there are no studies done to assess the long-term performance of self-healing systems. It is proven that glass capsules cannot survive the alkaline concrete environment more than few years, therefore their replacement by ceramic or cementitious capsules is advance in this direction. Moreover, there is no evidence of agent degradation as the polymer is stored for several years at the healing reservoir. Life-cycle assessment models should be developed in the near future to predict the potential use of self-healing technology.

4.4. Additive manufacturing of healing systems

The tubes vascular network remained science fiction since the tubes network is characterized by extend geometrical complexity. The recently developed additive manufacturing technology is called to overpass this limitation: 3D printed polyamide plastic agent reservoir and delivery elements were easily and cost-effectively manufactured [14]. A prototype delivery element for a 2D tubes network is presented in **Figure 9a** and **b**. The reservoir is attached to the tubes edges, is externally filled up with agent that is delivered through the tubes only in the presence of cracks. In preliminary studies, the reservoir appears to attract cracks formation due to greater than concrete material stiffness. Ongoing research investigates the potential additive manufacturing of the reservoir by cementitious media.

In plane healing vascular networks as discussed in Section 2.2, the short tubes are interconnected by additive manufactured circular, thin and light-weight PVC nodes (**Figure 9c**) [17]. Holes through which the agent circulates are drilled through the nodes. The tubes are attached to the holes. Additive manufacturing appears an essential tool for the design of real-scale concrete structural elements with healing ability.

4.5. Healing in the service of heritage conservation

Recently the interest on historic buildings conservation has arisen. Following the trend of smart materials of today, the composition of traditional mortars is re-invented in an attempt to optimally repair existed heritage structures. In a recent study, lime mortar casted as historical masonry filler is designed considering commercial crystalline admixtures and tailored encapsulated additives that permit early age autogenous and post-curing autonomous healing [29]. The self-healing lime mortar compatibly fits to the historic mortar and moreover is able to heal micro-cracks and develop superior mechanical capacity. This

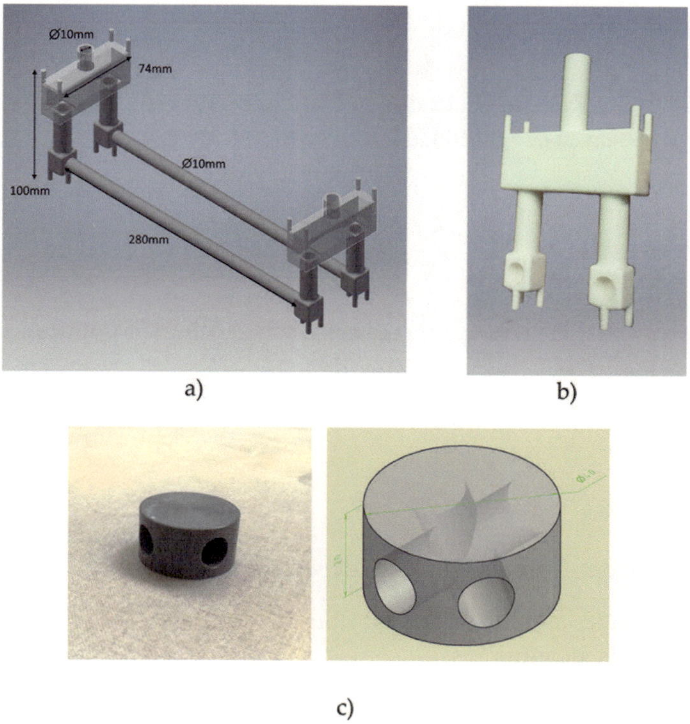

Figure 9. Additive manufactured components of healing agent vascular network: (a) the reservoirs-tubes model (design Case 3 [14]); (b) detail of the reservoir and (c) connection node in slab design case (Case 5 [17]).

pioneering work illustrates the potential application of self-healing technology for the repair and retrofitting of heritage infrastructure, a hot-topic research especially today since the architectural monuments of modernism are tremendously deteriorated and call for effective repair.

5. Conclusions

A retrospective and the current tendencies in self-healing concrete design are presented in this chapter. Self-healing concrete research introduced a new class of intelligent construction materials. The autonomous sensing of fracture and cracks repair feasibility is critically revised leading to the most promising material design: the embedment into concrete of a plane vascular healing agent carrying tubes network. The healing efficiency is assessed using a plethora of inspection experimental methods (AE, DIC, UPV, etc.) and it is concluded that each method contributes to sealing, healing, repair and mechanical restoration characterization. A decade ago, the self-healing concrete idea was abstractly projected in the long future. Recent scientific progress has made revolutionary jumps giving us the confidence that the futuristic self-healing concrete is a product of today.

Acknowledgements

Financial support of the Research Foundation Flanders (FWO-Vlaanderen, Project No 28976) and SIM Flanders (research project SECEMIN) for this study is gratefully acknowledged.

Conflict of interest

The author declares that there is no conflict of interest regarding the publication of this chapter.

Author details

Eleni Tsangouri

Address all correspondence to: eleni.tsangouri@vub.be

Department of Mechanics and Materials and Constructions (MeMC), Vrije Universiteit Brussel (VUB), Brussels, Belgium

References

[1] Jackson MD, Mulcahy SR, Chen H, Li Y, Li Q, Cappelletti P, et al. Phillipsite and Al-tobermorite mineral cements produced through low-temperature water-rock reactions in Roman marine concrete. American Mineralogist: Journal of Earth and Planetary Materials. 2017;**102**(7):1430-1450

[2] Hearn N. Self-sealing, autogenous healing and continued hydration: What is the difference? Materials and Structures. 1998;**31**(8):563

[3] Yang Y, Lepech MD, Yang E, Li VC. Autogenous healing of engineered cementitious composites under wet-dry cycles. Cement and Concrete Research. 2009;**39**(5):382-390

[4] Snoeck D, De Belie N. Repeated autogenous healing in strain-hardening cementitious composites by using superabsorbent polymers. Journal of Materials in Civil Engineering. 2015;**28**(1):04015086

[5] Li VC. On engineered cementitious composites (ECC). Journal of Advanced Concrete Technology. 2003;**1**(3):215-230

[6] Dry CM. Three designs for the internal release of sealants, adhesives, and waterproofing chemicals into concrete to reduce permeability. Cement and Concrete Research. 2000; **30**(12):1969-1977

[7] Dry C, McMillan W. Three-part ethylmethacrylate adhesive system as an internal delivery system for smart responsive concrete. Smart Materials and Structures. 1996;**5**(3):297

[8] White SR, Sottos NR, Geubelle PH, Moore JS, Ressler MR, Sriram SR, et al. Autonomic healing of polymer composites. Nature. 2001;**409**(6822):794

[9] Van Tittelboom K, Adesanya K, Dubruel P, Van Puyvelde P, De Belie N. Methyl methacrylate as a healing agent for self-healing cementitious materials. Smart Materials and Structures. 2011;**20**(12):125016

[10] Van Tittelboom K, De Belie N, Van Loo N, Jacobs P. Self-healing efficiency of cementitious materials containing tubular capsules filled with healing agent. Cement and Concrete Composites. 2011;**33**(4):497-505

[11] Van Tittelboom K, De Belie N, Lehman F, Grosse CU. Acoustic emission analysis for the quantification of autonomous crack healing in concretee. Construction and Building Materials. 2012;**28**(1):333-341

[12] Van Tittelboom K, Snoeck D, Vontobel F, Wittmann H, De Belie N. Use of neutron radiography and tomography to visualize the autonomous crack sealing efficiency in cementitious materials. Materials and Structures. 2013;**46**(1-2):105-121

[13] Pelletier M, Brown R, Shukla A, Bose A. Self-healing concrete with a microencapsulated healing agent. 2011. Available online: https://pdfs.semanticscholar.org/e48b/937bde7c53ce059d464822ebc837a31647dc.pdf (Accessed on 12 December 2018)

[14] Minnebo P, Thierens G, De Valck G, Van Tittelboom K, De Belie N, Van Hemelrijck D, et al. A novel design of autonomously healed concrete: Towards a vascular healing network. Materials. 2017;**10**(1):49

[15] Hilloulin B, Van Tittelboom K, Gruyaert E, De Belie N, Loukili A. Design of polymeric capsules for self-healing concrete. Cement and Concrete Composites. 2015;**55**:298-307

[16] Yang Z, Zhao N, Li W, Jiang Z, Yuan W. Self-healing efficiency of cementitious materials containing microcapsules filled with healing adhesive: Mechanical restoration and healing process monitored by water absorption. PLoS One. 2013;**8**(11):e81616

[17] Tsangouri E, Lelon J, Minnebo P, Aggelis D, Van Hemelrijck D. Ultrasound pulse velocity to measure repair efficiency of concrete containing a self-healing vascular network. In: 6th International Symposium on Life-Cycle Civil Engineering (IALCCE2018). Ghent; 2018

[18] Sangadji S, Schlangen E. Self healing of concrete structures—Novel approach using porous network concrete. Journal of Advanced Concrete Technology. 2012;**10**(5):185-194

[19] Kuang Y, Ou J. Passive smart self-repairing concrete beams by using shape memory alloy wires and fibers containing adhesives. Journal of Central South University of Technology. 2008;**15**(3):411-417

[20] Jefferson A, Joseph C, Lark R, Isaacs B, Dunn S, Weager B. A new system for crack closure of cementitious materials using shrinkable polymers. Cement and Concrete Research. 2010;**40**(5):795-801

[21] Teall O, Davies R, Pilegis M, Kanellopoulos A, Sharma T, Paine K, et al. Self-healing concrete full-scale site trials. In: 11th FIB International PhD Symposium in Civil Engineering. Tokyo; 2016

[22] Tsangouri E, Aggelis DG, Van Tittelboom K, De Belie N, Van Hemelrijck D. Detecting the activation of a self-healing mechanism in concrete by acoustic emission and digital image correlation. The Scientific World Journal. 2013;**2013**:10. Article ID: 424560

[23] Van Tittelboom K, Tsangouri E, Van Hemelrijck D, De Belie N. The efficiency of self-healing concrete using alternative manufacturing procedures and more realistic crack patterns. Cement and Concrete Composites. 2015;**57**:142-152

[24] Van Tittelboom K, Wang J, Araujo M, Snoeck D, Gruyaert E, Debbaut E, et al. Comparison of different approaches for self-healing concrete in a large-scale lab test. Construction and Building Materials. 2016;**107**:125-137

[25] Karaiskos G, Tsangouri E, Aggelis DG, Van Tittelboom K, De Belie N, Van Hemelrijck D. Performance monitoring of large-scale autonomously healed concrete beams under four-point bending through multiple non-destructive testing methods. Smart Materials and Structures. 2016;**25**(5):055003

[26] Feiteira J, Tsangouri E, Gruyaert E, Lors C, Louis G, De Belie N. Monitoring crack movement in polymer-based self-healing concrete through digital image correlation, acoustic emission analysis and SEM in-situ loading. Materials & Design. 2017;**115**:238-246

[27] Tsangouri E, Karaiskos G, Aggelis DG, Deraemaeker A, Van Hemelrijck D. Crack sealing and damage recovery monitoring of a concrete healing system using embedded piezoelectric transducers. Structural Health Monitoring. 2015;**14**(5):462-474

[28] Tsangouri E. Experimental assessment of fracture and autonomous healing of concrete and polymer systems [PhD dissertation]. Brussels: Vrije Universiteit Brussel (VUB); 2015

[29] De Nardi C, Bullo S, Ferrara L, Ronchin L, Vavasori A. Effectiveness of crystalline admixtures and lime/cement coated granules in engineered self-healing capacity of lime mortars. Materials and Structures. 2017;**50**(4):191

Chapter 3

Biocomposite Cement-Based Mortar

Xiaoniu Yu and Yan Gao

Additional information is available at the end of the chapter

http://dx.doi.org/10.5772/intechopen.79262

Abstract

This chapter presents the preparation of a new kind of biocement based on the biophosphate minerals, which have cementation, by the bacteria reacting with the substrate. Ammonia/ammonium can be changed into environment-friendly struvite when the soluble phosphate is added to biocarbonate cement. After that, struvite and carbonate, which can be considered as composite cements, are applied to cement loose particles. The biocement is environmentally friendly, which has important application prospects. Based on mixing-precipitation process, the injection process was adopted to bind loose sand particles. Permeability, porosity, compressive strength, and internal microstructure of the biosandstones cemented by composite cement were determined under different number of injections. Mixing-precipitation process was inferior to injection process according to compressive strength of the biosandstones caused by the particle size and morphology of composite cement. Permeability, porosity, compressive strength, and fixation ammonia ratio of the biosandstones were compared when three different formulations of composite cement (CJ1, CJ1.5, and CJ2) were adopted to bind sand columns. The results show that the CJ2 has the best overall performance. The molar ratio of $K_2HPO_4{\cdot}3H_2O$ and urea was 2:1 in the CJ2 formulation.

Keywords: biocement, struvite, carbonate, bind, sand particles, injection process, compressive strength, sand columns

1. Introduction

Portland cement is the most commonly used cementing material in the construction field and an important part of the building materials industry. The main raw material of cement is limestone. During the production process, limestone is burned with fossil fuels, and the greenhouse gas CO_2 is decomposed and released. When the fuel is burned, harmful gases such as SO_2 and NO_x are also emitted, causing adverse effects on the ecological and

IntechOpen

© 2018 The Author(s). Licensee IntechOpen. This chapter is distributed under the terms of the Creative Commons Attribution License (http://creativecommons.org/licenses/by/3.0), which permits unrestricted use, distribution, and reproduction in any medium, provided the original work is properly cited. (cc) BY

environmental systems. Based on microbial cement, it can consolidate loose particles with low energy consumption and less pollution. It can solve some disadvantages of traditional materials in the fields of ground reinforcement, desert treatment, dust control, etc. Biocement, based on microbial-induced mineralization, can consolidate the loose particles and would be a novel and sustainable cementing material which is applied to foundation reinforcement, dust control, and other fields.

The biocalcite cement, as one kind of biocement, has been investigated widely [1–6]. However, ammonia will be released in the cementation process of biocalcite cement, which has a negative impact on the ecological environment. It is a new type of sustainable development of cementitious materials. Biocarbonate cement can be used to cement loose grains into a whole with good mechanical properties through a grouting process. The main purpose of this chapter is to convert the ammonia/ammonium released from urease hydrolysis of urea into environmentally friendly struvite during the cementation process, and cemented loose particles together with carbonates. This can partially replace the chemical grouting materials commonly used in the treatment of foundations today, such as cement pastes, water glass, epoxy resins, methacrylates, polyurethanes, acrylic amines, lignin, and other chemical reinforcement materials. The biocomposite cement is injected into the prepared quartz sand mold through a peristaltic pump, and the loose particles can be cemented well into a whole with mechanical properties, and the ammonia/ammonium produced during the cementation of the biocarbonate cement can be converted into environmentally friendly struvite. The mineralization reaction of the cement in the pores between the loose particles produces struvite and hydromagnesite composite cementitious materials, which can improve the permeability, pore structure, and mechanical properties of the sand.

2. Biocomposite cement bind sand process

2.1. Preprecipitation mixing process molding sand columns

2.1.1. Biocomposite cement slurry with different standing time molding sand columns by preprecipitation mixing process

About 4 mol of $K_2HPO_4{\cdot}3H_2O$ was completely dissolved in the carbonate-mineralized bacteria (*Sporosarcina pasteurii*, 2 L), for CJ2. Divided into 18, 100 mL each, 1 group for 3 samples. Then, 100 mL of urea (1 mol/L) and $MgCl_2$ (3 mol/L) were sequentially added to obtain a precipitated solution. The settling solution was left standing every three groups for 0, 2, 6, 12, 24, and 40 h. The supernatant was removed, each group of sediment (30% of the total sand column) was mixed with quartz sand (particle size 425–212 μm), mechanically mixed until uniform, and the mixture was poured into a plastic mold (Φ 3cm × 6 cm), as shown in **Figure 1**. The molded specimens were placed in a 30 ± 2°C oven for drying. The molds were removed and the corresponding sand columns were obtained.

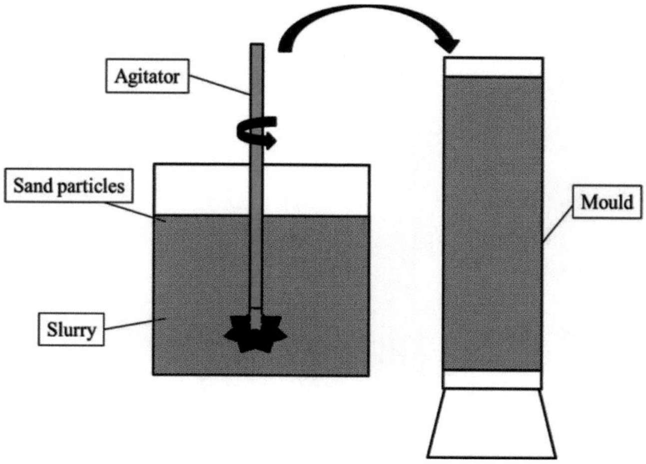

Figure 1. Schematic diagram of the preprecipitation mixing process.

Standing time (h)	0	2	6	12	24	40
Compressive strength (MPa)	0.18	0.42	0.38	0.34	0.34	0.21

Table 1. Effect of standing time of precipitated slurry on the average compressive strength of the biosandstones.

The average compressive strength of the sand column is shown in **Table 1**, which are 0.18, 0.42, 0.38, 0.34, 0.34, and 0.21 MPa, respectively. The average compressive strength of the bio-$BaHPO_4$ slurry (30% of the total sand column) cement sand column (Φ 3cm × 6 cm) is 0.90 MPa [1]. The results show that the average compressive strength of sand columns cemented by the composite cement slurry in mixing process is lower than that of the bio-$BaHPO_4$ slurry.

2.1.2. *Different contents of biocomposite cement slurry forming sand column by preprecipitation mixing process*

The raw file of the composite cement slurry (standing time for 6 h) was analyzed by the MDI Jade 5.0 program. The results showed that the composition of the product was the mixture of $MgNH_4PO_4{\cdot}6H_2O$ (JCPDS No. 03-0240) (JCPDS No. 03-0240) and $Mg_5(CO_3)_4(OH)_2(H_2O)_4$ (JCPDS No. 70-1177) (**Figure 2**). SEM images show that the morphology of the mixture is irregular, the surface is relatively rough, and the particle size is in the range of 150–500 μm, as shown in **Figure 3**.

The average compressive strength of the sand columns (Φ 3cm × 6 cm) cemented 10, 20, 30, 40, 50, and 60% for the composite cement slurry (standing time for 6 h) were 0.13, 0.25, 0.38, 0.36, 0.35, and 0.36 MPa, respectively, as shown in **Table 2**. Compared with the average compressive

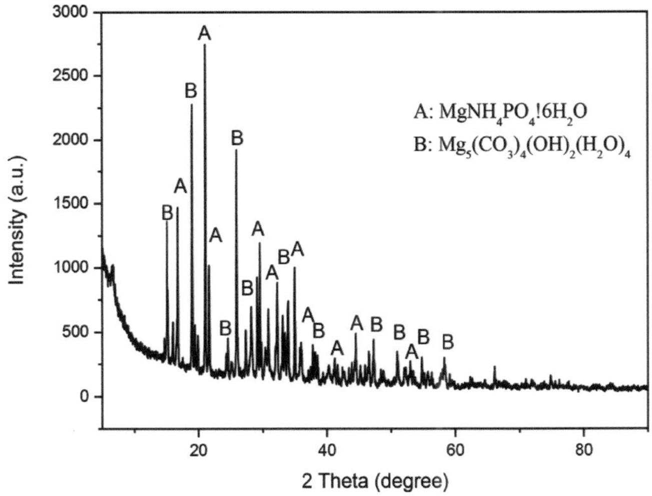

Figure 2. XRD patterns of the precipitated slurry.

Figure 3. SEM images of the precipitated slurry.

strength of sand columns (Φ 3cm × 6 cm) cemented bio-$BaHPO_4$ slurry at different dosages, the average compressive strength of sand columns cemented by composite cement slurry is lower [7]. Therefore, the grouting process was used to cement the loose sand.

2.2. Biogrouting process forming sand columns

Quartz sand with different particle sizes (particle diameter 425–212 μm) was mechanically mixed until uniform. CJ1 represents per liter of carbonate-mineralized bacteria containing 1 mol of $K_2HPO_4{\cdot}3H_2O$. Before adding loose sands, put a 1.0-mm high gauze on the bottom of the PVC pipe (Φ 5cm × 15 cm), then place sands in a PVC pipe and compact it to dense. Finally, put a 1.0-mm high gauze on top of PVC pipe. The bottom of the PVC tube (the bottom of the cylinder) was connected with a peristaltic pump that the flow rate can be regulated, and

Content (%)	10	20	30	40	50	60
Compressive strength (MPa)	0.13	0.25	0.38	0.36	0.35	0.36

Table 2. Effect of content of composite cement slurry on the average compressive strength of the biosandstones.

inject the solution from the bottom to top. The steps are as follows: (1) injecting tap water into the mold at a flow rate of 16 mL/min to eliminate air bubbles existing between the particles; (2) injecting 100 mL of CJ1; (3) injecting 100 mL of (1 mol/L) urea and (2 mol/L) $MgCl_2$ mixed solution and stand for 6 h; (4) injecting 100 mL of CJ1 and allowed to stand for 6 h; (5) 100 mL of (1 mol/L) urea and (2 mol/L) $MgCl_2$ mixed solution was injected and allowed to stand for 6 h; (6) 100 mL of CJ1 was injected and allowed to stand for 6 h, and so on, until unable to inject CJ1 and urea and $MgCl_2$ of mixed solution to sand columns. Then, the samples with the mold were placed in an oven (30 ± 2°C) to dry for 21 days, and then the mold was removed. All experiments were performed in triplicate and cemented at 30 ± 2°C. The number of injections of CJ1 was 2, 4, and 6, respectively (**Figure 4**).

2.2.1. XRD pattern of sand columns

The constituents of the sand column were analyzed through XRD, as shown in **Figure 5**. The sand column components were mainly a mixture of quartz sand (JCPDS No. 46-1045), $MgNH_4PO_4(H_2O)_6$ (JCPDS No. 71-2089), and $Mg_5(CO_3)_4(OH)_2{\cdot}4H_2O$ (JCPDS No. 25-0513). The XRD results indicated that the cementation material in the biosandstone was a mixture of $Mg_5(CO_3)_4(OH)_2{\cdot}4H_2O$ and $MgNH_4PO_4(H_2O)_6$. Ammonia and ammonium could be converted into magnesium ammonium phosphate. Therefore, the biocomposite cement can

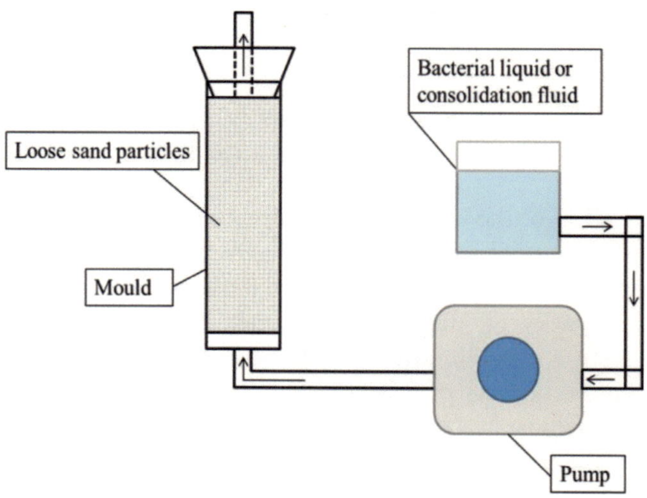

Figure 4. Schematic diagram of biogrouting process [8].

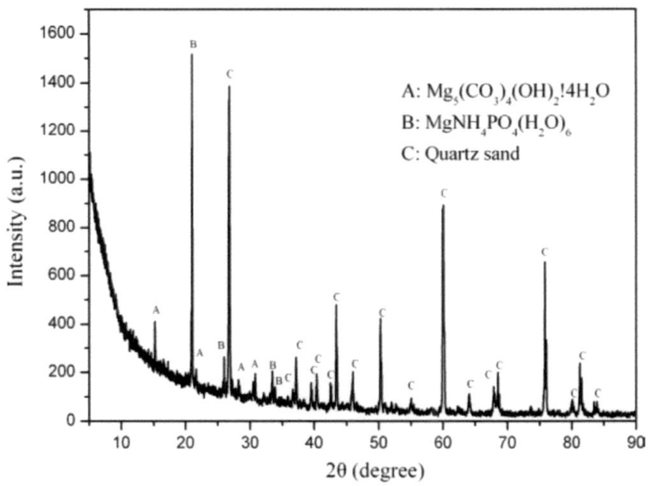

Figure 5. X-ray diffraction of the biosandstone.

be synthesized through *Sporosarcina pasteurii*-induced precipitation. $MgNH_4PO_4(H_2O)_6$ was mainly prepared by magnesium ions reacting with ammonium and HPO_4^{2-} ions. Meanwhile, $Mg_5(CO_3)_4(OH)_2{\cdot}4H_2O$ could also be prepared by magnesium ions reacting with carbonate ions in the alkaline solution.

2.2.2. *Influence of the number of injections on the hydraulic conductivity of sand columns*

Figure 6 shows the hydraulic conductivity of the biosandstone. The average hydraulic conductivity of the sand was 3.97×10^{-2} cm/s before cementation. The average hydraulic conductivity of the biosandstone was 2.72×10^{-2}, 2.22×10^{-2}, and 2.03×10^{-2} cm/s when the number of injections was 2, 4, and 6, respectively. The hydraulic conductivity of the biosandstone decreased as the number of injections increased.

2.2.3. *Influence of the number of injections on compressive strength and porosity of sand columns*

Figure 7 shows the effect of the number of injections on compressive strength of the sand columns. When the number of injections increases from 2, 4, and 6, compressive strength of the sand columns increases sequentially, and the average compressive strength is 0.37, 0.80, and 1.53 MPa, respectively. After injecting 6 times, it is difficult to inject the biocomposite cement. Therefore, the maximum number of injections was 6 times. **Figure 8** shows the relationship between the number of injections and porosity. The average porosity of sand columns decreased from the initial 45.01 to 34.62, 29.55, and 25.15%, when CJ1, the number of injections was 2, 4, and 6, respectively, and the decrease rates were 10.39, 15.46, and 19.86%, respectively. The average porosity of sand columns shows a decrease with the increase in the number of injections.

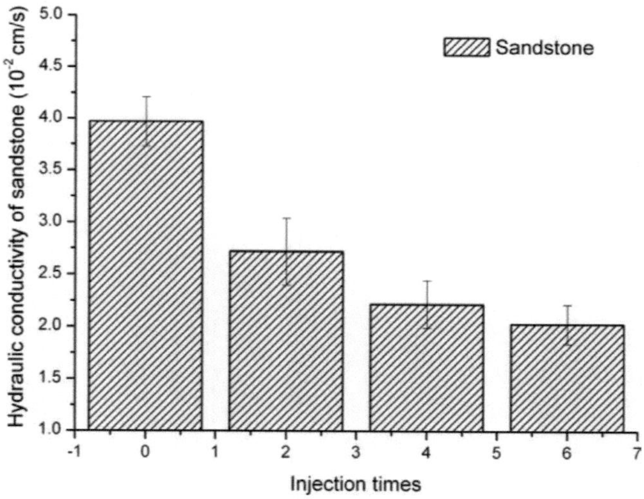

Figure 6. Effect of the number of injections on the hydraulic conductivity of the biosandstone.

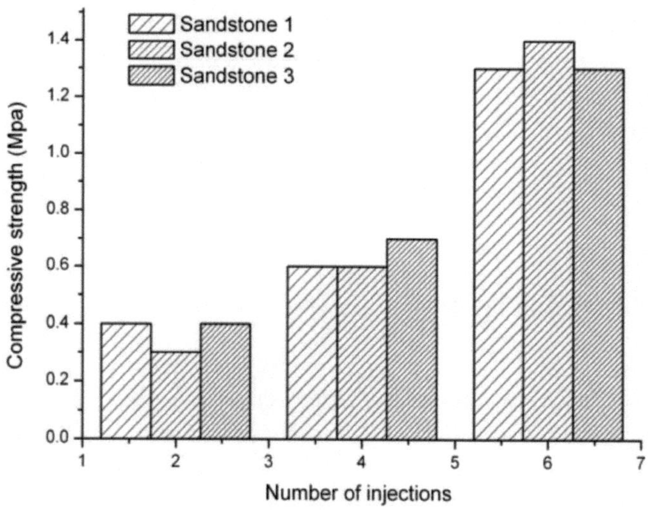

Figure 7. Effect of the number of injections on the compressive strength of sand columns.

2.2.4. Microstructure of sand columns under different number of injections

SEM images of sand columns with different number of injections are shown in **Figure 9**. **Figure 9(a–f)** shows the internal filling of the sand columns under numbers of injection 2, 4, and 6, respectively. The morphology of the particles is mainly an irregular sheet structure. **Figure 9(a, b)** shows that the internal filling of the sand column results in a small amount of composite cementitious material and cannot be filled in between the sand particles. **Figure 9(c, d)** shows that the interior of the sand column is filled, and the amount of the

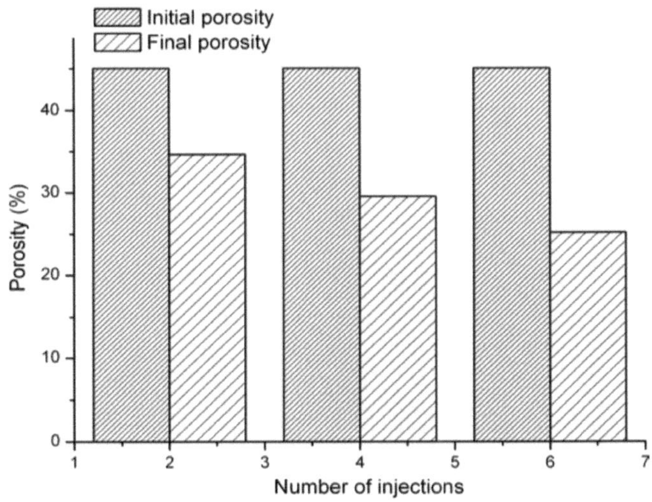

Figure 8. Effect of the number of injections on the porosity of sand columns.

composite cementitious material produced is more than **Figure 9(a, b)** and can be partly filled between the sand particles. **Figure 9(e, f)** shows that the interior of the sand column is well filled, and the amount of the composite cementitious material produced is more than that of 2 and 4 times. The loose sand grains could well be cemented into a whole with good mechanical properties. The above results also indicate that compressive strength, permeability, and porosity of the sand columns decrease with the increase in the numbers of injection.

2.2.5. 3D pore structure evolution of biosandstone produced using different number of injections

The XCT was Y. CT Precision S series high-precision computed tomography system from YXLON, Germany, with a resolution of micrometers. The test method was described in the literature [8]. XCT in situ trace tests were performed on sand columns with different cementation times (2, 4, and 6 times). The test results were analyzed using the defect analysis module in the VG Studio Max 2.0 software on the XCT device. According to the literature [9], VG Studio Max 2.0 software's defect analysis module and gray threshold algorithm are used to extract hole information, and a color scale is used to represent the size of the hole, in which blue to red represent from small holes to large holes. **Figure 10** shows the three-dimensional (3-D) pore structure evolution of the sand columns at different number of injections. When the number of injections increased, the defect volume in the sand columns gradually decreases. The maximum defect volume of the sand column with injecting 2 times was 1401 mm^3. After injecting 6 times, it decreased to 738 mm^3. This is due to the increase in the number of injections, resulting in more and more microbial cement product filling defects in the sand column. With the increase of the number of cementation, the size of the defect gradually decreased. This is also due to the increase in the number of

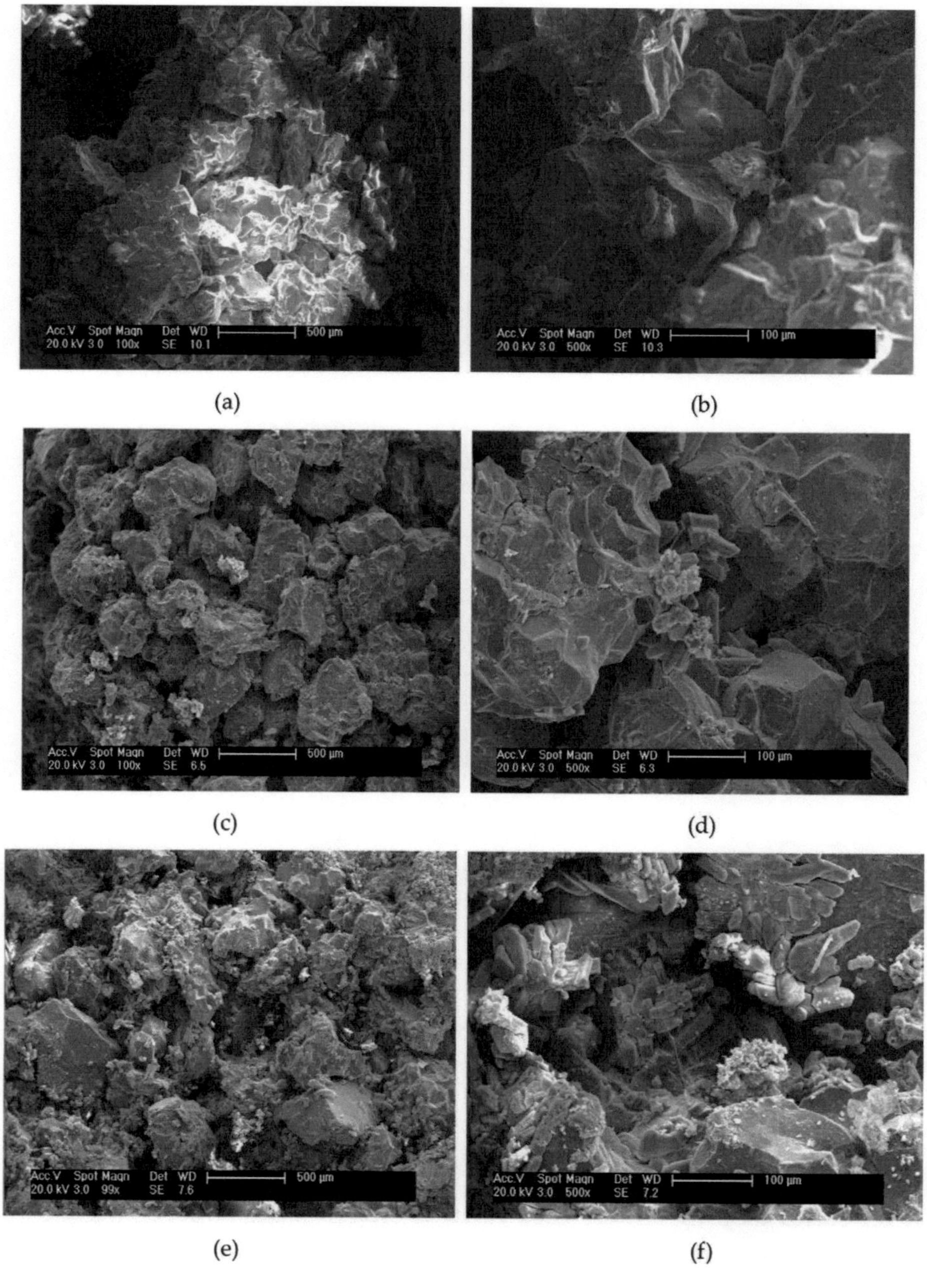

Figure 9. SEM images of sand columns: (a, b) 2 injections; (c, d) 4 injections; (e, f) 6 injections.

injections, resulting in the continuous filling of microbial cement in the defects, thereby reducing the size of defects and the number of defects. The overall defect distribution of the sand columns is not uniform. This is due to the inhomogeneity of defect distribution during the formation of loose sand grains, i.e., uneven distribution of internal defects, as

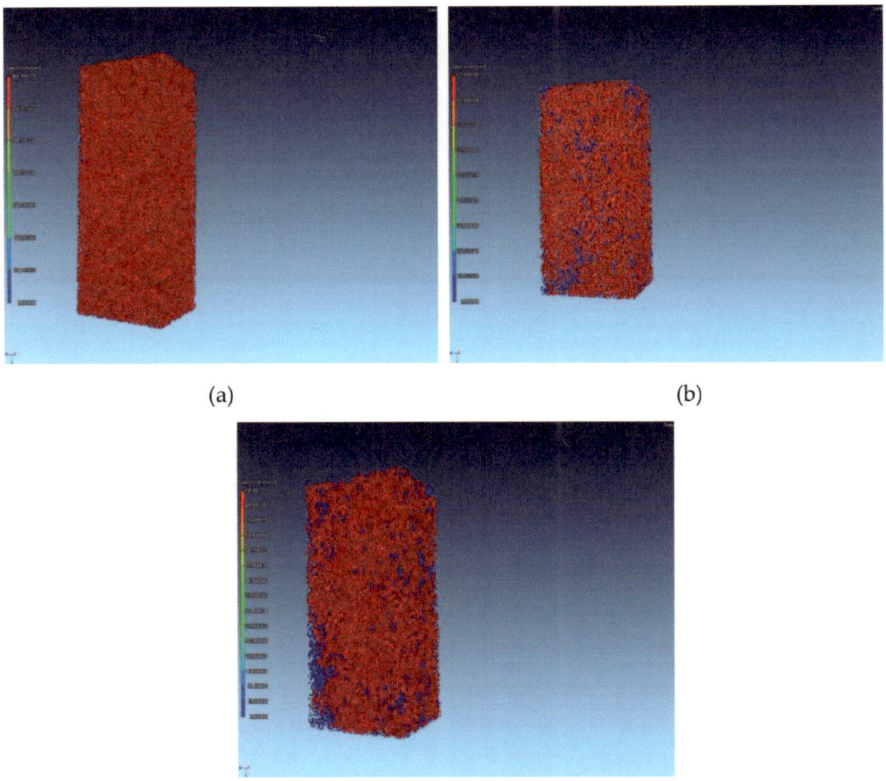

(a) (b) (c)

Figure 10. 3D pore structure evolution of biosandstone following: (a) 2 injections, (b) 4 injections, and (c) 6 injections.

shown in **Figure 10**. The average porosity of the sand columns was 36.20, 32.21, and 20.21% with corresponding injection times 2, 4, and 6, respectively. This result is similar to average porosity of 34.62, 29.55, and 25.15% for the sand columns measured at 2, 4, and 6 with paraffin drainage method.

3. Comparison of performance of sand columns cemented by biocomposite cement with preprecipitation mixing process and biogrouting process

By comparing the preprecipitation mixing process and the biogrouting process, the compressive strength of sand columns cemented by the preprecipitation mixing process sand column is poor, and the biogrouting process achieves a certain number of cementation can significantly improve the compressive strength, permeability, and porosity of sand columns. The reason for the compressive strength of sand columns cemented by preprecipitation mixing process is inferior to the biogrouting process caused by the particle size and structure. For example, **Figure 3** shows that the particle morphology of the biocomposite cement is an irregular massive structure with a size in the range of 150–500 μm, while the pore size under tightly packed loose grains is less than 100 μm, resulting in the increase in the distance between sand grains under the preprecipitation mixing process. This is one of the reasons for the low compressive strength of the sand columns. Secondly, through the particle structure of the biocomposite cement, it can be judged that the bonding force between the particles is poor, which is also the reason why the compressive strength of sand columns is low. However, the particle size of the biocomposite cement formed by the grouting process is less than 100 μm, and the particles are tightly bound together in the sand columns, as shown in **Figure 9**. Therefore, it is better consolidate loose sand grains into a whole with good mechanical properties. In the next experiment, different formulations of biocomposite cement were used to cement loose grains under the grouting process, and the porosity, permeability, compressive strength, and internal microstructure of the sand columns were studied.

4. Different formulations of biocomposite cement binding loose sand

Each liter of solution for *Sporosarcina pasteurii* and $K_2HPO_4{\cdot}3H_2O$ (1, 1.5, and 2 mol/L) are named as CJ1, CJ1.5, and CJ2, respectively. Different types of magnesium ammonium phosphate and carbonate are synthesized by CJ1, CJ1.5, and CJ2 reacting with the mixture solution of $MgCl_2$ and urea (equimolar). Biocomposite cement was then obtained.

All sand columns were pared according to Section 2.2. All injection experiments were performed at room temperature of 25–30°C. PVC pipe (bottom of the cylinder) was connected with a peristaltic pump, and the CJ1, CJ1.5, and CJ2 and mixture solution of $MgCl_2$ and urea were injected from the bottom to top. The steps are as follows: (1) tap water was injected into three PVC pipes at a flow rate of 16 mL/min to exclude bubbles; (2) 100 mL of CJ1, CJ1.5, and CJ2 were injected to three PVC pipes, respectively; (3) injecting 100 mL of mixed solution of urea and $MgCl_2$ to PVC pipes standing for 6 h. The next steps are the same as in Section 2.2. Until CJ1, CJ1.5, and CJ2, mixed solution of urea and $MgCl_2$ could not be injected, and number of injections was 6, 4, and 3, respectively. Then, the specimens were placed in a 30 ± 2°C oven for curing for 15 days. The molds were removed and the corresponding sand columns were obtained. This method can well cement loose sand grains into a whole with mechanical properties, as shown in **Figure 11**. All tests were performed in triplicate and cemented at 30 ± 2°C. The initial porosity is 45.01 ± 2%.

4.1. XRD patterns of sand columns

The XRD diffraction peak of the sand column cemented by CJ1 was quartz (JCPDS No. 46-1045), $MgNH_4PO_4(H_2O)_6$ (JCPDS No. 71-2089), and $Mg_5(CO_3)_4(OH)_2{\cdot}4H_2O$ (JCPDS No.

25-0513) (**Figure 12(a)**). The XRD results show that the cementitious materials formed by CJ1 cementation sand column are mainly $MgNH_4PO_4(H_2O)_6$ and $Mg_5(CO_3)_4(OH)_2{\cdot}4H_2O$ complexes.

The XRD diffraction peak of the sand column cemented by CJ1.5 was quartz (JCPDS No. 86-2237), $MgNH_4PO_4(H_2O)_6$ (JCPDS No. 71-2089), $Mg_3(PO_4)_2(H_2O)_8$ (JCPDS No. 84-1148), and $MgCO_3(OH)_2(H_2O)_3$ (JCPDS No. 70-0591), as shown in **Figure 12(b)**. The XRD results show that the cementitious materials formed by CJ1.5 cement sand column are mainly $MgNH_4PO_4(H_2O)_6$ and $MgCO_3(OH)_2(H_2O)_3$ composites.

The XRD diffraction peak of the sand column cemented by CJ2 was quartz (JCPDS No. 89-1961), $MgNH_4PO_4{\cdot}6H_2O$ (JCPDS No. 15-0762), and $Mg_5(CO_3)_4(OH)_2{\cdot}5H_2O$ (JCPDS No. 23-1218) (**Figure 12(c)**). The XRD results showed that the cementitious materials formed by CJ2 cemented sand column were mainly $MgNH_4PO_4{\cdot}6H_2O$ and $Mg_5(CO_3)_4(OH)_2{\cdot}5H_2O$ complexes. The above results indicate that the main components of cementation product in sand columns are mainly magnesium ammonium phosphate (struvite) and hydromagnesite composites. Therefore, ammonia/ammonium (NH_3 /NH_4^+) can also be changed into struvite when *Sporosarcina pasteurii* contained $K_2HPO_4{\cdot}3H_2O$ in all cementation process.

4.2. Influence of different biocomposite cement on hydraulic conductivity of sand columns

Figure 13 indicates the influence on hydraulic conductivity of biosandstones by three different biophosphate and carbonate composite cements. The average hydraulic conductivity of the sand was 39.7×10^{-3} cm/s before cementation, as shown in **Figure 13(a)**. **Figure 13(b–d)** indicates that the average hydraulic conductivity of biosandstones cemented by CJ1, CJ1.5, and CJ2 is 20.3×10^{-3} cm/s, 2.52×10^{-3} cm/s, and 3.59×10^{-3} cm/s, respectively. Therefore, permeability of biosandstones can be retained when loose particles are cemented by three different biocomposite cements.

Figure 11. Loose sand particles to sand column cemented by biocomposite cement.

http://dx.doi.org/10.5772/intechopen.79262

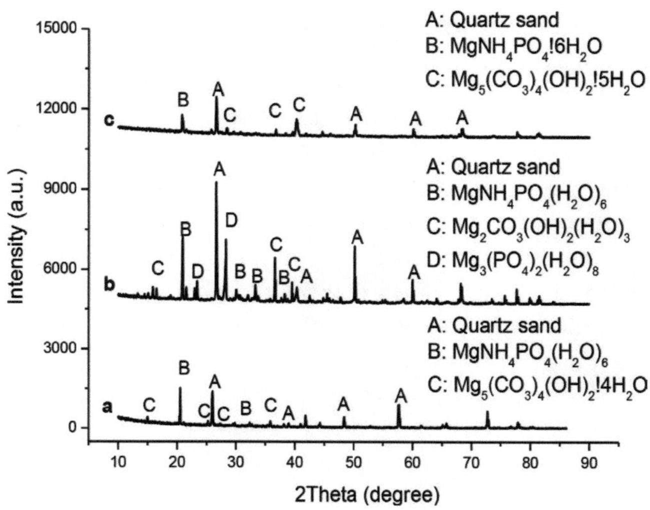

Figure 12. X-ray diffraction of the sand columns cemented by three different biocomposite cements.

4.3. Influence of different biocomposite cement on compressive strength and porosity of sand columns

The effects of different biophosphate and carbonate composite cement on compressive strength are presented in **Figure 14**. **Figure 14(a–c)** shows that the average compressive strength of biosandstones cemented by CJ1, CJ1.5, and CJ2 are 1.53, 1.42, and 1.47 MPa, respectively. Results show that the maximum compressive strength of biosandstone cemented by CJ1 cementation is 1.59 MPa.

The relationship between average porosity and different biophosphate and carbonate composite cement is presented in **Figure 15**. The average porosity of biosandstones cemented by CJ1, CJ1.5, and CJ2 is effectively reduced from initial 45.01% down to 25.15%, 26.08%, and 25.87%, respectively, as shown in **Figure 15(a–c)**. The reductions were 19.86%, 18.93%, and 19.14%, respectively.

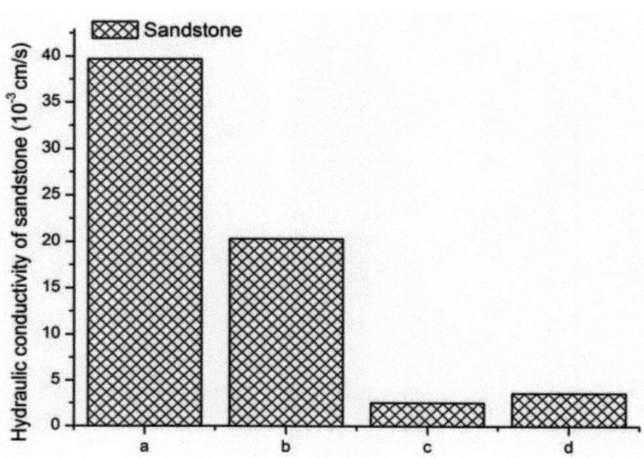

Figure 13. Effect of different biocomposite cement on the average hydraulic conductivity of sand columns.

4.4. Microstructure of sand columns cemented by different biocomposite cements

The SEM images of different formulations of biocement cemented sand columns are shown in **Figure 16**. **Figure 16(a, b)** shows the internal state of the CJ1 cemented sand column. The morphology of the composite cement product was mainly an irregular sheet structure and could be well filled between sand grains. The internal state of the CJ1.5 cemented sand column shows that the biocomposite cement product of morphology is an irregular particle cluster structure and can also be well filled between sand grains (**Figure 16(c, d)**). The internal state of the CJ2 cemented sand column shows that the morphology of the biocomposite cement product is also an irregular particle cluster structure, and can also be well filled between sand grains, as shown in **Figure 16(e, f)**. SEM images show that the porosity of sand columns can well be filled by three formulations of biocomposite cement.

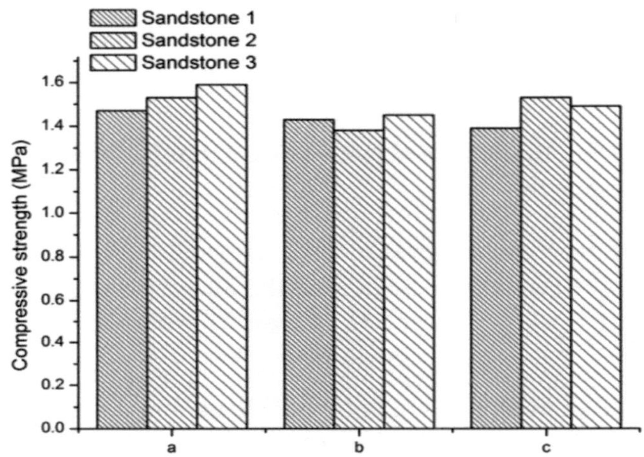

Figure 14. Effect of different biocomposite cements on compressive strength of sand columns.

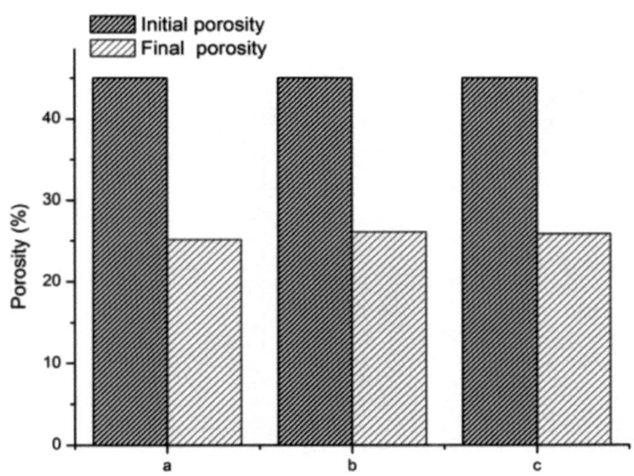

Figure 15. Effect of different biocomposite cements on the average porosity of sand columns.

4.5. Comparison of performance of three types of biocomposite cement

Above results show that the strength of sand columns cemented by three bio-composite cements is significantly lower than that of the biocarbonate cement bind sand columns. Similarly, three biocomposite cements can replace biocarbonate cement and can be applied to desert or dust treatment, sandy soil foundation, etc. The average permeability coefficients of sand columns cemented by CJ1, CJ1.5, and CJ2 were 2.03×10^{-3}, 2.52×10^{-2}, and 2.59×10^{-2} cm/s, respectively, and the average compressive strengths were 1.53, 1.42, and 1.47 MPa, respectively (**Table 2**). The content of biocomposite cement and compressive strength of sand columns are similar. The ammonia is released for 9.4, 7.5, and 5.7 g/L in CJ1, CJ1.5, and CJ2 cementation process, respectively. Therefore, CJ2 can effectively reduce ammonia emissions, and cementation numbers are least. Therefore, overall performance of CJ2 is optimum.

Figure 16. SEM images of the sand columns: (a, b) CJ1 cementation, (c, d) CJ1.5 cementation, and (e, f) CJ2 cementation.

5. Conclusions

Using biocomposite cement, loose sand grains can be cemented into sand columns with good mechanical strength by the biogrouting process. Ammonia was produced during the formation of biocarbonate cement and could be effectively become into struvite by phosphate. The number of injections has an important influence on the mechanical properties of sand columns. XRCT analysis showed that content of the biocomposite cement increased, the pores between sand grains gradually filled, and finally reducing the defect volume. Loose sand particles can well be cemented into biosandstone by three biocomposite cements. The performance of the biosandstones cemented by three formulations of composite cement was compared. Experiment results show that the compressive strength of sand columns cemented by three types of biocomposite cement is greater than 1.0 MPa. The weight of ammonia produced by CJ2 is less than CJ1 and CJ1.5.

Conflict of interest

The authors declare no conflict of interest.

Notes/thanks/other declarations

This work was supported by the National Nature Science Foundation of China (Grant No. 51702238).

Author details

Xiaoniu Yu[1]* and Yan Gao[2]

*Address all correspondence to: xnyu@wzu.edu.cn

1 College of Civil Engineering and Architecture, Wenzhou University, Wenzhou, China

2 School of Continuing Education, Northeast Normal University, Changchun, China

References

[1] Qian CX, Yu XN, Wang X. A study on the cementation interface of bio-cement. Materials Characterization. 2018;**59**:1186-1193. DOI: 10.1016/j.matchar.2017.12.011

[2] Xiao P, Liu HL, Xiao Y, Stuedlein AW, Evans TM. Liquefaction resistance of bio-cemented calcareous sand. Soil Dynamics and Earthquake Engineering. 2018;**107**:9-19. DOI: 10.1016/j.soildyn.2018.01.008

[3] Terzis D, Laloui L. 3-D micro-architecture and mechanical response of soil cemented via microbial-induced calcite precipitation. Scientific Reports. 2018;**8**:1416. DOI: 10.1038/s41598-018-19895-w

[4] Azadi M, Ghayoomi M, Shamskia N, Kalantari H. Physical and mechanical properties of reconstructed bio-cemented sand. Soils and Foundations. 2017;**57**:698-706. DOI: 10.1016/j.sandf.2017.08.002

[5] Sari YD. Soil strength improvement by microbial cementation. Marine Georesources and Geotechnology. 2015;**33**:567-571. DOI: 10.1080/1064119X.2014.953234

[6] Salifu E, MacLachlan E, Iyer KR, Knapp CW, Tarantino A. Application of microbially induced calcite precipitation in erosion mitigation and stabilisation of sandy soil foreshore slopes: A preliminary investigation. Engineering Geology. 2016;**201**:96-105. DOI: 10.1016/j.enggeo.2015.12.027

[7] Yu XN, Qian CX, Xue B, Wang X. The influence of standing time and content of the slurry on bio-sandstone cemented by biological phosphates. Construction and Building Materials. 2015;**82**:167-172. DOI: 10.1016/j.conbuildmat.2015.02.038

[8] Yu XN, Qian CX, Sun LZ. The influence of the number of injections of bio-composite cement on the properties of bio-sandstone cemented by bio-composite cement. Construction and Building Materials. 2018;**164**:682-687. DOI: 10.1016/j.conbuildmat.2018.01.014

[9] Rong H, Qian CX, Li LZ. Study on microstructure and properties of sandstone cemented by microbe cement. Construction and Building Materials. 2012;**36**:687-694. DOI: 10.1016/j.conbuildmat.2012.06.063

Chapter 4

Development of Sustainable Building Materials from Agro-Industrial Wastes in Nigeria

Abel Olajide Olorunnisola

Additional information is available at the end of the chapter

http://dx.doi.org/10.5772/intechopen.81662

Abstract

One of the continuing challenges posed by unprecedented urbanisation in Nigeria, estimated at about 5% per annum, is the provision of adequate and affordable housing. The shortage in housing, due in part to the ever-rising prices of construction materials, makes it logical to consider alternative building materials. Paradoxically, Nigeria is grappling with the challenges of managing solid waste, many of which could find suitable applications in the production of cement-, concrete- and clay-based walling, roofing and ceiling products as well as pozzolans for partial replacement for ordinary Portland cement. The objective of this chapter is to present information on the development, experimental investigations and practical application of sustainable building materials from agro-industrial wastes in Nigeria. Agroforestry residues such as bagasse and corn cob ashes have been found suitable as pozzolans; cement- and clay-bonded reinforced composite roofing tiles, hollow concrete blocks and stabilised clay bricks have been developed using a variety of lignocelluloses as sources of fibre reinforcement, while biomaterial substitutes for steel reinforcement in concrete have been tested. However, for these products to become widely acceptable, greater awareness has to be created among all stakeholders in the building construction industry, coupled with the development of appropriate building codes.

Keywords: housing, construction materials, agro-industrial wastes, rattan, bamboo

1. Introduction

Housing is one of the three basic needs of mankind, and it is the most important for the physical survival of man after the provision of food. Housing, either in units or in multiple forms, is a significant component of the physical form and structure of a community. In other words, adequate housing contributes to the attainment of physical and moral health of a nation and

IntechOpen

© 2019 The Author(s). Licensee IntechOpen. This chapter is distributed under the terms of the Creative Commons Attribution License (http://creativecommons.org/licenses/by/3.0), which permits unrestricted use, distribution, and reproduction in any medium, provided the original work is properly cited.

stimulates social stability and work efficiency. It is also an indicator of a person's standard of living and of his place in the society [1].

One of the continuing challenges posed by unprecedented urbanisation in Nigeria and many other Sub-Saharan African countries is the provision of adequate and *affordable housing*. Affordable housing is a term used to describe dwelling units whose total housing costs are deemed "affordable" to a group of people within a specified income range. Although the term is often applied to rental housing that is within the financial means of those in the lower-income ranges of a geographical area, the concept is applicable to both renters and purchasers in all income ranges. In the United States and Canada, a commonly accepted guideline for housing affordability is a housing cost that does not exceed 30% of a household's gross income. Housing costs considered in this guideline generally include taxes and insurance for owners and usually include utility costs. When the monthly carrying costs of a home exceed 30–35% of household income, then the housing is considered unaffordable for that household [2].

The challenges of urbanisation and the attendant consequences on housing provision are probably more widespread in Nigeria than anywhere else in Sub-Saharan Africa. Approximately 50% of the Nigerian population lives in urban cities with predictions that the urban population will hit the 65% mark by the year 2020. Rapid urban growth has resulted in problems of urban congestion or overcrowding and poor housing, among other challenges. Typical manifestations of this unmet demand include proliferation of slums in the cities with Nigeria having the fourth largest number of slum dwellers in the world (**Figure 1**) and the menace of skyrocketing house rents. Estimates show that Nigeria housing deficit is over 16 million units. An average of 1 million housing units per year is required not only to replenish decaying housing stock but also to meet rising demand [3, 4].

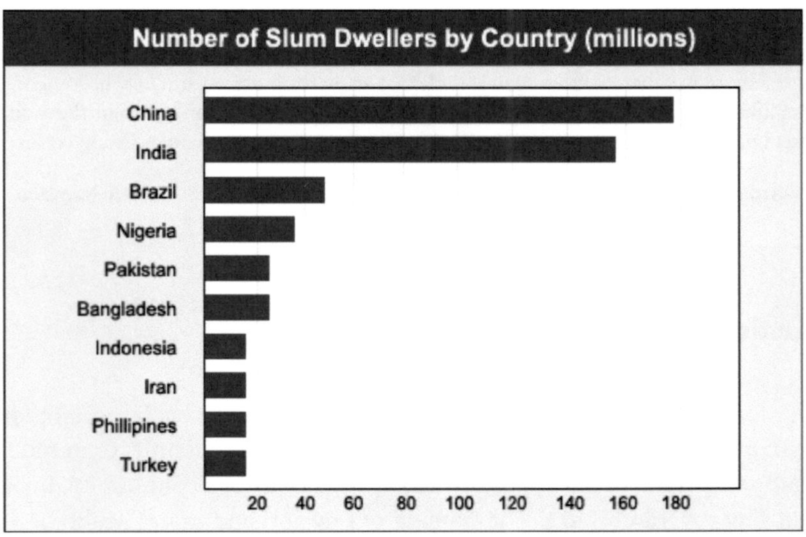

Figure 1. Number of slum dwellers in different countries. Source: [3].

Unfortunately, the twin concepts of *Public housing*, a form of housing tenure in which the property is owned by a central or local government authority, and *Social housing*, an umbrella term referring to rental housing owned and managed by the state, not-for-profit organisations, or a combination of the two, are not popular yet in Nigeria, even though the common goal of both concepts is to provide affordable housing. The consequence is that a vast majority of individuals engage in the construction of personal houses financed through different means including savings and loans, excluding in most cases long-term financing—mortgage financing and mortgage-backed securities—which is still at the rudimentary state of existence in the country at the moment [4].

The housing problem in Nigeria is more serious for the low-income groups whose challenges are complicated by several factors including the ever-increasing cost of construction materials. The aim of this chapter, therefore, is to discuss the potentials of nonconventional building materials derived from agroforestry and municipal wastes in addressing the challenges of affordable housing construction in Nigeria.

2. A brief review of conventional building materials in common use in Nigeria

Rocks and its derivables including stones, granite, gravel, sand and clay and wood, twigs and leaves have been used to construct buildings across Nigeria over the ages. Rock is the longest lasting building material available and is readily available in the country. Wood, a product of trees and sometimes other fibrous plants, is also used for construction purposes when cut or pressed into lumber and timber, such as boards, planks and similar materials. Apart from the aforementioned naturally occurring construction materials, many man-made products are in use including steel used as structural framework for larger buildings such as skyscrapers or as an external surface covering, glass and concrete. The most common form of concrete is Portland cement concrete, which consists of mineral aggregate (generally gravel and sand), Portland cement and water [4].

Concrete hollow blocks (**Figure 2**) and unfired clay bricks (**Figure 3**) are two of the predominant conventional materials for the construction of houses in Nigeria. Concrete hollow blocks manufactured in a factory or on-site made of Portland cement and sand in a ratio of 1:8, and more commonly used in the urban and peri-urban areas, are usually rectangular, often 45 cm

Figure 2. Freshly manufactured concrete hollow blocks.

Figure 3. A typical building constructed using clay bricks.

Figure 4. A typical building under construction using concrete hollow blocks.

wide, 15 cm thick and 30 cm high and yellow/whitish in colour. The hollows tend to run from top to bottom and occupy about two-thirds of the volume of the block [5]. The blocks are usually joined together with mortar in building construction as shown in **Figure 4**.

Clay is relatively cheap, environmental friendly and abundantly available, and there are large deposits of laterite clay in the six geopolitical zones of Nigeria [6]. Hence, clay is widely used for building construction across the country. Unfired clay bricks are particularly prominent in rural housing construction. Such buildings come in different forms such as when the walls are made directly with the mud mixture, when walls are built by stacking air-dried mud bricks, when the walls are made with clay combined with straws to create light clay and when brick walls are made from cement-stabilised bricks, i.e. laterite stabilised with just 5% cement or cement-stabilised interlocking blocks [7].

3. Recent advances in the production of sustainable building materials in Nigeria

One of the challenges associated with the use of concrete hollow blocks in building construction is the relatively high cost of Portland cement. While Nigeria does not produce enough cement domestically to meet demand, imports have been restricted, leading to sharp increases in the price of cement. Also, concrete blocks and unfired clay bricks tend to exhibit brittleness in failure. One way of addressing the challenge of escalating cost of cement is the partial replacement of Portland cement with either with pozzolana or a lime in the production of concrete hollow blocks. The challenge of cost and brittle failure could both be addressed simultaneously by the addition of fibrous materials to delay and control tensile cracking of the matrix [8]. Incidentally, vast quantities of agroforestry, industrial and municipal solid waste are generated which constitute a source of health hazards and environmental pollution in Nigeria. One way of reducing the cost of housing construction is by recycling some of the agroforestry and industrial municipal waste materials either as pozzolanas, fillers or reinforcement materials in innovative cement- and clay-bonded low-cost building materials. Doing so should help in conserving energy and preserving the environment even as the products are expected to exhibit acceptable strength, sound, thermal and durability properties [8]. Examples of such sustainable building materials investigated for diverse applications in Nigeria within the last two decades include the following paragraphs.

3.1. Pozzolans and blended cement

In Nigeria, many agricultural residues are freely available and often treated as waste. Examples include bagasse (waste obtained from sugar cane (*Saccharum officinarum*) processing) and corn (*Zea mays*) cobs. The pozzolanic activities of the ashes of these two residues have been tested and proven [9, 10]. The chemical analysis of bagasse ash is presented in **Table 1**. The combined percentage composition of silica, alumina and ferric oxide exceeds the minimum requirement of 70% for a good pozzolan for the manufacture of blended cement in both bagasse and corn cob ashes. It has been further reported that a corn cob ash (CCA)-blended cement containing not more than 15% of CCA satisfies the NIS 439:2000 and ASTM C 150 requirements for cement. The workability and compressive strength of CCA-blended cement concrete were also investigated [11]. Experimental results showed that CCA-blended cement containing not more than 8% of CCA is suitable for structural concrete works.

Calcium carbide residue is a by-product of oxyacetylene gas welding, a very toxic material whose toxicity can be contained by incorporating it in construction materials [12]. However, it is mostly sent to landfills in Nigeria. It is generally believed that since calcium

Element	Quantity (%)
SiO_2	57.95
Al_2O_3	8.23
Fe_2O_3	3.96
K_2O	2.41
CaO	1.17
Loss in ignition (LOI)	5.00

Table 1. Chemical analysis of bagasse ash.

carbide residue is rich in calcium hydroxide, it behaves like hydrated lime. Hence, calcium carbide residue has also been analysed for potential use as partial replacement for cement in concrete works [13]. A comparison of the results of chemical analysis of the calcium carbide samples tested in Nigeria with those reported by other researchers elsewhere [14, 15] and presented in **Table 2** shows that there were little but insignificant variations in the chemical compositions.

3.2. Natural fibre-reinforced roofing tiles

The major types of roofing materials presently in use in Nigeria include corrugated iron and aluminium sheets, slates and asbestos sheets [16]. While corrugated iron sheets are prone to rusting and can be noisy when it is raining, asbestos roofing sheets are relatively expensive and have been outlawed in many countries due to carcinogenic nature of asbestos fibres. Investigations on the development of alternative roofing materials from wood fibre-cement composites have been going globally on for over 30 years now [17].

A major source of fibre for cement-bonded roofing tile production is rattan, a specialised group of scaly fruited, spiny, climbing palms with flexible stem generally found near water courses in no less than 20 African countries including Nigeria. The stem popularly referred to as 'cane' is generally considered an 'open-access' resource that is readily harvested from wild forests largely for furniture and handicraft production. However, in many rural areas, the cane is split to make ropes used to tie bamboo and stick frame of houses before the frame is plastered with mud [18, 19]. To its advantage, rattan is much easier to harvest, requires simpler tools and is much easier to transport and grows much faster than most tropical wood [19]. The bonding of whole rattan canes, fibres and splits with ordinary Portland cement has been reported [20] as well as the development of methodologies for rattan fibre, particle and

Element	Sources of calcium carbide residue		
	Nigeria [9]	Bahrain [13]	Thailand [14]
	Quantity (%)		
SiO_2	2.69	<0.10	3.4
Al_2O_3	1.78	<1.22	2.6
Fe_2O_3	0.17	0.02	0.3
K_2O	0.10		0.0
CaO	61.41	65.05	51.9
MgO	0.80	0.97	0.5
Na_2O	0.0	Not indicated	0.0
SO_3	Not indicated	0.64	0.2
Loss in ignition (LOI)	32.51	27.92	41.7

Table 2. Chemical analyses of calcium carbide residues of various origins.

strand application in cement-bonded composite manufacture [21–23]. Also, the production of relatively strong and dimensionally stable cement-bonded roofing tiles reinforced with rattan fibres in which cement was partially replaced with carbide waste has been reported [24, 25], and samples are shown in **Figure 5**.

Banana (*Musa acuminata*) fibre has also been investigated as a reinforcement material in cement-bonded composite roofing tile production in Nigeria [26]. The researchers studied the effects of partial replacement of cement with calcium carbide residue and the addition of calcium chloride ($CaCl_2$) on the properties of banana fibre-reinforced roofing tiles. The fibre content was fixed at 3%, while lime replacement levels by mass of cement were 0 (control), 10, 20 and 30%. Two percent of Iron II oxide was added for colouring, while $CaCl_2$ was added at 0 (control) and 3% levels. The density of the roofing tiles ranged between 1.63 and 2.0 g/cm^3. Calcium carbide residue and $Cacl_2$ reduced the density of the composites. There was also a decrease in impact strength as the calcium carbide residue content increased.

Other fibrous materials already investigated and found suitable for cement-bonded composite roofing tile production in Nigeria include bamboo (*Bambusa vulgaris*), coconut husk (*Cocos nucifera*), sugar cane bagasse (*Saccharum officinarum*), raffia palm (*Raphia africana*) and luffa (*Luffa cylindrica*) [27–31]. Samples of bamboo and coconut husk roofing tiles are shown in **Figures 6** and **7a,b**.

In a departure from the use of ordinary Portland cement alone as the binder, the production of clay-cement-sawdust composite roofing tiles using sawdust derived from teak (*Tectona grandis*) and wood ash as a partial replacement for cement was investigated [32]. The basic properties of the lateritic clay material used are presented in **Table 3**, while the properties of the tiles produced are presented in **Table 4**. Partial replacement of ordinary Portland cement with about 10% of wood ash was found very acceptable in producing composite roofing tiles with relatively good bending strength, while partial replacement of cement with between 20 and 30% of wood ash reduced the thermal conductivity of the composite roofing tiles to acceptable levels.

Figure 5. A prototype roof fabricated from rattan-cement composite tiles. Source: [16].

Figure 6. Fibre-reinforced composite roofing tiles installed on a gable-roofed building in Ibadan, Nigeria. Source: [23].

(a)

(b)

Figure 7. (a) Samples of cured coconut husk fibre-reinforced roofing tiles, (b) coconut husk fibre-reinforced roofing tiles installed on a flat roof building. Source: [30].

Properties	Quantity
Natural moisture content (%)	3.18
Liquid limit (%)	71.5
Plastic limit (%)	59.63
Plasticity index (%)	11.87
Percentage passing BS.NO.200 sieve	233.5
Specific gravity	2.78
AASHTO classification	A-7-6
USCS classification	CH
Maximum dry density, MDD (M/m^3)	1.45
Optimum moisture content, OMC (%)	25.25
pH value	6.7
Colour	Reddish brown

Table 3. Basic properties of the laterite used for composite tile production.

Sample composition	Mean density (Kg/m^3)	Mean impact energy (J)	Mean thermal conductivity (W/m.k)	Mean 24-h water absorption (%)
Clay + cement + sawdust (2:3:1) (control)	850	0.95	1.44	32.2
Clay + 90% cement + 10% wood ash + sawdust	960	1.14	1.26	23.8
Clay + 90% cement + 20% wood ash + sawdust	940	0.68	1.18	38.3
Clay + 90% cement + 30% wood ash + sawdust	980	0.64	1.18	43.2

Table 4. Selected properties of laterite-cement composite roofing tiles.

3.3. Cement-bonded composite ceiling boards

Cement-bonded particleboard is a generic term for a panel product manufactured from lignocelluloses primarily in the form of discrete pieces or particles, combined with cement and compacted. Some of the admirable properties of cement-bonded particleboards include relatively high strength-to-weight ratio and durability; high resistance to moisture uptake, nailability and ease of sawing; excellent insulation against noise and heat; and high resistance against fire, insect and fungus attack. The panels do not emit gases or leak harmful chemicals [33]. One of the common nonstructural uses of cement-bonded particleboard in housing construction is for ceiling, i.e. as an overhead interior surface that bounds the upper limit of a

Figure 8. Made-in-Nigeria ceiling board-making machine. Source: [16].

Figure 9. Made-in-Nigeria wood-cement composite ceiling boards after installation. Source: [16].

room, a finished surface concealing the underside of the roof structure. Ceiling boards serve the purposes of thermal insulation, noise reduction/absorption and fire protection.

Ceiling boards have been manufactured from a number of lignocellulosic fibres and wastes in Nigeria including sawdust, waste paper, rattan cane, coconut husk, maize husk, elephant grass (*Pennisetum purpureum*), pawpaw (*Carica papaya*) pseudostem, okra (*Abelmoschus esculentus* L. Moench) and *Cissus populnea*, among others. In many of these investigations, ordinary Portland cement was partially replaced either with rice husk ash (RHA) or calcium carbide residue [34–39]. The development of a low-cost motorised ceiling board-making machine (**Figure 8**) has also been reported [16]. Samples of selected made-in-Nigeria ceiling boards are shown in **Figures 9** and **10**, while a comparison of the basic properties of some of the ceiling boards with asbestos cement ceiling board is presented in **Table 5**.

3.4. Sawdust-reinforced concrete hollow blocks and clay bricks

As earlier noted, concrete hollow blocks and clay bricks remain popular construction materials in Nigeria for walls of single-storey buildings in urban and rural areas, respectively.

Figure 10. Made-in-Nigeria wastepaper-cement composite ceiling boards. Source: [16].

Board properties	Asbestos cement	Sawdust cement	Maize stalk cement	Coconut husk cement	Rattan cement
Density (kg/m^3)	1200	1200	1200	990–1200	1360
Modulus of rupture (N/mm^2)	7.7	2.0–4.0^a 8.7–11.2^b	3.1–5.4	4.3–7.4	7.0
Modulus of elasticity (N/mm^2)	3142	1250–3000^a 3000–4000^b	6409	4253	3350
Water absorption at 24 h (%)	13.8–17.8	18.0^a 28.0^b	25.7	22.1	2.5
Thickness swelling at 24 h (%)	0.21–0.29	0.16^a 0.43^b	0.88	0.6	0.5
Sound absorption ratio (%)	35.0	35.0^c	35.0^c	35.0–40.3^c	
Thermal conductivity (W/Km)	0.38	0.38^c	0.38^c	0.30–0.38^c	

^{a}100% sawdust cement panel
^{b}Three-layered boards incorporating flakes
^{c}Theoretical computed value

Table 5. A comparison between asbestos cement and cement-bonded composite ceiling boards manufactured with selected agroforestry materials.

Though the strength of a concrete hollow block is less than that of fired clay bricks, it is considerably cheaper. In a bid to further reduce the cost of concrete hollow blocks, the possibility of incorporating Nigerian-grown teak (*Tectona grandis*) sawdust and partial replacement of cement with calcium carbide residue and poultry egg shell powder in the mix for the production of low-cost concrete hollow blocks was investigated [30]. Representative samples of 100 × 100 × 100 mm sawdust-reinforced concrete hollow blocks were produced using different percentages of sawdust (20, 25, 30%) and carbide calcium carbide residue (30, 25, 20%) in the mix. These were cured for 28 days, after which 24-h water absorption and thickness swelling as well as compressive strength were determined. The water absorption (6.1–10.3%)

and thickness swelling (1.2–1.9%) were quite acceptable. The maximum compressive strength obtained in the mix containing 20% sawdust and 30% of carbide waste fell within the measured strength values (0.5–1 N/mm^2) of commercially available concrete hollow blocks in Nigeria. The preliminary cost analysis indicated the possibility of a 20% reduction in the production of concrete hollow blocks with the incorporation of sawdust and calcium carbide residue in the production mix.

In another investigation, the possibilities of using sawdust of *Cordia millenii* timber species, calcium carbide residue and poultry eggshell powder for the production of stabilised laterite clay bricks were explored [40]. It is well known that poultry egg contains about 95% calcium carbonate and hence can be used as a supplement for lime. The index properties of the laterite were first determined. A set of 100 mm × 100 mm × 100 mm bricks (**Figure 11**) was then produced using laterite stabilised with 50% of sawdust and 10% ordinary Portland cement (w/w). Based on the combined results of the liquid limit and the particle size analysis, the clay was classified as A-7-6 and CH in accordance with AASHTO and the Unified Soil Classification System, respectively. Its specific gravity of 2.8 fell within the range of 2.6 and 3.4 reported for lateritic soils, while its pH of 6.7 showed that it was slightly acidic. In some of the samples, ordinary Portland cement was partially replaced (w/w) with calcium carbide residue (50%) and eggshell powder (30%). The density, compressive strength and 24-h water absorption of the bricks were determined after 28 days of air drying under a shade and shown in **Table 6.** Both the calcium carbide residue and egg shell powder significantly lowered the density and compressive strength ($p \leq 0.05$). Incorporation of the poultry egg shell powder led to an increase in water absorption of the bricks. Only bricks produced from a mixture of laterite, sawdust and cement met the minimum compressive strength requirement of 1.65 N/mm^2 specified by the Nigerian Building and Road Research Institute for building construction.

3.5. Natural fibre-reinforced floor and wall tiles

Tiles are hard-wearing thin, flat slabs or blocks typically made from porcelain, fired clay or ceramic with a hard glaze or other materials such as glass, metal, cork and stone. They

Figure 11. Samples of nonconventional clay bricks. Source: [33].

http://dx.doi.org/10.5772/intechopen.81662

Brick composition	Density (Kg/m³)	Compressive strength (N/mm²)	24-h water absorption (%)
Sawdust + cement + laterite (control)	950	1.73	54.9
Sawdust + cement + calcium carbide residue + laterite	940	1.59	59.2
Sawdust + cement + egg shell + laterite	930	1.53	51.9
Sawdust + calcium carbide residue + laterite	860	1.39	62.1
Sawdust + egg shell + laterite	830	1.13	63.5

Table 6. Selected properties of the modified clay bricks.

are usually used to form wall and floor coverings and can range from simple square tiles to complex mosaics. Modern concrete tiles are made from the mixture of Portland cement and quarry sand as an aggregate. The major advantage of fibre reinforcement of concrete tiles is to impart additional energy-absorbing capability and to transform a brittle material into a pseudo-ductile material [41]. One of the sources of natural fibres found throughout West Africa and readily available in Nigeria is *Cissus populnea* (Guill. and Perr.) shown in **Figure 12**, a strong woody climbing shrub, typically 8–10 m long and 7.5 cm in diameter. Although greater attention has been paid to the binding properties of *Cissus populnea* [42], less attention has been paid to the fibrous leftover once the gum is extracted, a potential material for fibre-reinforced cementitious composite production.

A manually operated machine for the manufacture of *Cissus* fibre-reinforced cement-bonded composite floor tiles (**Figure 13**) was developed and used to produce 200 (length) × 100 (width) × 10.5 mm (thickness) tiles of different colours—white, red, yellow, blue, green, black and grey (**Figure 14**) [41]. The average mass per unit area was 25.5 kg/m². The impact energy absorption capacity of the composite tiles (2943 Nmm) was relatively lower than that of a typical ceramic tile (3746 Nmm). However, the fibre-reinforced tiles had much lower water absorption of 7% than 12% for a typical ceramic tile at after 24 h of soaking in cold water, while both types of tiles had the same thickness swelling of 3%.

Figure 12. *Cissus populnea* shrubs.

Figure 13. Side view of the tile-making machine. Source: [41].

Figure 14. Samples of cement-bonded *Cissus populnea* fibre-reinforced floor tiles produced. Source: [41].

3.6. Concrete beam reinforcement

Portland cement concrete is relatively strong in compression but weak in tension. The weakness in tension is typically overcome by the use of the conventional rod (steel bar) reinforcement. However, the permeability of concrete allows the entry of other aggressive elements which leads to carbonation and chloride ion attack resulting in corrosion problems, hence the search for alternative reinforcing materials.

The bond strength of the canes of selected rattan species in concrete was investigated [43]. It was reported that the strength could be up to roughly 30% of that of mild steel with concrete. The use of whole canes, fibres and splits of rattan canes as reinforcement materials in concrete slabs was also investigated [20]. It was noted that rattan canes remained stable in strength and dimensions after 36 months of embedment in concrete and that canes could be used as reinforcement in lightly loaded structures. In another study, the suitability of bamboo (*Bambusa vulgaris*) as an alternative reinforcement material in concrete slabs was also investigated and confirmed [44], as well as the suitability of bamboo and rattan cane as reinforcement in concrete struts [45]. It was reported that the average compressive strength of the bamboo- and rattan-reinforced struts was about 78 and 64% of the equivalent steel reinforcement, indicating that both materials could be used as replacements for steel in struts of low load-bearing structures.

(a) (b) (c)

Figure 15. (a) Rattan cane reinforcement, (b) lintel beam casting in progress, (c) the cured concrete lintel beam after 28 days. Source: [23].

The potential use of oil palm (*Elaeis guineensis*) stem as reinforcement in concrete was also confirmed [46] as well as the suitability of rattan cane in reinforced concrete lintel beam fabrication in rural building construction in Nigeria [30]. The concrete lintel beam dimensions were 0.23 m × 0.254 m × 1.37 m, while the mean diameter of the rattan cane stems was 30 mm. They were soaked in water for 24 h to ensure flexibility, manually straightened and cut into 4 long (1300 mm) and 12 short (187 mm) pieces. The short pieces were tied to the long ones as stirrup with binding wires, sun-dried for 2 days to about 15% moisture content and then placed in the formwork for casting three 0.23 m × 0.254 m × 1.37 m concrete lintel beams (**Figures 15a,b**). The cement content of the concrete was partially replaced with 20% RHA to reduce the alkalinity. After 28 days post-casting operation, the formwork was removed, and it was observed that the lintel was straight with no sign of deformation (**Figure 15c**). Hence, it was successfully demonstrated that rattan canes with diameters ranging from 26 to 31.5 mm could be used as alternative to steel in lintel construction with up to 54% reduction in production cost.

4. Conclusion

Article 25 of the Universal Declaration of Human Rights states that "Everyone has the right to a standard of living adequate for the health and well-being of himself and of his family, including food, clothing, housing …". In confronting the daunting challenges of inadequate housing in Nigeria, research efforts in recent years have been geared towards using locally available materials in developing sustainable wall construction, roofing, flooring and ceiling products, as well as concrete slab and strut reinforcement products in residential building construction. Some efforts have also been geared towards developing appropriate facilities for the small-scale manufacturing of fibre-reinforced roofing, flooring and ceiling tiles. However, for these products to become widely acceptable as building components, greater awareness has to be created among all stakeholders in the building construction industry, coupled with the development of appropriate design codes.

Conflict of interest

There is no conflict of interest.

Author details

Abel Olajide Olorunnisola

Address all correspondence to: abelolorunnisola@yahoo.com

Department of Wood Products Engineering, University of Ibdan, Ibadan, Nigeria

References

[1] Adeniyi E. The provision of housing a challenge to urban planning and development in Africa. Ibadan, Nigeria: Nigeria Institute for Social & Economic Research (NISER), reprint series 96. 1996. pp. 701-710

[2] Adejumo AA. Social Housing in Nigeria—An Imminent Mass Housing Revolution? Available from: http://www.nigeriansinamerica.com/articles/3098/1/Understanding-the-Concepts-of-Affordable-and-Social-Housing-in-Nigeria/Page1.html [Accessed: 29-10-2011]

[3] Peterside CS. Policy Foundation for Affordable Housing in Nigeria. Role of the Secondary Mortgage Market. Available from: Nigeriaworld.com [Accessed: 14-10-2003]

[4] Olorunnisola AO. Looking beyond the challenges of affordable housing development in Nigeria: Capitalizing on the engineering opportunities. In: The 18th Engr. Lawrence Oluwemimo Arokodare Memorial Lecture. Ibadan Branch: Nigerian Society of Engineers. p. 40

[5] Mohammed M, Anwar AR. Assessment of structural strength of commercial Sandcrete blocks in Kano state. Nigerian Journal of Technological Development. 2014;**11**(2):39-43

[6] Kolawole FO, Adeniji SA, Idowu AT, Owoseni TO, Ngasoh OF, Soboyejo WO. Corrugated laterite based ceramic roof tile stabilized with cement. International Journal of Engineering and Technology. 2014;**4**(3):2049-3444

[7] Osinubi KJ, Mu'Azu MA. Use of cement and bagasse ash in the modification of lateritic soil. Journal of Applied Science, Engineering and Technology. 2010;**10**:60-65

[8] Ganiron TU. Effect of sawdust as fine aggregate in concrete mixture for building construction. International Journal of Advanced Science & Technology. 2014;**63**:73-82

[9] Osinubi KJ. Lime modification of black cotton soil. Spectrum Journal. 1995:112-122

[10] Adesanya DA, Raheem AA. Development of corn cob ash blended cement. Construction and Building Materials. 2007;**23**(1):347-352

[11] Adesanya DA, Raheem AA. A study of the workability and compressive strength characteristics of corn cob ash blended cement concrete. Construction and Building Materials. 2007;**23**(1):311-317

[12] Semikolennykh AA, Rahleeva AA, Poputnikova TB. Spent carbide waste retains toxicity long term after disposal in caves and mines. Acta Carsologica. 2012;**41**(1):129-137

[13] Agbede IO, Joel M. Effect of waste carbide on the properties of Makurdi shale and burnt bricks made from the admixture. American Journal of Scientific & Industrial Research. 2011;**2**(4):670-673

[14] Al-Khaja WA. Potential use of carbide lime waste as an alternative material to conventional hydrated lime of cement-lime mortars. Engineering Journal of Qatar University. 1992;**5**:57-67

[15] Makarata N, Jaturapitakkul C, Namarak C, Sata V. Effects of binder and $CaCl_2$ on the strength of calcium carbide residue-fly ash concrete. Cement & Concrete Composites. 2011;**33**:436-443

[16] Olorunnisola AO. Harnessing the Forester's harvest for sustainable development. An Inaugural Lecture delivered at the University of Ibadan; 16 May 2013; Publishing House, University of Ibadan. 2013. pp. 102

[17] Savastano H Jr. The use of coir Fibres as reinforcement to Portland cement mortars. In: Sobral HS, editor. Vegetable Plants and Their Fibres as Building Materials: Proceedings of the 2nd International Symposium Sponsored by RILEM and CIB (International Council for Building Research Studies). London: Chapman and Hall; 1990. pp. 150-157

[18] Morakinyo AB. Profiles and Pan-Africa distribution of the rattan species (Calamoideae) recorded in Nigeria. Principe. 1995;**1**(4):197-209

[19] Oteng-Amoako A, Obiri-Darko B. Rattan as sustainable industry in Africa: The need for technological interventions. In: Rattan: Current Research Issues and Prospects for Conservation and Sustainable Development. Non Wood Forest Products, No. 14. Rome: FAO; 2002

[20] Lucas EB, Dahunsi BIO. Short and medium duration tests on rattan cane-reinforced concrete slabs. Lautech Journal of Engineering and Technology. 2004;**2**(1):21-29

[21] Olorunnisola AO, Adefisan OO. Trial production and testing of cement-bonded particleboard from rattan furniture waste. Wood and Fiber Science. 2002;**34**(1):116-124

[22] Olorunnisola AO. Effects of pre-treatment of rattan (*Laccosperma secundiflorum*) on hydration of Portland cement and the development of a new compatibility index. Cement and Concrete Composites. 2008;**30**(1):37-43

[23] Olorunnisola AO, Agrawal SP. Effects of NaOH concentration and fibre content on physico-mechanical properties of cement-bonded rattan fibre composites. Pro Ligno—International Journal of Wood Engineering. 2015;**11**(4):192-198

[24] Olorunnisola AO. Ogundipe: Effects of rice husk and calcium chloride on impact strength and water resistance of rattan-cement composite roofing sheets. In: Proceedings of the 11th Annual Congress of Materials Science and Engineering Society of Nigeria; 22-24 November 2012. Ile-Ife: Centre for Energy Research & Development, Obafemi Awolowo University. pp. 153-155

[25] Olorunnisola AO. Experimental investigations on cement-bonded rattan cane composites. In: Savastano H Jr, Fiorelli J, Francisco Dos Santos S, editors. Sustainable and Nonconventional Construction Materials using Inorganic Bonded Fiber Composites. United Kingdom. eBook ISBN: 9780081020029, Hardcover ISBN: 9780081020012: Elsevier Woodhead Publishing; 2017. pp. 429-444

[26] Olorunnisola AO, Ope-Ogunseitan F. Effects of lime and $CaCl_2$ on impact strength and dimensional stability of banana fibre-reinforced composite roofing tiles. In: Proceedings of the 15th Inorganic-Bonded Fiber Composites Conference (IIBCC); 8-11 November 2016. China: Fuzhou. pp. 76-84

[27] Alade GA, Olutoge FA. Bamboo fibre-reinforce cement used as a roofing sheet. Journal of Civil Engineering Research and Practice. 2004;**1**(2):107-118

[28] Omoniyi TE. Development and evaluation of roofing sheets from bagasse-cement composite [PhD thesis]. Nigeria: Faculty of Technology, University of Ibadan. 2009

[29] Odera RS, Onukwuli OD, Osoka EC. Optimization of the flexural strength of raffia palm fibre-cement composites. Journal of Emerging Trends in Engineering and Applied Sciences. 2011;**2**(2):294-297

[30] Oludele OO. Production and installation of rattan cane-reinforced lintel beam and coconut fibre composite roofing sheets on a sawdust-reinforced sandcrete block building [project report]. Nigeria: Department of Agricultural & Environmental Engineering, University of Ibadan; 2015

[31] Olorunnisola AO, Oshifeso S. Dimensional stability of conventionally cured and CO_2-injected Luffa fibre-reinforced composite roofing tiles. In: Proceedings of the 15th Inorganic-Bonded Fiber Composites Conference (IIBCC); 8-11 November 2016. China: Fuzhou. pp. 153-161

[32] Fadipe EA. Effects of partial replacement of cement with wood ash on physico-mechanical properties of clay-cement-sawdust composite roofing tiles [project report]. Nigeria: Department of Agricultural & Environmental Engineering, University of Ibadan; 2016

[33] Ramirez-Coretti AC, Eckelman CC, Wolfe RW. Inorganic bonded composite wood panel Systems for low-cost housing: A Central American perspective. Forest Products Journal. 1998;**48**(4):62-68

[34] Owonubi JJ, Badejo SO. Industrial wood waste conversion into building materials at FRIN, Ibadan. In: 38th Annual Conference of Science Association of Nigeria; 10-14 December, 2000

[35] Ajayi B. Properties of maize-stalk-based cement-bonded composites. Forest Products Journal. 2004;**56**(6):51-55

[36] Olorunnisola AO, Nzor SN. Production and evaluation of eco-friendly composite ceiling boards from Elephant Grass (*Pennisetum purpureum*) fibre and carbide waste mixed with cement. In: International Conference on "Eco-Materials for Construction: A Pillar to the Green Growth in Africa?"; 10-12 June, 2013; organized by 2iE, at Laico Hotel Ouaga 2000, Ouagadougou, Burkina Faso

[37] Babatunde AO. Production and evaluation of cement-bonded composite ceiling boards reinforced with Pawpaw (*Carica papaya)* fibre [project report]. Nigeria: Department of Agricultural & Environmental Engineering, University of Ibadan; 2014

[38] Lazeez MO. Production and evaluation of cement-bonded composite using okra fibre as a potential reinforcement material [project report]. Nigeria: Department of Agricultural & Environmental Engineering, University of Ibadan; 2014

[39] Akinbosoye TBS. Trial production and evaluation of *Cissus Populnea* and pulp fibre-reinforced composite panels [project report]. Nigeria: Department of Agricultural & Environmental Engineering, University of Ibadan; 2015

[40] Nnate FN. Effects of partial replacement of cement with carbide waste and eggshell on physico-mechanical properties of sawdust-reinforced stabilised bricks [project report]. Nigeria: Department of Agricultural & Environmental Engineering, University of Ibadan; 2015

[41] Amoo K, Adefisan OO, Olorunnisola AO. Development and evaluation of cement-bonded composite tiles reinforced with *Cissus populnea* fibres. International Journal of Composite Materials. 2016;**6**(4):133-139. DOI: 10.5923/j.cmaterials.20160604.06

[42] Iwe MO, Obaje PO, Akpapunam MA. Physicochemical properties of Cissus gum powder extracted with the aid of edible starches. Plant Foods for Human Nutrition. 2004; **59**:161-168

[43] Lucas EB, Dahunsi BIO. Bond strength in concrete of canes from three rattan species. Journal of Applied Science, Engineering and Technology. 2004;**4**(1):1-5

[44] Alade GA, Olutoge FA, Alade A. The durability and mechanical strength properties of bamboo in reinforced concrete. Journal of Applied Science, Engineering and Technology. 2004;**4**(1):35-40

[45] Akinyele JO, Aresa S. Structural characteristics of bamboo and rattan cane reinforced concrete struts. Physical Review & Research International. 2013;**3**(4):602-611

[46] Olutoge FA. Bonding characteristics of oil palm (*Elaeis guineensis*) stem in concrete. Global Journal of Engineering & Technology. 2009;**2**(3):451-458

Chapter 5

Oil Shale Ash Addition Effect in Concrete to Freezing-Thawing

Sabit Oymael, Alper Bideci and Özlem Sallı Bideci

Additional information is available at the end of the chapter

http://dx.doi.org/10.5772/intechopen.79285

Abstract

Modifying the properties of cement by including mineral-based additives into cement ensures benefits such as preventing environmental hazards of waste additives and providing increase in cement amount. Based on these main considerations, as a part of the study in which contribution of ash, obtained as a result of the burning of oil shale rocks in fluidized bed thermal plants, to Portland cement (PC) in 15% ratio is chemically proven, this study is performed in order to understand the direction and severity of the effects that oil shale ash (OSA) might have on the properties of cements. For this purpose, freezing-thawing experiments were performed on cement samples produced with PC 42.5 cement, which has 0, 10, 20, and 30% of OSA additions. It is desired to make a conclusion by finding weight loss factor (AF_w) and relative elasticity module (E) loss related with press and ultrasonic test methods based on given experiments. As a result, it was observed that durability of concretes obtained by addition of oil shale rocks in 15% ratio to PC 42.5 cements against freezing/thawing effects is greater than that of PC 42.5 control sample.

Keywords: Portland cement, oil shale ash, concretes, freezing-thawing, resistance

1. Introduction

Concrete should not be damaged by decomposing due to the compressive strength of the ice formed in it. If compression stress in ice that is the resultant of freezing, reaches greater values than original compressive strength of ice, the ice turns to water by decomposing and the negative effect of freezing are removed [1]. Depending on the freezing temperature, there can be increases in the volume of water frozen inside the concrete by up to 12%, and this causes wearing in concrete by creating internal stresses.

IntechOpen

© 2019 The Author(s). Licensee IntechOpen. This chapter is distributed under the terms of the Creative Commons Attribution License (http://creativecommons.org/licenses/by/3.0), which permits unrestricted use, distribution, and reproduction in any medium, provided the original work is properly cited.

Frost occurrence initially starts at big pores in the concrete and then builds up at small pores. The freezing degree of water decreases as the size of gaps inside the material is reduced. Finally the water inside capillary gel gaps gets frozen. This happens as salt crystals cannot be formed in gel gaps that are in microns size, and the freezing of water in these zones cannot be prevented. Temperature value should be under −78°C for water to be frozen inside the mentioned small capillary pores. Here, the major effects are the methods and techniques and materials used during the concrete manufacture and also the fact of adding or not adding additive substances inside cement qualification and the type and amount of the additive substance, if added.

Oil shale is formed by the simultaneous sedimentation of granule mineral fragments and the rotting organisms of low-ranked animals and plants and also is a kind of marlite containing combustible organisms. Several researchers studied about the addition of oil shale ash into cement. Xiang-penga et al. studied the effects of burning temperature on the reactivity of oil shale ash [2]. Al-Hasan [3] searched to replace cement with ash, even with small amounts, and he found it to be an effective way to improve thermal conductivity of concrete mixtures. Further, the higher was the level of cement replacement by oil shale ash, the lower was the compressive strength [3]. Oymael [4] obtained at 700°C by burning oil shale and then added to cement at the ratio of 15 and 30% by its weight. The optimum pozzolanic characteristics and performances were provided by the mixture containing 15% ashes [4]. Smadi and Haddad [5] observed that oil shale ash replacement of cement, sand, or both by about 10% (by weight) yielded the optimum compressive strength. Moreover, its replacement of cement by up to 30% was not reducing its compressive strength [5]. Raado et al. [6] focused on the use of oil shale ash for low-strength concrete. Two main types of oil shale ash and their mixes were tested, and they discovered that expansion and water resistance tests showed the content of CFB ash in OSA binders to increase and water resistance was developed and expansion disappeared [6].

The purpose of this study is to reveal the attribute of concretes made up with pozzolanic oil shale ash added Portland cements (PC) against freezing-thawing effects. Based on these considerations, this study investigates the relationship between compressive strength values and material loss that would happen due to frost under freezing-thawing forcing in air-entrained and non-air-entrained (10 cm^3) concrete samples that are made with PC 42.5 cements as per TSE CEN/TS12390-9 [7].

2. Materials and method

2.1. Cement

PC 42.5 cement manufactured in Gaziantep Cement Factory and brought as bulk to Elazığ El-Beton Prefabricated Concrete Building Elements Factory was used. A chemical and physical property of the mentioned cement is given in **Table 1**.

2.2. Oil shale ash

Oil shale rocks of Ankara-Çayırhan region are broken to aggregate size and grinded to fineness of cement. By firing this powdered material in laboratory-type ovens, at the rate of 125

Chemical properties (%)		Physical properties		
CaO	63.92	Specific gravity (g/cm³)		3.11
SiO_2	19.57	Blaine fineness (cm²/g)		3510
Al_2O_3	5.72	Compressive strength (N/mm²)	7 days	41.9 (lim: 31.5) N/mm²
Fe_2O_3	3.69			
MgO	1.17			
SO_3	3.19		28 days	54 N/mm²
Unidentified	0.24			
(Na_2O + K_2O)	0.15 + 0.62			
Chloride	≤0.1			
Loss on ignition	1.73			
C_3S	58.68			
C_2S	11.83			
C_3A	8.91			
C_4AF	11.23			

Table 1. Properties of Portland cement.

capsules for 1.5 h at 500, 600, 700, 800, 900, and 1030°C, oil shale ash is obtained. Obtaining oil shale ash as waste substance in industry can be made by firing/combusting oil shale rocks as combustion of low calorie lignite in fluidized bed thermal plants. If the most suitable values of ash were obtained by firing at 700°C [8] in the study, this temperature at fluidized bed thermal plants would be needed to determine and control the rate of material to be transferred into the oven and air circulation (**Table 2**).

The values obtained in activity experiments performed according to TS25/T1 [9] and TS EN 197-1 [10] are given in **Table 3**. In the activity experiment performed by lime, the amount of ash is found by the formula of (specific weight of ash/specific weight of lime) × 300 and that

Firing temperature (°C)	SiO_2 (%)	Al_2O_3 (%)	Fe_2O_3 (%)	CaO (%)	MgO (%)	SO_3 (%)	Loss on ignition (%)	Insoluble residue (%)	Unidentified (%)
500	29.26	10.15	4.65	36.13	6.82	2.30	3.24	4.21	3.24
600	28.6	8.97	4.47	32.80	8.04	3.79	6.47	3.88	2.98
700	39.32	7.80	4.20	26.40	9.26	5.21	3.82	1.28	1.11
800	28.4	9.40	4.40	31.18	7.70	2.27	8.82	4.30	3.53
900	20.25	13.00	4.35	29.00	6.63	4.25	15.4	5.37	1.87
1030	14.54	17.45	4.90	26.82	5.35	6.23	19.84	3.94	0.93

Table 2. Chemical change of oil shale ash depending on firing temperature.

Firing temperature (°C)	Lime activity (TS25/T1) [9]		Cement activity (TS EN 197-1) [10]					
	Compressive strength (N/mm²)	Flexural strength (N/mm²)	Compressive strength (N/mm²)			Flexural strength (N/mm²)		
	7 days		2 days	7 days	28 days	2 days	7 days	28 days
500	—	—	—	—	—	—	—	—
600	9.1	3.0	—	—	—	—	—	—
700	13.0	3.6	15.3	23.9	29.5	3	4.5	8.1
800	10.1	3.0	11.1	22.2	23.3	2.7	4.3	5.8
900	8.7	2.6	—	—	—	—	—	—
1030	4.2	1.3	—	—	—	—	—	—

Table 3. Activity experiments values and results of OSAs.

amount of ash is mixed with 150 g of lime, 239 g of water, and 1350 g of standard sand and standard mortar is obtained. However, in cement experiment, mortar is obtained by mixing the ash, for which the amount is firstly found by the formula of (*147 × specific weight of ash)/ specific weight of cement, with 293 g of Portland cement, 1350 g of standard sand and water in an amount of 0.5 (Portland cement + ash). For experiment of activity with lime, after the mortar mixtures were given shape in 40 × 40 × 160 mm of standard molds and kept under room temperature for 24 h, the open surfaces of the molds were covered with glass and their surroundings were made airtight with paraffin, and they were kept in an oven for 6 days at 55 ± 2°C. Those mortar samples were cured in water for 2, 7, and 28 days for experiment of activity with cement. Activity experiment values and results are given in **Table 3**.

The mean numeric relationship between addition ratios of OSA at the temperatures is obtained, and its fineness is given in **Table 4**. The ash at 700°C, which was sieved through mentioned sieves, was seen to give the optimum fineness values. Blaine fineness for ashes obtained as a result of firing was found to be 3180–3500 cm²/g without performing any additional grinding process.

Firing temperature (°C)	200 μm Sieve (%)	90 μm Sieve (%)	Blaine fineness (cm²/g) (average)
PC-Control	0.5	6.7	3180–3500
500	0.7	5.9	
600	0.9	4.0	
700	0.6	4.0	
800	1.0	9.0	
900	1.8	9.8	
1030	1.1	4.7	

Table 4. Relationship between firing temperature and fineness.

Particle group (mm)	Volume spec, weight (DYK)	Water absorption (%)	Present humidity (%)	Density (kg/m³)		Abrasion (%) (Tuvennan)	
				Loose	Tight	100 cycles	500 cycles
0–8	2.69 (2.59)	1.15	0.7	1650	1755	6.98	18.63
8–16	2.69	1.2	0.62	1580	1720		
16–19	2.69	1.2	0.44	1455	1650		

DYK, saturated dried surface.

Table 5. Granulometry of Tuvennan aggregate.

The specific weight of oil shale ash had shown a relative increase from 2.26 to 2.77 g/cm^3 based on the firing temperatures at 500, 600, 700, 800, 900, and 1030°C. As this is lower than the specific weight of Portland cement, this causes a partial increase in the volume of the cement.

2.3. Standard sand

All of the mortar mixtures in the study were used in the standard sand produced by the Pınarhisar Cement Plant which is in compliance with TS EN 196-21 [11].

2.4. Aggregate

In cement experiments, washed Palu aggregate used in Fırat University Elazığ Vocational School Laboratories was used. The maximum aggregate size was selected by using one-fifth of the experiment sample molds (19 mm) as a basis. Three sizes (0–8, 8–16, 16–19 mm) of aggregates were used. Properties of granulometric structured aggregate in accordance with TS 706 EN 12620 + A1 [12] are given in **Table 5**.

3. Experiment method and results

In the study, PC 42.5 Portland cement and aggregate were used. In the design of concrete mixture, water amounts were specified based on DYK (saturated dried surface) aggregate (**Table 7**). It was taken as a basis to have 105–110% dispersion in mixtures and to add liquid air-entraining admixture chemical substance specified in TS EN 934-4 [13] to water amount in air-entrained samples. In the experiment, OSA in 0, 10, 20, and 30% ratios were added to PC 42.5 cement, and in order to meet the predicted dispersion values, corrections in water amounts were made. It was observed that in concrete mixtures as OSA, percentage ratio is increased, the water requirement is also increased, and dispersion value is decreased. This fact reflects on the water amount in concrete mixture design and accordingly on the fresh concrete density. Non-air-entrained concrete mixture design is given in **Table 6**, and air-entrained concrete mixture design is given in **Table 7**.

For the study, cubic samples of 10 cm size were prepared in two groups. The first group includes air-entrained concrete samples, whereas the second group includes non-air-entrained (normal) concrete samples. Each group of samples was exposed to three experiments. First of

Mixture	0% PC-Control	10%	20%	30%
Cement (kg)	360	324	288	252
OSA (kg)	—	36	72	108
Fine aggregate (0–8 mm)	663	663	663	663
Coarse aggregate (8–16 mm)	807	807	807	807
Coarse aggregate (16–19 mm)	346	346	346	346
*Water + HK** (kg)	202 + 1.2	209 + 1.2	216 + 1.3	245 + 1.5
Water/binder (%)	0.52	0.54	0.6	0.68
Dispersion (mm)	115	116	114	116
Fresh concrete density (kg/m^3)	2378	2385	2392	2421

*Net water (aggregate DYK).
**Air is predicted, included to the fresh concrete weight.

Table 6. Non-air-entrained concrete mixture design.

them was the experiment on determination of weight loss factor (AF_w), the second one was the experiment on press compressive strength loss factor, and the third one was the experiment on determination of dynamic elasticity module.

In the experiment on determination of weight loss factor (AF_w) in freezing/thawing experiment, following 28 days of cure period in water at average (+22°C) temperature, the samples for compressive strength were broken by press. Other samples were exposed to freezing-thawing experiments at Elazığ El-Beton Concrete Prefabricated Building Elements Factory Laboratory.

Experiment samples were dried at 110 ± 5°C in an oven until they reached constant weight, then they were taken from the oven and cooled in a desiccator until they reached room temperature. Then, they were weighed with 0.1 g precision, and W_0 is found. Those experiment

Mixture	0% PC-Control	10%	20%	30%
Cement (kg)	360	324	288	252
OSA (kg)	—	36	72	108
Fine aggregate (0–8 mm)	663	663	663	663
Coarse aggregate (8–16 mm)	807	807	807	807
Coarse aggregate (16–19 mm)	346	346	346	346
*Water + HK** (kg)	207 + 1.8	214 + 1.8	236 + 1.8	243 + 1.8
Water/binder (%)	0.58	0.6	0.66	0.68
Dispersion (mm)	105	111	115	116
Fresh concrete density (kg/m^3)	2384	2391	2413	2441

*Net water (aggregate DYK).
**HK: air entraining ratio 0.5%.

Table 7. Air-entrained concrete mixture design.

samples were saturated with water under normal atmospheric conditions and were frozen in the freezer. The cooling rate of the freezer was adjusted as it would decrease to −20°C within 4 h. After it was observed that the temperature of the freezer decreased to −20°C, the experiment samples that were kept at this temperature for about 2 h were taken out at the end of the period. By immersing them in water as they would be completely under water and keeping them as such for 2 h, it was ensured that the ice completely melted. At the end of the freezing-thawing processes repeated for 25 times like this, after the experiment samples that reached constant weight in the oven at 110 ± 5°C were cooled in a desiccator, they were weighed with 0.1 g precision (W_n). As can be seen in **Tables 8** and **9**, the decrease in mass, weight loss factor (AF_w), caused by the portions that break and leave from the samples due to freezing-thawing effect is calculated by the following formula (Eq. (1)).

$$AF_w = 1 - W_n/W_0 \times 100 \tag{1}$$

Here AF_w: weight loss factor (%); W_0: mean compressive strength of samples before frost experiment (N/mm²); W_n: arithmetic mean of compressive strength of concrete after freezing (N/mm²).

According to weight loss and compressive strength

Mixture (air addition (HK))	Number of samples	According to weight loss			According to compressive strength		
		Mean initial weight (g) (W_0)	Mean final weight (g) (W_n)	$AF_w = 1 - W_n/W_0 \times 100$(%)	Initial compressive strength (N/mm²) (f_0)	Final compressive strength (N/mm²) (f_n)	$DF_f = 1 - f_n/f_0 \times 100$ (%)
0% HK, PC-Control	6	2.32	2.303	0.7	20.1	16	20.3
10% + HK	6	2.28	2.254	1	13.5	10.1	25.1
20% + HK	6	2.27	2.238	1.4	12.7	9.5	23.6
30% + HK	6	2.18	2.139	2	11.3	8.3	26.5

According to relative dynamic elasticity module

Mixture (air addition (HK))	Number of samples	Unit weight (kg/dm³)	Number of freezing–thawing cycles and ultrasonic measurements						$E = 10^6 \times v^2 \times \Delta/g$	$E = 10^6 \times v^2 \times \Delta/g$	$E = 10^6 \times v^2 \times \Delta/g$
			V_0 (m/s)	V_n (m/s)				V_n (m/s)			
			0	5	10	15	20	25	E_0 (N/mm²)	E_n (N/mm²)	EMF (%)
0% HK, PC-Control	6	2.32	4.44	4.16	4.14	4.11	4.08	3.9	466,213	35,976	23
10% + HK	6	2.28	4.30	4.06	4.06	4.0	3.86	3.61	429,737	302,886	29
20% + HK	6	2.27	4.20	3.93	3.92	3.79	3.63	3.45	408,183	275,419	33
30% + HK	6	2.18	4.00	3.87	3.86	3.86	3.80	3.15	355,555	225,000	38

Table 8. Calculation table of freezing-thawing experiment for air-entrained samples.

According to weight loss and compressive strength

Mixture	Number of samples	According to weight loss			According to compressive strength		
		Mean initial weight (g) (W_0)	Mean final weight (g) (W_n)	$AF_w = 1 - W_n/W_0 \times 100(\%)$	Initial compressive strength (N/mm²) (f_0)	Final compressive strength (N/mm²) (f_n)	$DF_f = 1 - f_n/f_0 \times 100$ (%)
0% PC-Control	6	2.408	2.383	1	22.5	16.3	27.5
10%	6	2.443	2.411	1.3	20.6	14.3	30.5
20%	6	2.369	2.319	2.1	15.6	9.1	41.6
30%	6	2.293	2.233	2.6	11	6.1	39

According to relative dynamic elasticity module

Mixture	Number of samples	Unit weight (kg/dm³)	Number of freezing–thawing cycles and ultrasonic measurements						$E = 10^6 \times v^2 \times \Delta/g$	$E = 10^6 \times v^2 \times \Delta/g$	$E = 10^6 \times v^2 \times \Delta/g$
			V_0 (m/s)	V_n (m/s)				V_n (m/s)			
			0	5	10	15	20	25	E_0 (N/mm²)	E_n (N/mm²)	EMF (%)
0% PC-Control	6	2.408	4.58	4.32	4.32	4.25	4.18	4.13	514,894	418,685	19
10%	6	2.443	4.46	4.27	4.25	4.06	3.86	3.83	495,363	365,301	26
20%	6	2.369	4.40	4.23	4.11	4.04	3.47	3.38	467,521	275,885	41
30%	6	2.293	4.31	4.03	4.00	3.93	3.27	3.11	434,199	226,076	48

Table 9. Calculation table of freezing-thawing experiment for non-air-entrained samples.

In the second part of the experiment, a study taking losses at compressive strength as a basis was performed. Decrease factor of compressive strength (DF_f) is calculated by the following formula by finding mean compressive strength of samples after experiment (f_n) for which mean compressive strength values at press (f_0) for equivalents of original samples of experiment were found before freezing-thawing experiment (Eq. (2)).

$$DF_f = 1 - f_n/f_0 \times 100 \quad (2)$$

Here DF_f = decrease factor of compressive strength (%); f_0 = mean compressive strength of sample before experiment (N/mm²); f_n = mean compressive strength of sample after freezing (N/mm²).

In the third part of the experiment, ultrasonic pulse velocity (v = km/s) of samples exposed to freezing-thawing were found before experiment (DYK) and during experiment at the end of 5, 10, 15, 20, and 25 cycles. Based on these values, the dynamic elasticity modules (E) of samples are calculated by the following formula (Eq. (3)).

$$E = 106 \times v^2 \times \Delta/g. \quad (3)$$

Here E = relative dynamic elasticity module (N/mm^2); v = ultrasonic pulse velocity (km/s) (V_0 = values before experiment; V_n = values after experiment); Δ = unit weight of concrete (kg/dm^3); g = gravitational acceleration (g/cm^2).

In experiments conducted as per TSE CEN/TS 12390-9 [7] principles, comparative values of after frost losses of weight and compressive strength and relative dynamic elasticity module of air-entrained and non-air-entrained samples which were exposed to freezing-thawing in water are shown in **Figures 1–3** and **Tables 8** and **9**.

When AF_w values after 25 cycles of freezing–thawing in air-entrained concrete samples are examined, it is observed that for those with 10% addition, partial frost loss is 0.30 times less than the control sample and for those with 20 and 30% addition, partial frost loss is 2.42 and 2.85 times less than the control sample, respectively. In general, those values are 27% less than those with no air entrainment.

For air-entrained and non-air-entrained samples, decrease factor of compressive strength (DFf) as given in **Tables 8** and **9** (if pressure loss value at 20% OSA ratio is excepted), it is seen that there is a linear relation between pressure loss value and addition ratio, and as the addition ratio increases, the pressure loss also increases. Both in added or not added samples, considering that the pressure loss due to freezing-thawing experiments (DF_f) cannot be more than 20% [14], it is understood that 15% of the addition ratio would be more suitable.

In evaluations performed according to relative elasticity module (E) in air-entrained samples, the value for 0% added (control) sample was 23%, whereas it was 29, 33, and 38% for those with 0, 20, and 30% addition, respectively. After 25 cycles, if maximum 0.30 [14] limit is considered for decrease in relative dynamic elasticity module of concrete samples, it can be suggested that 15% added one would be suitable. This result is also valid for non-air-entrained samples. Relative dynamic elasticity module values are given in **Figure 3**.

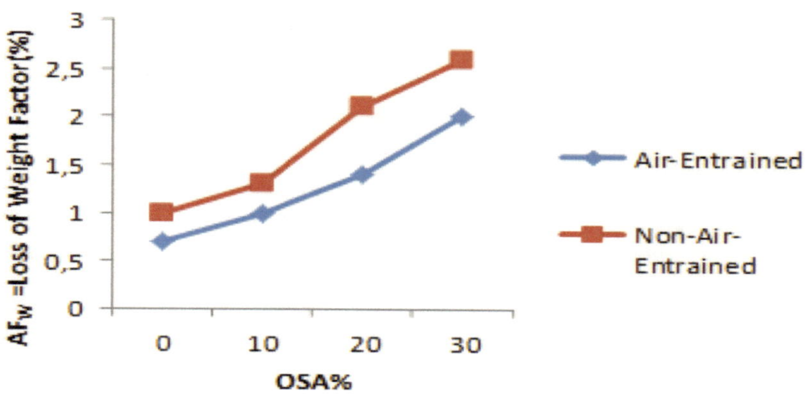

Figure 1. Relationship between weight loss and OSA percentage in samples.

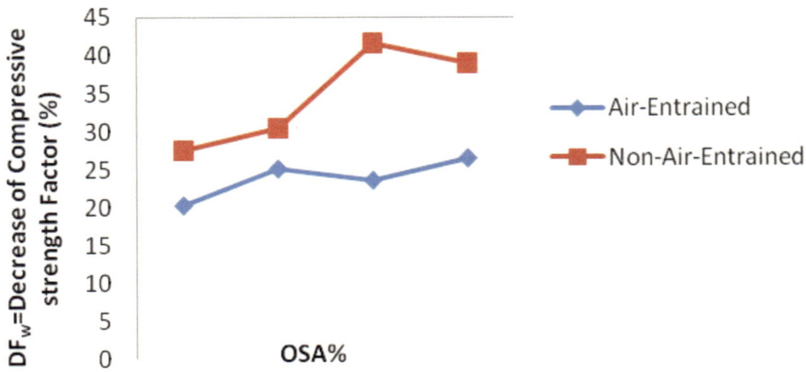

Figure 2. Relationship between compressive strength factor and addition percentage in air-entrained and non-air-entrained samples.

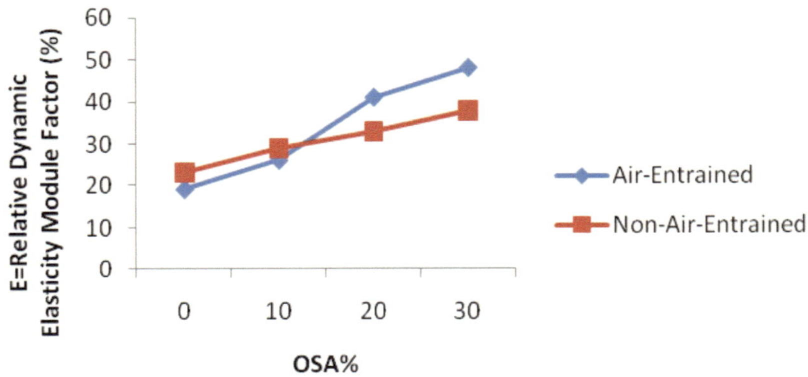

Figure 3. Relationship between relative dynamic elasticity module and addition percentage in air-entrained and non-air-entrained samples.

4. Conclusion and recommendations

By firing oil shale rocks at different temperatures, oil shale ash (OSA) was obtained. With a series of physical and chemical tests performed, in order to obtain the maximum performance, the ratio of the ash to be added to PC (Portland cement) was suggested. As a conclusion, it was determined that all chemical test values of the cement obtained as a result of addition of OSA, which was obtained by firing at 700 °C, to PC (Portland cement) with 15% ratio (when ignition is considered separately, which was also 0.65% greater) were within the limits of TS EN 197-1 [15].

In this study which was performed to determine the resistance of concrete produced with 15% OSA-added cement to freezing-thawing effects, air-entrained and non-air-entrained concretes were produced and weight loss factor, press strength factor, and relative dynamic elasticity module was investigated. As a result of the performed studies, it was observed that optimum results could be taken from concrete samples produced by addition of 15% of OSA to PC 42.5 cement.

In examinations about weight loss of air-entrained and non-air-entrained samples, it was seen that by freezing-thawing cycles there occurred more breaks and weight decreases for air-entrained samples. This fact demonstrates the protection of concrete against frost by air-entraining substance.

In experiments of press compressive strength, the compressive strength (loss) factor (DF_f) found by interpolation for those with 15% addition was 24.3% for non-air-entrained samples and 20.7% for air-entrained samples. Therefore, for resistance to freezing–thawing, the air entraining (HK) chemical substance that is added as 0.5% of the cement weight decreased the strength (loss) factor (DF_f) by about 14%. These values comply with the condition that compressive strength loss in freezing-thawing experiments should be not more than 20%.

Although frost resistances of air-entrained concrete samples are more than that of non-air-entrained concrete samples, their compressive strength is less. The alterations in compressive strength might change based on amount and type of air entrainment.

According to relative dynamic elasticity module (E) examination, for those with 15% addition the value found by interpolation is 33% for non-air-entrained samples and 31% for air-entrained samples. When this result is compared with the limit value for which the relative dynamic elasticity module loss would be maximum 30%, the suitability of air-entrained 15% OSA + HK samples (although the limit is a little bit forced) was observed. Relative dynamic elasticity module (E) is an indicator of durability in concretes.

In air-entrained samples, there is an inverse relation between OSA addition ratios and ultrasonic measurement values that are basis for compressive strength. As pozzolanic addition ratio increases in concrete structure, compactness increases, but pressure loss does not increase because there is a relation between water/binder ratio and compressive strength of air-entraining substance. As air-entraining substance makes closed air gaps in the inner structure of the concrete, the compressive strength is negatively affected. By decreasing Water/Binder ratio in concrete, the number and size of capillary small channels might be decreased. So, less water would enter inside of concrete by capillaries. This fact also shows itself by decreasing permeability.

In places where freezing–thawing is frequently happening, concrete with 15% OSA addition can be used. It is preferred to use mentioned concrete in detail elements, not in structural building elements that are exposed to flexural strength. This includes construction of pavement concrete and so on.

As a conclusion, as 15% OSA-added concretes show performance over desired limit values against freezing-thawing, they can be used in engineering structures of airports, highways, railways and in building elements in buildings as walls, floorings, columns and beams, and in every kind of coatings. So, by adding OSA, which was a waste product, to Portland cement, an economic benefit would be obtained.

Acknowledgements

This manuscript was presented in the International Oil Shale Symposium 2016 in Tallinn, Estonia, but it was not issued.

Author details

Sabit Oymael[1], Alper Bideci[2*] and Özlem Sallı Bideci[2]

*Address all correspondence to: alperbideci@duzce.edu.tr

1 Faculty of Engineering and Architecture, Architecture Department, İstanbul Arel University, İstanbul, Turkey

2 Faculty of Art, Design and Architecture, Architecture Department, Düzce University, Düzce, Turkey

References

[1] Postacıoğlu B. Beton: Agregalar, Beton (Cilt:II). İstanbul: Matbaa Teknisyenleri Basımevi; 1987 (in Turkish)

[2] Xiang-penga F, Xue-lianb N, Xueb B, Xiao-mingb L, Heng-hua S. Cementing properties of oil shale ash. Journal of China University of Mining and Technology. 2007;**17**(4):498-502. DOI: 10.1016/S1006-1266(07)60133-3

[3] Al-Hasan M. Behavior of concrete made using oil shale ash and cement mixtures. Estonian Academy Publishers. 2006;**23**(2):135-143

[4] Oymael S. Behavior of concrete made using oil shale ash and cement mixtures. Estonian Academy Publishers. 2007;**24**(1):45-58

[5] Smadi MM, Haddad RM. The use of oil shale ash in portland cement concrete. Cement & Concrete Composites. 2003;**25**:43-50. DOI: 10.1016/S0958-9465(01)00054-3

[6] Raado L, Hain T, Liisma E, Kuusik R. Composition and properties of oil shale ash concrete. Estonian Academy Publishers. 2014;**31**(2):147-160. DOI: 10.3176/oil.2014.2.05

[7] TSE CEN/TS12390-9. Test Hardened Concrete—Part 9: Freeze-Thaw Resistance—Scaling, Turkish Standard, Ankara. 2012

[8] Oymael S. The Suitability of oil shale ash as an admixture in cement and concrete [PhD thesis]. Graudate School of Natural and Applied Sciences Department of Construction Education, Fırat University; 1995

[9] TS25/T1. Naturel Pozzalan (Trass) for Use in Cement and Concrete-Definitions, Requirements and Confirmity Criteria. Ankara: Turkish Standard; 2011

[10] TS EN 197-1. Cement—Part 1: Compositions and Conformity Criteria For Common Cements. 2002

[11] TS EN 196-2. Rilem-Cembureau Standards Sand. Ankara: Turkish Standard; 2002

http://dx.doi.org/10.5772/intechopen.79285

[12] TS706 EN 12620+A1. Aggregates for Concrete. Ankara: Turkish Standard; 2009

[13] TS EN 934-4. Air Entraining Admixtures for Concrete. Ankara: Turkish Standard; 2002

[14] Yılmaz K. "Yapı Malzemesi ve Beton Teknolojisi", Sakarya Mühendislik Mimarlık Matbası Ders Kitapları Sayı: 67/3. İstanbul (in Turkish)

[15] TS EN 197-1. Cement-Blended Cement. Ankara: Turkish Standard; 2002

Chapter 6

Connection Node Design and Performance Optimization of Girder Truss

Ze-li Que, Liu-liu Zhang, Fei-bin Wang, Yi-fan Gao, Wei-zhen Cai, Xin-meng Wang and Chang-tong Mei

Additional information is available at the end of the chapter

http://dx.doi.org/10.5772/intechopen.80992

Abstract

Girder truss is a kind of high-performance truss, which is combined with some single trusses by connectors. It is the common structural form of the key-bearing node in the modern wood structure floor and roof structure system. With the development of the sponge city and green building in China, girder truss is widely used in wood structure buildings and re-roofing project for its lightweight, high strength, good seismic performance, simple construction, design flexibility, and other excellent characteristics. Since the stress environment of girder truss is more complicated than single wood truss, the wood girder truss needs higher bearing capacity. This chapter emphatically provides a theoretical basis for practical engineering and mainly introduces a new type of girder truss connected with different diameters of wood dowels. The deformation of each node in the static loading process is measured in situ and continuously by using the self-designed loading device and the advanced measuring system. Research contents include the increasing effect of girder truss than single truss and influence of different connection modes on the mechanical properties of girder trusses. We can restore the mechanical properties and failure mechanism from the two aspects of phenomena and mechanism by comparing the test results.

Keywords: girder truss, static load test, carrying capacity, anti-deformation capability, connection node design

1. Introduction

For many years, the construction industry has been called the 'big energy-consuming households' in China with the industry and transportation. Building energy consumption accounts

IntechOpen

© 2018 The Author(s). Licensee IntechOpen. This chapter is distributed under the terms of the Creative Commons Attribution License (http://creativecommons.org/licenses/by/3.0), which permits unrestricted use, distribution, and reproduction in any medium, provided the original work is properly cited.

for one-third of the total energy consumption of the whole society, which is 2–3 times higher than other countries under the same climatic conditions (see Ref. [1]). This is mainly due to the traditional Chinese construction materials, such as steel, cement, clay bricks, etc. These materials not only waste a lot of natural resources but also cause pollution to the ecological environment. Therefore, the use of green building materials has become the key to energy conservation and emission reduction in the construction industry. On March 5, 2016, Premier Li Keqiang clearly stated in the "Government Work Report" of the Fourth Session of the 12th National People's Congress that the focus of the work in the field of housing construction is to further promote new urbanization and vigorously develop green buildings and building materials (see Ref. [2]). A very important direction for the development of green buildings is wood structure architecture (see Ref. [3]). A large number of studies have also shown that wood structure is better able to save energy and reduce emissions than other structural forms (see Ref. [4–6]). As one of the major trends in modern architecture, building energy efficiency can be beneficial to the growth of national economy as well as help protect the ecologic environment (see Ref. [6, 7]). Besides, timber structure building has a strong prefabrication because most of its components are processed in the factory. The study on the components is very crucial because the components are closely related to the safety and energy efficiency of timber structure.

As the important parts of timber structure building, floor system and roof are usually divided into two kinds of systems, the traditional grille-rafter system and light wood truss structure system, and the latter is more widely used. With the development of light wood structure in China, the application prospect of light wood truss in modern wood structure in China will be more and more broad. The girder truss is composed of several pieces of single light wood truss by connectors and commonly used in key parts of the roof or floor system in modern wood structure buildings and re-roofing projects. For the floor and roof system of modern timber structure, the key joints have suffered both the upper uniform load and the concentrated load from other trusses that connected with them. So the force circumstance is so complicated that the ordinary single wood truss can hardly bear (see Ref. [8–10]). A common solution in practical engineering is to increase the cross-sectional area of the member by combining a plurality of ordinary light wood trusses as a structural member to obtain a greater load carrying capacity (as shown in **Figure 1**). The form of the girder truss can be easily obtained and conforms to the developing trend of industrialization and modularization of buildings. Besides, some long-span and cantilevered structure is emerged with the development of modern timber structure building, which needs the wood truss with a higher carrying capacity. Emerging as the times require, the girder truss appears with a higher carrying capacity, greater span, and wider range of use, compared to the single wood truss. At present, the study on the single truss is very mature (see Ref. [11–13]), but few studies have been done on the girder truss. In most practical engineering projects, many builders work mostly depending on their experiences without any reliable standard, which will bring some potential safety issues. Girder truss is usually connected with nail and bolt, which is easy corrosion, and the mechanical property will reduce under the fire resistance circumstance. Therefore, this chapter has designed a new type of connection method that is used for girder truss (as shown in **Figure 2**). Wood dowel connector is not easy to be rusted and its mechanical property will not be reduced rapidly under fire resistance circumstance. Besides, wood-made connectors

Figure 1. Application of girder truss in building structure.

can increase the ductility of the connected components. Thus, the performance of girder truss is improved. The connection node of the wood structure is also related to the bearing capacity and normal use of the whole building in the future. Therefore, it is very meaningful to study the connection nodes of wood structure (see Ref. [14]).

2. New connection node design of girder truss

The connection modes of girder truss are currently relatively simple. The domestic connection method is recommended in the technical specification for light wood trusses (JGJ/T 265-2012), but the following problems remain when it is connected with nails.

1. Processing is complicated. The girder truss needs to turn over the truss constantly during processing. Nailing in different parts is not conducive to industrialized line processing.

2. Poor fire resistance. When subjected to fire, the steel will soften and its mechanical properties will rapidly decrease. The failure of the girder truss node affects its overall bearing performance, resulting in transient failure of the structure.

3. Easy to rust. Steel or iron nails are prone to rusting when exposed to air, which are more pronounced under conditions of high humidity and high salt, thus reducing the durability of the entire timber structure.

4. Poor energy dissipation. Nails are the fastening type joints, which constrain the relative rotation between the truss and the truss, and cannot consume the energy generated by the lateral force, which leads to the lateral resistance of the whole building becoming weak.

In response to the problems with connection modes of girder truss, this chapter proposes a new type of connection modes of girder truss, which replaces traditional iron connectors with wooden connectors. The specific scheme is as follows: all the single trusses that make up the girder truss are preassembled and temporarily fixed, then predrilled at specific positions of all the trusses and finally inserted into a wood or bamboo round dowel, which is a wood dowel connector (see Ref. [15]) (as shown in **Figure 2**).

The use of wood or bamboo connectors is mainly due to the fact that wood or bamboo joints are less susceptible to corrosion than iron joints (see Ref. [16, 17]). There is also no problem of a sharp drop in mechanical properties under fire-resistant conditions. In addition, the wood or bamboo joints can greatly improve the ductility of the connected members (see Ref. [18]), thereby improving the performance of girder truss when resisting lateral forces.

The selection of the position of the wood dowel connection is determined by the force characteristics of the parallel chord truss. Parallel chord truss can be considered as a simply supported beam when subjected to an upper uniform load. The force is mainly borne by the

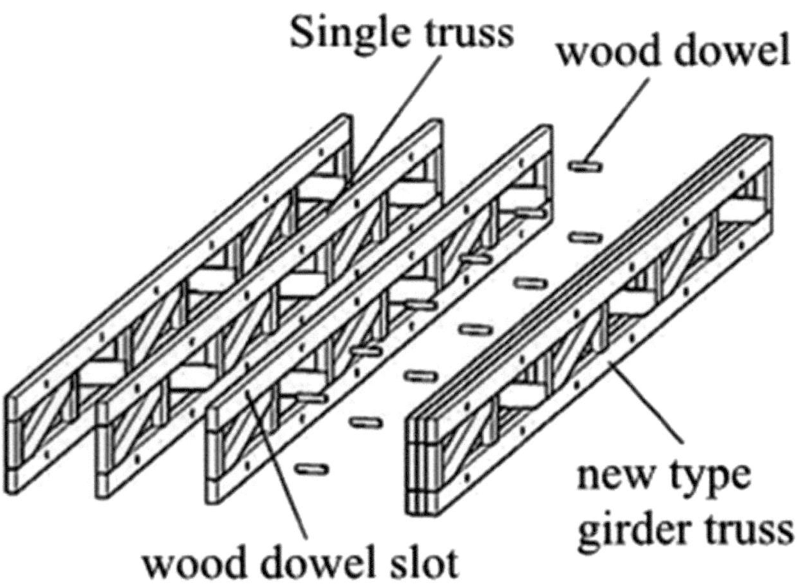

Figure 2. A new type of girder truss.

upper and lower chords of the truss. The upper chord is under pressure and the lower chord is subjected to tension, but the web only plays a supporting role. **Figure 3** shows the internal force diagram of the light wood truss supported by the upper uniform unit. It can be seen from the internal force diagram that if the parallel chord truss is regarded as a static combination structure, which means the chord is broken and both ends are hinged. Under the uniform load, the middle bending moment value of each chord is the largest and the shear force is at least zero. The use of wood dowel connectors requires predrilling the upper and lower chords of the truss, thus reducing the net cross-sectional dimensions of the chord. The shear force calculation formula of the structural member is:

$$\tau = \frac{Q}{A} \tag{1}$$

A represents the sheared net cross-section of the sheared member. The decrease in A means an increase in the shear stress in the member. Therefore, the position of the connector must be placed in where the chord shear force is the smallest, which is the middle of every two nodes of the chord.

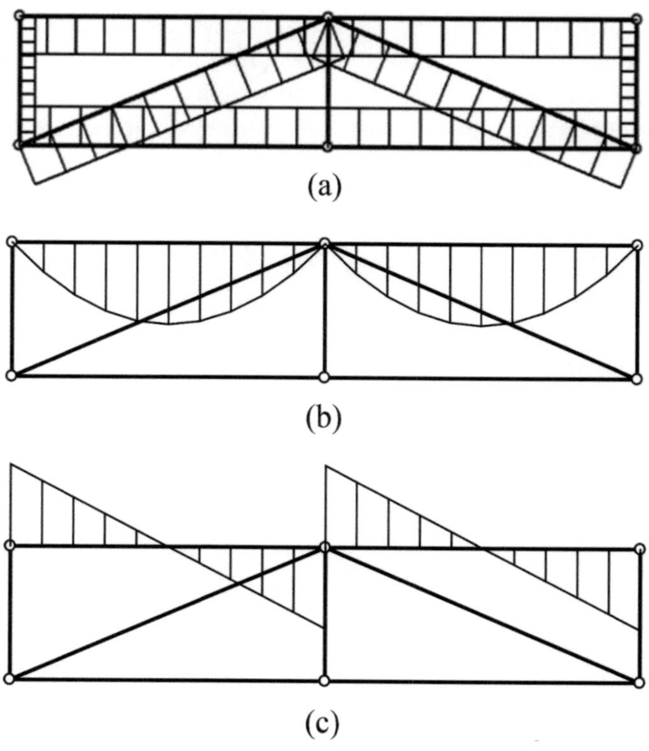

Figure 3. Internal force diagram of parallel chord truss. (a) Parallel chord truss axial force diagram; (b) parallel chord truss bending moment diagram; (c) parallel chord truss shear force diagram.

3. Experiment overview

3.1. Experimental design

The material used in the test is the Larch (*Larix gmelinii)* specification material imported from Russia. The material grade is grade II and the density is 0.657 g/cm^3. The moisture content is 17.4%, according to the general requirements for physical and mechanical tests of wood (GB/T 1928–2009).

According to the method of continuous loading of trusses in the standard for test methods of timber structures (GB20329-2012), the static load test of six types of small-span trusses was carried out, and the test piece number is expressed as S.

In order to explore the influence of different diameter wood dowels on the girder truss performance, the experiment in this chapter contains girder truss of three different diameter wood dowels. The wood dowels are 12, 16, and 20 mm in diameter. The performance evaluation of the three girder truss is still considered from the two aspects of ultimate bearing capacity and deformation resistance. Among them, the anti-deformation ability includes creep resistance and elastic recovery performance.

In addition, the experiment also set up a girder truss composed of three single trusses to explore the enhancement effect of girder truss with the increase of the number of single truss. The diameter of the wood dowels connecting the girder truss depended on the experimental results of the girder trusses with two single trusses. In order to distinguish other girder trusses with two single trusses, girder trusses with three single trusses are denoted by G3, while other girder trusses are denoted by G2.

Figure 4 shows the structural form and specific dimensions of the test piece used in this test. The girder trusses used in the experiment are all composed of this single truss.

The specific test piece composition is shown in **Table 1**.

3.2. Theoretical calculation

Calculation of standard load P_k

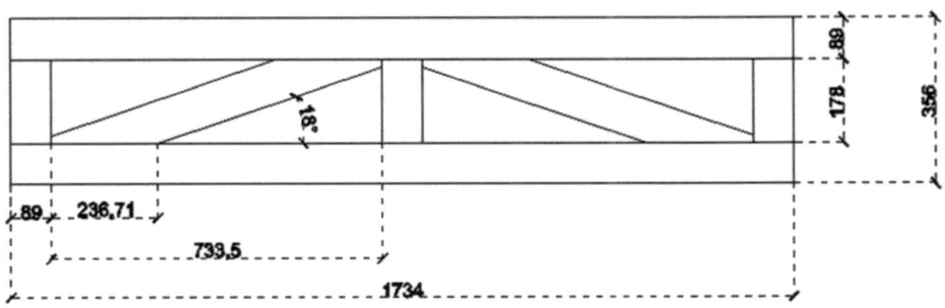

Figure 4. Module's size of girder truss (unit: mm).

http://dx.doi.org/10.5772/intechopen.80992

Truss number	Description	Quantity
SPT-S	Normal single truss	1
SPT-G2-N	Girder truss made by two SPT-S nails	1
SPT-G2-12	Girder truss made of two SPT-S connected by a 12-mm diameter beech wood dowel	2
SPT-G2-16	Girder truss made of two SPT-S connected by a 16-mm diameter beech wood dowel	2
SPT-G2-20	Girder truss made of two SPT-S connected by a 20-mm diameter beech wood dowel	2
SPT-G3	Girder truss made of three SPT-S connected by a 16-mm diameter beech wood dowel	1

Table 1. Sample number and description.

Assume that the truss spacing is 406 mm and the building life is 50 years.

According to the 2012 edition of load code for the design of building structures (GB5009-2012):

Standard value of constant load: $0.885 \times 0.406 = 0.359$ kN/m

Truss weight: $0.106 \times 0.406 = 0.043$ kN/m

Snow load standard value: $0.5 \times 0.406 = 0.203$ kN/m

Standard value of live load: $2.0 \times 0.406 = 0.812$ kN/m

Load design value: $(0.359 + 0.043) \times 1.2 + (0.203 + 0.812) \times 1.4 = 1.9$ kN/m

Load on the node: $1.9 \times 1.734 \approx 3.3$ kN.

3.3. Load program and device

According to the truss grading loading in the standard for test methods of timber structures (GB50329-2012), the truss static load test added a first-order load every 10 minutes during the failure phase, with a load of $0.2P_k$ per stage. This test used a mechanical testing machine to load. So the loading procedure could be carried out in a continuous loading mode, which is 0.2 P_k every 10 minutes. The loading per minute was 0.02 P_k. After the above theoretical calculation, P_k was 3.3 kN, and the loaded force per minute was 0.066 kN. However, in the

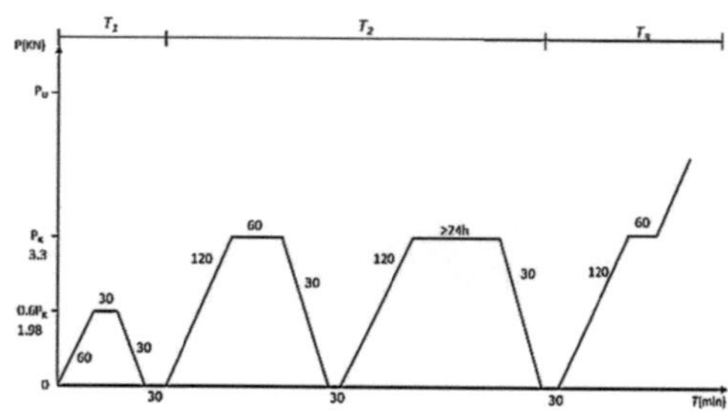

Figure 5. Loading system.

preliminary experiment, we found that the girder truss had more than twice the ultimate bearing capacity of the single truss. So during the loading process, the load of each stage was also doubled to 0.132 kN. If the girder truss is composed of three single trusses, the load per stage was also tripled to 0.198 kN. The specific loading system diagram is shown in **Figure 5**.

3.4. Evaluation index and measurement point arrangement

The purpose of this experiment is to explore the effect of different dowel diameters on the mechanical properties of the new dowel-connected girder truss. The performance evaluation

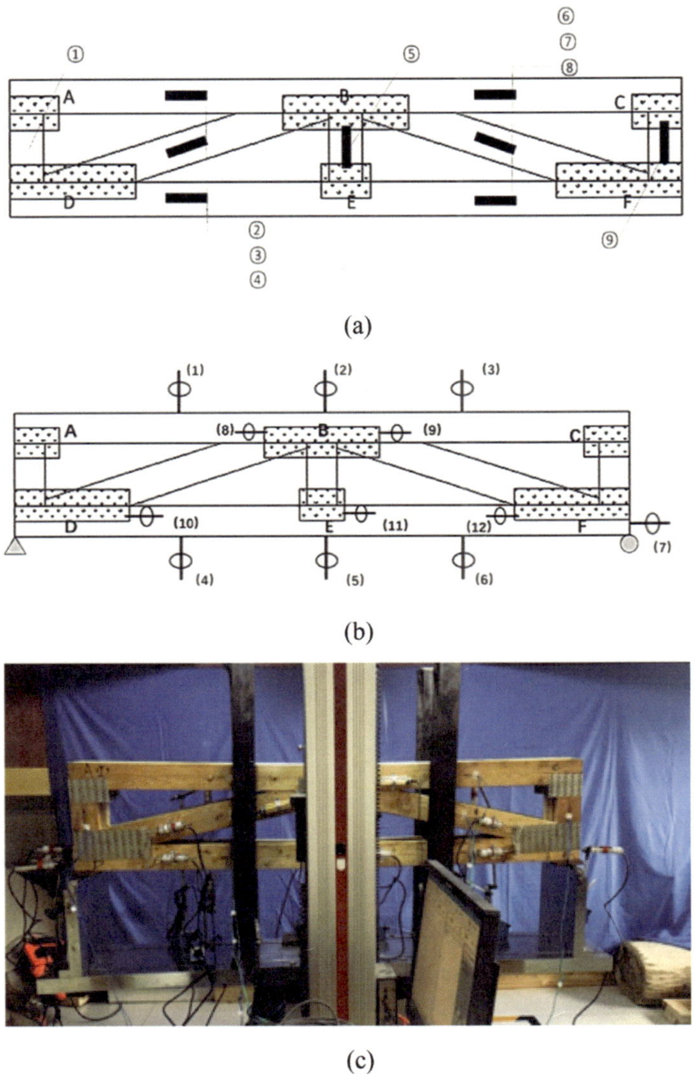

Figure 6. Layout of strain gauges and displacement gauges. (a) Layout of strain gauge; (b) displacement gauge; (c) universal mechanical testing machine.

http://dx.doi.org/10.5772/intechopen.80992

of girder truss for different dowel diameters should also start from the aspects of ultimate bearing capacity, deformation resistance, failure form and mechanism. Therefore, similar to the static load test of large-span wood trusses, it is necessary to track the displacement changes of various nodes of different types of wood trusses continuously. In this experiment, a small-range displacement sensor was also arranged between the chord and tooth plate for measuring the relative slip of the tooth plate relative to the chord. In addition, strain gauges were arranged at the important chords to measure the strain at various stages of the chords. The specific measuring point layout is shown in **Figure 6**.

4. Phenomenon description

4.1. Overall destruction

This chapter performs a static load test on one single truss and nine girder trusses, including a girder truss composed of three single trusses. There is a big difference in the ultimate bearing capacity and deformation of various types of trusses. However, the overall form and process of destruction of the truss are roughly the same. The damage form of the connector taken out after the girder truss test is also quite different. This also fully illustrates the different connection between girder trusses, which will have a greater impact on its performance.

First, during the preloading of the T_1 stage, the truss did not produce significant changes. After 30 minutes of loading, all types of trusses produced very small residual deformations. Especially, the girder truss could achieve full elastic recovery. From the load-displacement curve of the T_1 stage in **Figure 7**, a certain creep value appeared in the single truss during the preloading phase. The creep variables of other girder trusses were negligible. The use of dowel connectors of different diameters had little effect on the performance of the girder truss.

As the test progressed, there was no significant test phenomenon for each truss from the 24-hour holding load to the initial T_3 stage. However, when loading to 5 P_k, the test phenomenon began to occur in the span of the truss and there was no obvious phenomenon at other nodes. For example, a slight truss plate bulge occurred in the upper B node of the SPT-S and the lower chord appeared cracking near the knot (as shown in **Figure 8**). At other stages, other girder trusses were similar to the test phenomenon of single trusses, and the phenomena of destruction were also concentrated in these two places. In particular, the truss plate of the central B node of the upper chord bulged (as shown in **Figure 9**). This is mainly related to the force mechanism of parallel chord truss. When the parallel chord truss is subjected to the upper concentrated load, the upper chord is under compression and the lower chord is under tension. Combined with the analysis of structural mechanics, the diagonal web of the truss will generate a lateral force at the B node in order to resist the upper concentrated load. Therefore, the B node was subjected to the shear stress, and the stress environment was very complicated. Combined with the final failure form of the truss plate, the truss plate at the B node eventually appeared as a form of shear compression failure.

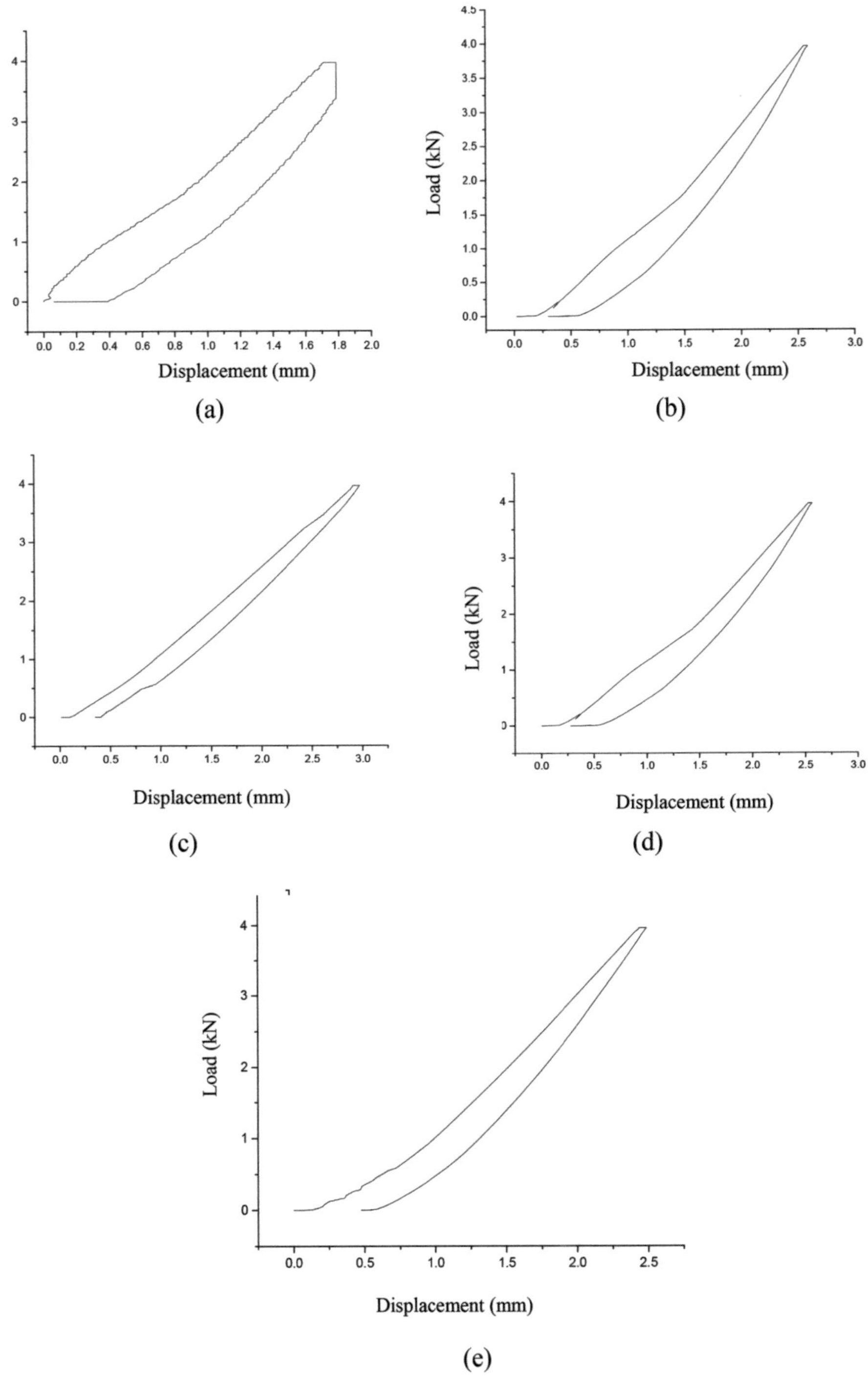

Figure 7. Load-displacement in T_1 stage. (a) SPT-S; (b) SPT-N; (C) SPT-G2-12 (d) SPT-G2-16; (e) SPT-G2-20.

Figure 8. The lower chord cracks accompanied by tooth extraction test process of SPT-S1.

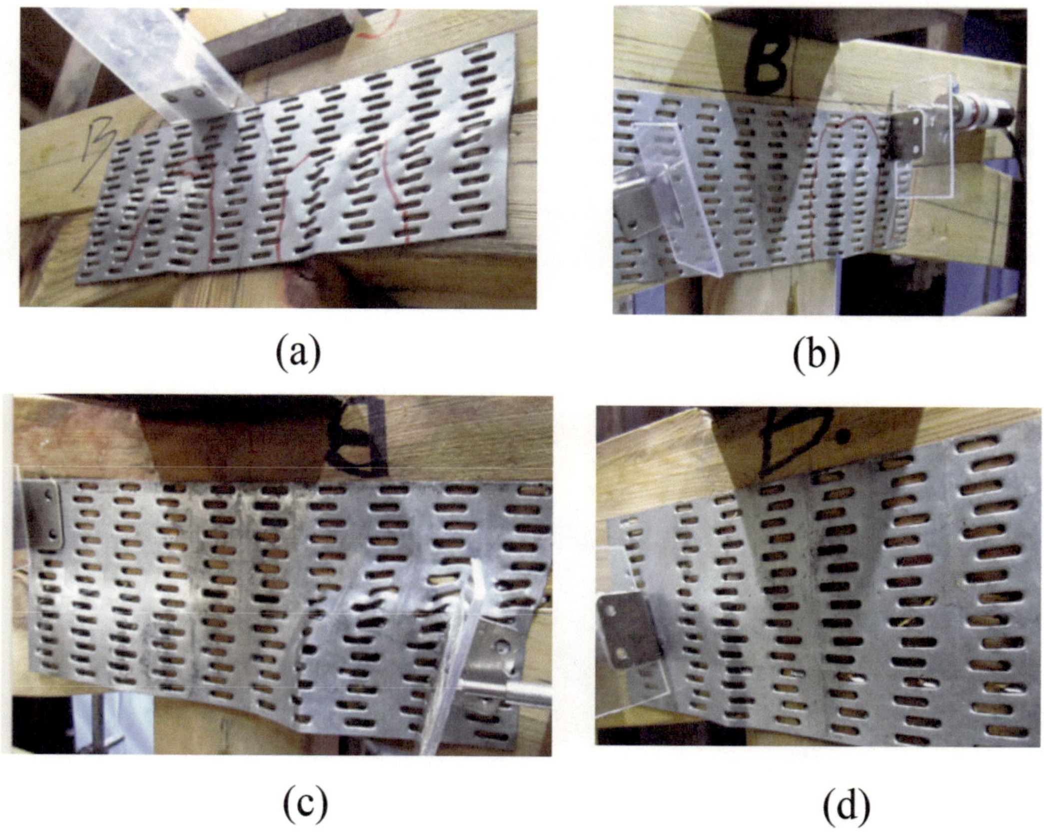

Figure 9. Shear compression failure of mate plate in B node. (a) SPT-G2-N; (b) SPT-G2-12-1; (c) SPT-G2-16-2; (d) SPT-G2-20-2.

In addition, many experiments have found that the overall damage of the truss is destroyed by the destruction of the lower chord. The knot of the lower chord also directly affects the force performance. **Figure 10** shows the actual photo of destruction in lower chords of trusses. When processing the truss, the experimenters must pay attention to the selection of the lower chord and try to avoid too many specifications with the knots. However, the upper chord and the web of the truss had obvious shear damage and the chord did not have obvious damage. Therefore, when the wood truss is processed, the grade of the processing material can be appropriately reduced.

The destruction of the lower chord of the girder truss SPT-G2-20 was due to different reasons. The SPT-G2-20 had a 19.5-mm diameter hole in its upper and lower chords. The opening of the lower chord was too large, destroying the fibers in the direction of the wood and also reducing the net cross-sectional area of the truss chord. Under the condition of constant force, reducing the net cross-sectional area of the rod will increase the stress on the chord. The tensile strength of large-sized wood is less than the compressive strength, so the lower chord is easily damaged. **Figure 11** shows the real photo of the girder truss SPT-G2-20 chord

(a) (b) (c)

Figure 10. Destruction in lower chords of trusses. (a) SPT-G2-N; (b) SPT-G2-12-2; (c) SPT-G2-16-1.

Figure 11. Failure phenomenon of SPT-G2-20-1.

failure. The lower chord crack started from the wood dowel joint and run through the entire chord. This ultimately led to the overall destruction of the truss, yet there was little damage to dowel.

4.2. Connection node destruction

The previous section described the node failure mode of a girder truss connected by a 20-mm diameter dowel between the single trusses of girder truss. The final damage was caused by the destruction of the lower chord, but the dowels showed almost no deformation. The size of the dowels and the nail connectors was damaged to varying degrees. **Figure 12** shows the connector removed from the truss after the final destruction of each girder truss connector.

From **Figure 12**, the deformation caused by the nail connection was large. Similar to the long-span nail connection girder truss, a plastic hinge appeared in the middle of the nail. When the truss was loaded to the later stage, a more obvious dislocation occured between the single trusses that make up the girder truss. Different diameters of dowels produced different forms of deformation or damage. First, similar to the nail, a 12-mm diameter wood dowel also produced a plastic hinge. However, the amount of deformation was less than the nail connection. The diameter of the wood dowels affected its stiffness. The dowel deformation of a large diameter was small. The dowels with a diameter of 20 mm showed almost no deformation. Wood dowels were almost unaffected by truss damage. The deformation of the wood dowels with 16 mm was also not obvious. The cross-sectional loss of the chord was reduced while providing sufficient joint strength. The relationship between the diameter of hole and dowel of the truss member also had an effect on the connection performance. The black color at the end of the 20-mm diameter dowel in **Figure 13** is the result of carbonization when dowel was screwed into the slot. When the hole diameter of the chord was less than 0.5 mm of the diameter of the

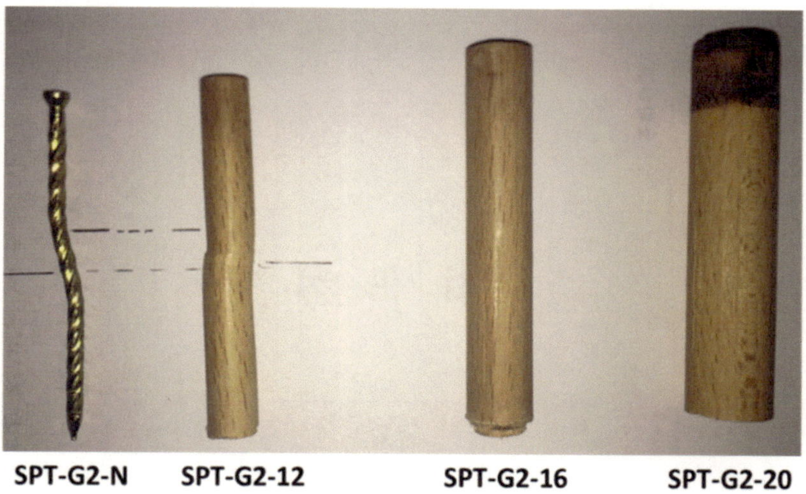

Figure 12. Failure form of connectors.

wood dowel, the wood dowel screwed in the rod was carbonized by the high-speed rotation heat generation and the carbonized layer formed on the surface of the dowel, which protected the surface of the wood dowel. The surface strength was improved. Therefore, it is necessary to select the connectors well and fit the appropriate size of predrilling from the perspective the durability of girder truss.

5. Ultimate bearing capacity

Figure 13 is a comparison of the ultimate bearing capacity of a category five trusses. Among them, the new wood dowel-joined girder truss takes the average of two tests. It can be seen from the figure that the ultimate bearing capacities of all kinds of wood trusses are much higher than its theoretical calculation value, so reducing the span of the truss will effectively improve its carrying capacity. In addition, the ultimate bearing capacities of various girder trusses are much larger than that of single truss, but the ultimate bearing capacities of the various girder trusses have little different from each other. The 12-mm and 16-mm wood dowel-joined girder trusses have a relatively high ultimate bearing capacity. The nail connection girder truss affected the synergy of girder truss due to the mutual dislocation between its single trusses, thus reducing its bearing performance. The girder truss with a wood dowel diameter of 20 mm had a large opening area at the lower chord of the truss, which reduced the net cross-sectional area of the tension member, thereby reducing the load-bearing performance of the truss.

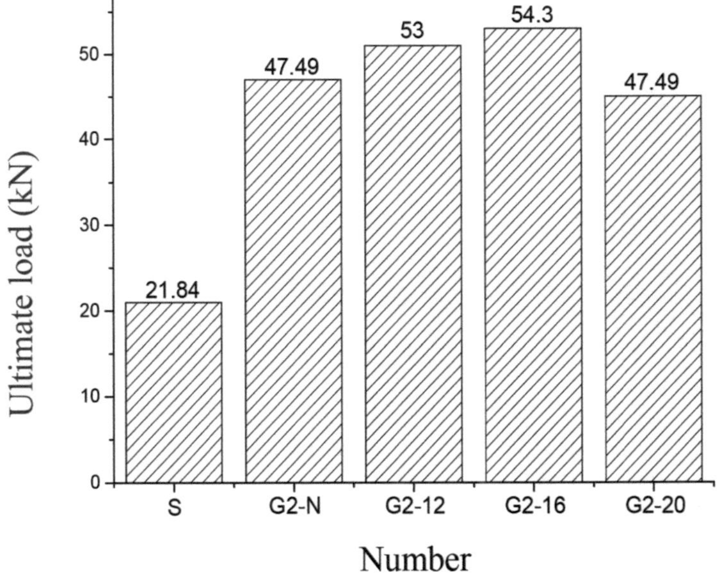

Figure 13. Ultimate bearing capacity of trusses.

http://dx.doi.org/10.5772/intechopen.80992

6. Analysis of node deflection test results

Figure 14 is a deflection diagram of the lower chords of three girder trusses using two different diameter wood dowel connections. It can be seen from the figure that the three types of trusses show good consistency in the first two stages of loading. The truss enters the nonlinear stage when it enters the failure stage, and the results of the two tests will vary due to the variability of the wood. **Figure 17** shows the variation of the overall deformation of the truss during the loading process. The image shows that it is difficult to distinguish the influence of different connection methods on the creep resistance, elastic recovery performance and deformation resistance of the truss on small-span truss specimens. Only in the stage of truss failure, the deflection curve can be distinguished and different failure modes and mechanisms of various trusses can be analyzed.

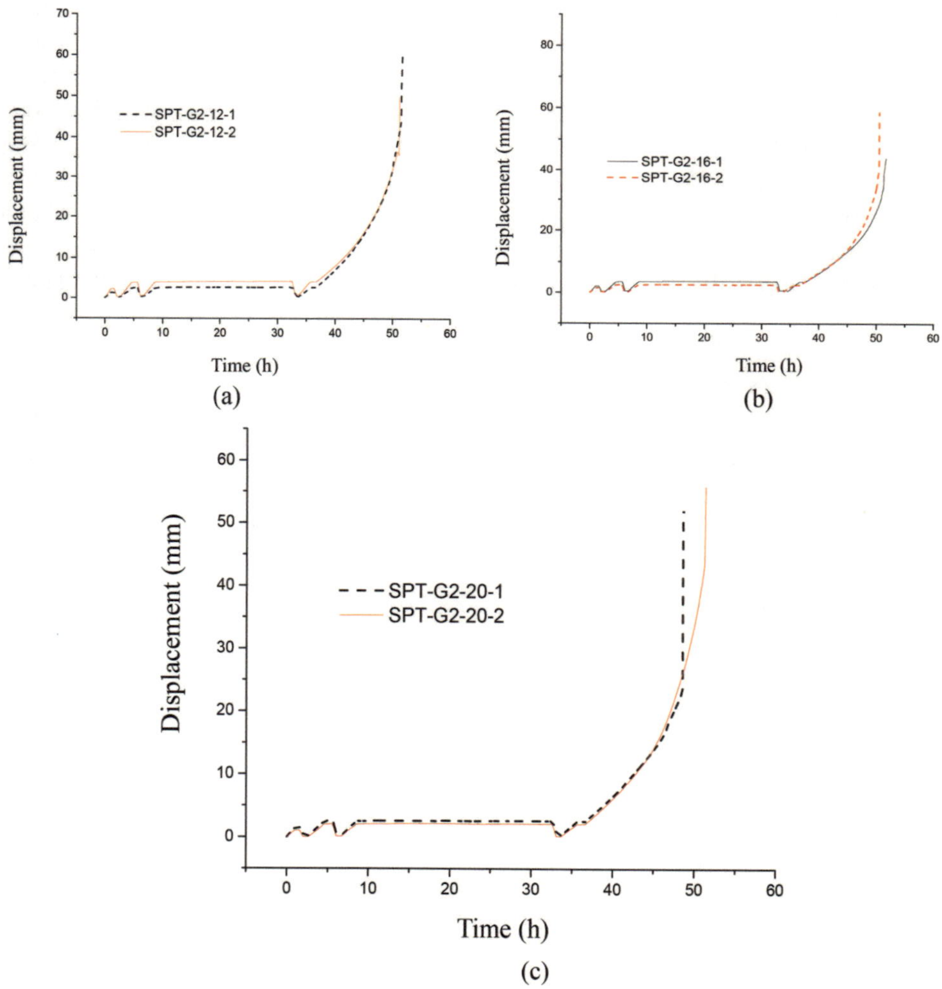

Figure 14. Time-deflection of new girder trusses from start to finish. (a) SPT-G2-12; (b) SPT-G2-16; (c) SPT-G2-20.

As shown in **Figure 15**, the load-deflection curves of the single truss, the girder truss of the nail connection, and the girder truss of the wood dowel connection are selected in the failure stage. It can be seen from the figure that different types of trusses exhibit different failure modes and mechanisms. Single truss showed obvious characteristics of brittle failure. There was no obvious sign in the vicinity of the failure. A crack occurred near the knot of the lower chord (as shown in **Figure 8**). Then, the crack continued to increase. Eventually, the overall failure of the truss was caused by the sudden fracture of the lower chord.

The ductility of the two girder trusses is significantly better than that of single truss. In the middle and later stages of truss failure, the load-displacement curve often shows a twist at one end. The reason for the twists and turns is that one single truss in the girder trusses was destroyed first. Since the other truss still had the carrying capacity, it would quickly bear the upper load. However, it would also be destroyed quickly because only one single truss was stressed. Due to the different connections between the selected single trusses of girder trusses, the above situation would be different. Although the shear span ratio was reduced, the girder truss of the nail connection still exhibited in-plane instability at the later stage of loading. Larger span trusses were not very obvious. There was mutual dislocation between the upper trusses. The girder truss would be obvious that one single truss were destroyed first and then another truss would be destroyed quickly. Therefore, the nail connection girder truss did not produce the expected effect of "one plus one is greater than two." The girder truss connected by wood dowel could still maintain good synergy between the loaded trusses. Therefore, SPT-G2-16 also had its first wave where SPT-G2-N had its twists and turns. However, it can be seen that the drop in displacement was very limited, indicating that the truss had not been

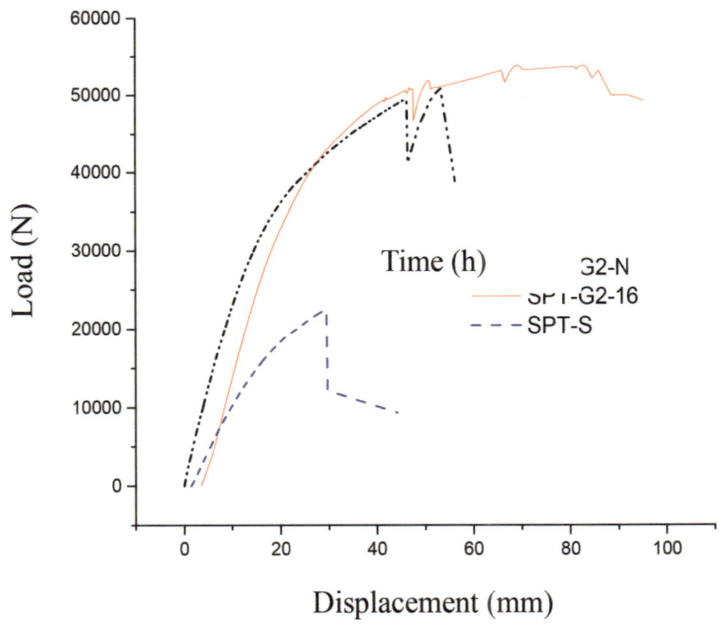

Figure 15. Load-deflection curve of trusses.

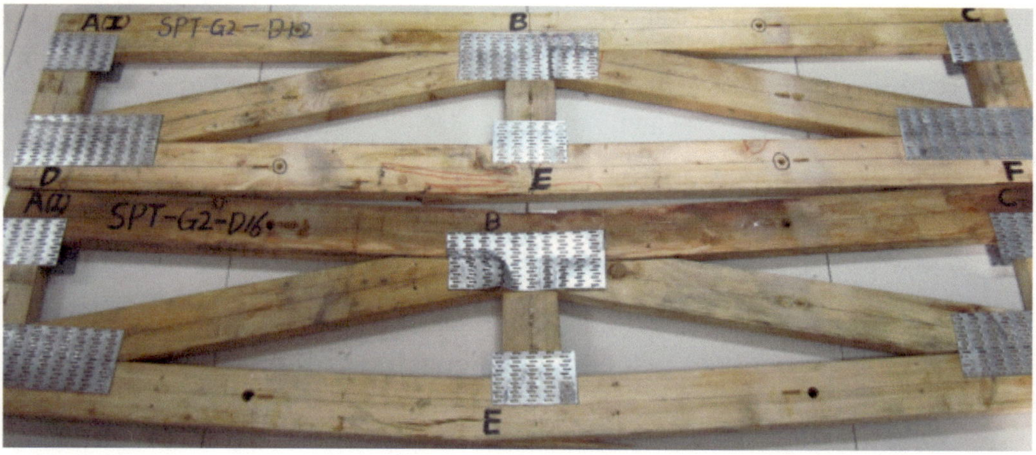

Figure 16. Failure phenomenon of SPT-G2-16.

completely destroyed. As the load continuing to increase, the curve appeared three or four small twists. Eventually, the cracks that was generated at the bottom of the two lower chords of the truss were excessive and completely penetrated (as shown in **Figure 16**), resulting in failure of the truss.

Figure 17 shows the load-deflection curves of three new girder trusses with different diameters of wood dowel as the connection between single trusses. It can be seen from the

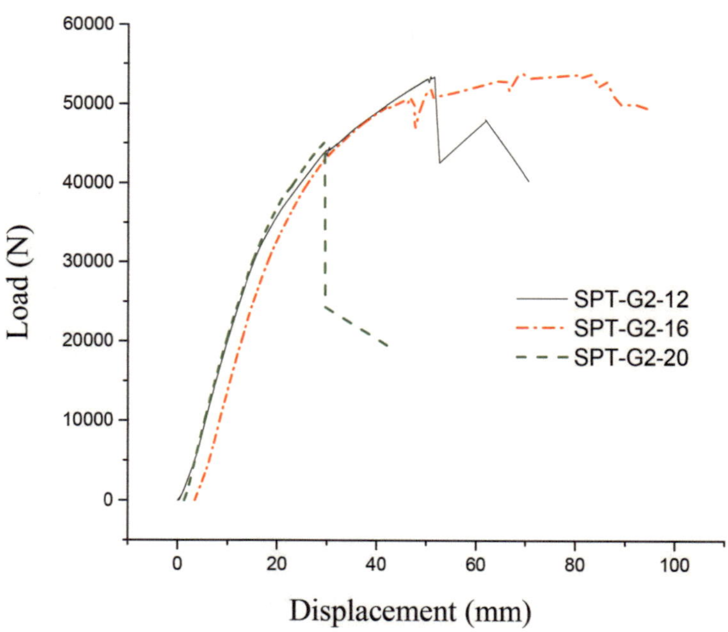

Figure 17. Load-deflection curve of new girder trusses.

figure that the load-deflection curves of the three girder trusses show different shapes. The girder truss with a wood dowel diameter of 20 mm had a similar failure mode of single truss, which was brittle failure (as shown in **Figure 10**). The lower chord was broken under the combined action of tension and shear. The 12-mm diameter dowel connected girder truss was similar to the nail-connected girder truss. Although the load-displacement curve had undergone a twist, the truss did not show good synergy. Finally, a plastic hinge appeared similar to the wood dowel and the nail. Therefore, the dowel connected girder truss showed a better mechanical performance with a 16-mm diameter wood dowel.

7. Further experiments of girder truss composed of three single trusses

In the last section, the girder trusses connected by three different diameter wood dowels were tested for static load, and the girder truss with 16-mm diameter wood dowels was the best. All previous experiments were performed on girder truss composed of two single trusses, but girder truss composed of three or more single trusses were not tested. For the girder trusses to be widely used in larger span structures and more complicated bearing environments, they cannot be composed of only two single trusses. It must consider more forms of single truss combinations. Combined with the test results of the previous section, this section performs a static load test on a girder truss composed of three single trusses connected by 16-mm diameter wood dowels. The enhancement effect of girder truss was explored by comparison with girder truss composed of two single trusses connected by the same diameter wood dowels.

In terms of bearing capacity, the girder truss composed of two single trusses was 53 kN and the girder truss composed of three single trusses had a bearing capacity of 77 kN, increasing 45%. Thus, the more quantities of the single trusses that make up the girder truss are, the more obvious enhancement effects have from the perspective of the failure mode, the two girder trusses were similar and the chord was destroyed under the tensile

Figure 18. The failure of SPT-G3-16.

shear environment, resulting in the destruction of the truss. As shown in **Figure 18**, the middle truss first appeared as cracks at the lower chord. The increase in force resulted in an overall failure of the intermediate truss, with only two trusses being stressed, but girder truss had lost synergy at this time. Then, the wood dowels of the lower chord were also destroyed (as shown in **Figure 19**). The lower chord of the single truss on the outside of girder truss was completely cracked in the direction of the grain. The truss failed as a whole. As shown in **Figure 20**, the load-displacement curves of the two girder trusses in the final failure stage, it can also be found that the two girder trusses have very similar failure modes and both exhibit good ductility.

Figure 19. Failure of connectors in SPT-G3-16.

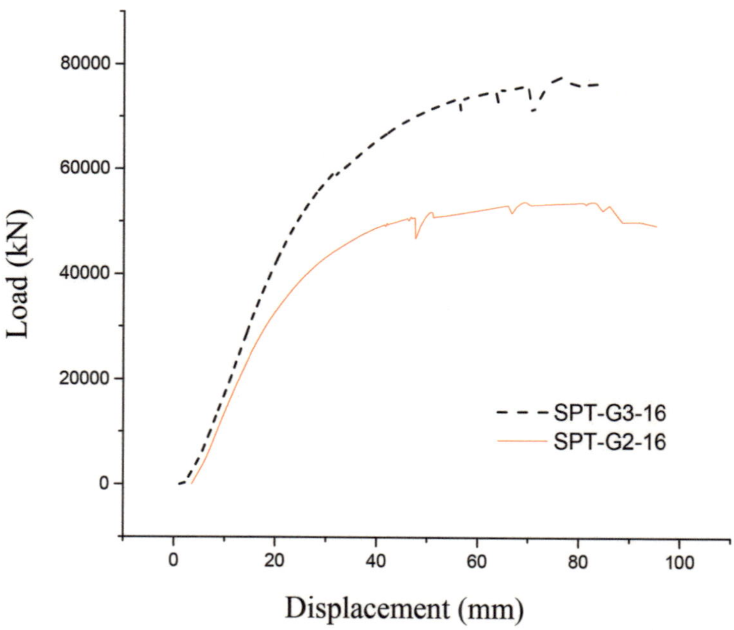

Figure 20. Load-displacement curve in failure stage.

8. Conclusion

In this chapter, the static load test of wood trusses was carried out to investigate the influence of different joints on the mechanical properties of girder truss between single trusses of girder truss, especially the influence of different dowel diameters on girder truss. The results showed that:

1. The girder truss with wood dowel connection should be connected to the girder truss as a whole, but the diameter of the wood dowel should be chosen reasonably.
2. From the perspective of the load capacity, failure mechanism, and mode of the truss, the girder truss has the best performance when the wood dowel diameter is 16 mm.
3. Under serviceability limit states, the use of dowel connectors of different diameters has little effect on the resistance to deformation of girder truss.
4. On the upper chord connected to the truss plate, the truss plate is prone to shear damage due to the combined action of pressure and shear. In the actual project, it should try to do partial reinforcement.
5. As the number of single trusses that make up the girder truss increases, there will also be a significant improvement in its mechanical properties.
6. Wood knots, especially dead knots, have a strong weakening effect on the carrying capacity of the chords of wood trusses. The overall failure of the truss is often due to the presence of the knot, so the selection of the truss should be done. The chord of the truss should avoid the use of materials with knots. If necessary, It can be chosen that using steel instead of wood, developing steel-wood composite structure.

Acknowledgements

This work is supported by the National Natural Science Foundation of China (grant No. 31670566) and National "Twelfth Five-Year" Plan for Science & Technology Support (2015BAD14B0503).

Author details

Ze-li Que[1]*, Liu-liu Zhang[1], Fei-bin Wang[2], Yi-fan Gao[1], Wei-zhen Cai[1], Xin-meng Wang[1] and Chang-tong Mei[1]

*Address all correspondence to: zelique@njfu.edu.com

1 College of Materials Science and Engineering, Nanjing Forestry University, Nanjing, China

2 College of Landscape Architecture, Nanjing Forestry University, Nanjing, China

References

[1] Li ZJ, Zhao XJ. Difficulties and countermeasures of building energy saving and emission reduction in China. Jianghuai Tribune. 2014;(2). (in Chinese)

[2] Engineering Construction Standardization Editorial Department. Interpretation of government work report, Four key points in the field of engineering construction. Standardization of Engineering Construction. 2016;(03):12-17. (in Chinese)

[3] Ministry of Industry and Information Technology and Ministry of Housing and Urban-Rural Development joint development. Action plan to promote green building materials production and application' release. Development Guide to Building Materials. 2015;**13**(20):11-14. (in Chinese)

[4] Huang DM. Life Cycle Assessment of Bamboo/Wood Structure Houses. Nanjing: Nanjing Forestry University; 2012. (in Chinese)

[5] Shang JC, Chu CL, Zhang ZH. Quantitative Assessment on Carbon Emission of Different Structures in Building Life Cycle. Building Science. 2011;**27**(12):66-70. (in Chinese)

[6] Liu BN, He Q. Building energy-saving technology. Advanced Materials Research. 2012;**472-475**:469-472

[7] Bank TW. China–Energy conservation study. Africa series. No. AF 34; 2010

[8] Wang GR, Li ZZ, Zhang WZ. On sponge city theory and its application ideas in urban planning. Shanxi Architecture. 2014;**40**(36):5-7. (in Chinese)

[9] Chen WD. Roof conversion in Shanghai from fiat to sloping. China Building Waterproofing. 2002;(3):4-5. (in Chinese)

[10] Burdzik WMG. Analysis of timber girder trusses for eccentric loading: technical paper. Journal of the South African Institution of Civil Engineering. 2004;**46**(2):15-22

[11] Liu XQ, Lam F. Reliability analysis of lateral bracing forces in metal-plated wood trusses. Journal of Structural Engineering. 2017;**143**(12):04017171-1-10

[12] Arya B, Jain A. Prefabricated wood truss roof failure under wind induced loads. Forensic Engineering. 2013;**2012**:1228-1237

[13] Arya B, Kargahi M, Hart GC. Load duration effect on failure of an overloaded wood truss structure. Forensic Engineering. 2010;**2009**:465-476

[14] Que ZL, Yang L, Fu Q, et al. Modes and causes of structural failure of timber structures. China Wood-Based Panels. 2012;(11):15-18. (in Chinese)

[15] Que ZL, Gao YF, Ge XY. A review of the research of wood girder truss. China Forestry Science and Technology. 2017;**2**(2):138-144. (in Chinese)

[16] Yang L, Que ZL. The new cooperation project between Suzhou Royal and Nanjing Forestry University was established–the research of the effect of high temperature, high

humidity and high salt environment on the performance of wood structural timber. International Wood Industry. 2013;(9):8-9. (in Chinese)

[17] Que ZL, Li ZR, Wang FB, et al. Influence of high salinity environment on shear strength of wood frame structures. Industrial Buildings. 2015;**45**(9):81-85. (in Chinese)

[18] He SJ, Shang P, Yang BS, et al. Influence of corrosion inducing treatment on shear behaviour of bamboo nail and steel nail. China Forestry Science and Technology. 2015;**29**(2):90-94. (in Chinese)

Chapter 7

Functionally Graded Concrete Structure

Ning Zhang, Aizhong Lu and Xuguang Chen

Additional information is available at the end of the chapter

http://dx.doi.org/10.5772/intechopen.81435

Abstract

By inverse analysis, the concept, functionally gradient, is introduced into thick-walled lining to improve the supporting performance. Theoretical results show that for two linings with the same sizes, an ideal functionally graded lining (IFGL) has higher elastic ultimate bearing capacity than a traditional single-layered lining (TSL). But the IFGL model requires that the Young's modulus should be a continuously monotonically increasing function in the radial direction, which, obviously, cannot be achieved currently for the concrete materials. In order to apply this idea to real lining, we use a simplest multilayered lining, a double-layered functionally graded lining (DFGL), as an approximate simulation of the IFGL. Then, we carried out elastoplastic analysis on IDFL and DFGL and model test on DFGL to assess the support performance. Results of elastoplastic analysis show that the elastic ultimate bearing capacities of both the IDFL and DFGL are higher than the traditional single-layered lining. Model tests also verify the conclusion.

Keywords: functionally graded structure, concrete, rock support, lining

1. Introduction

Concrete thick-walled lining is an important supporting structure widely used in many areas like mining, hydraulic power station, tunnel, underground storage project, and so on. Under the action of the outer loading, usually the in situ stresses, stress distributions are nonuniform, especially in the radial direction. For example, as a thick-walled hollow cylinder is subjected to hydraulic pressure, tangential stress concentration will occur along the inner boundary (see **Figure 1**). Nowadays, some underground engineering disciplines, like mining, are being excavated in depth more than 1000 m and the in situ stresses will be very large, maybe more than 27 MPa [1]. Under such a stress, if the concrete strength is not high enough, failure will probably initiate from the inner lining boundary. In order to improve the bearing capacity

IntechOpen

© 2018 The Author(s). Licensee IntechOpen. This chapter is distributed under the terms of the Creative Commons Attribution License (http://creativecommons.org/licenses/by/3.0), which permits unrestricted use, distribution, and reproduction in any medium, provided the original work is properly cited.

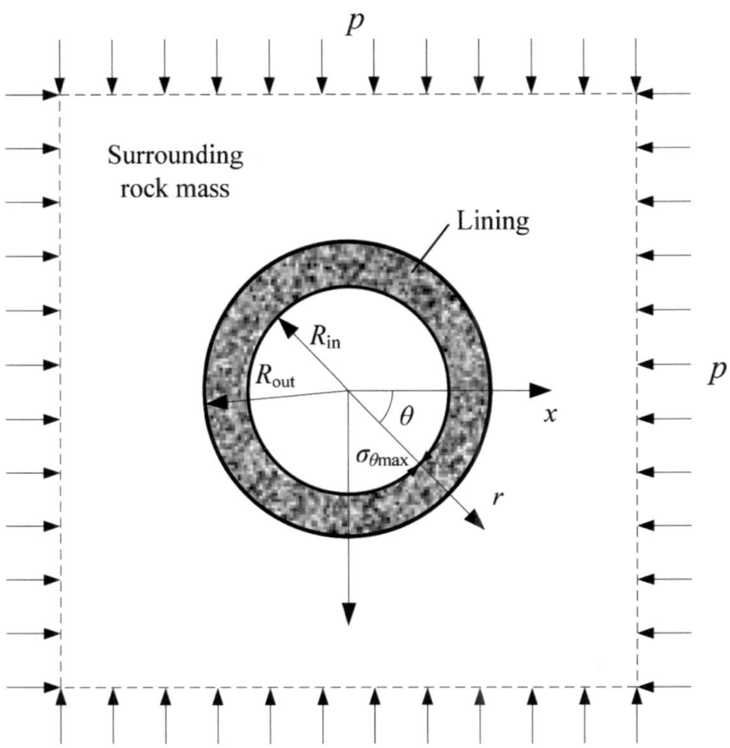

Figure 1. Traditional single-layered lining under hydrostatic in situ stress.

and ensure the lining can work safely, two means are usually used: (i) increase the thickness of lining and (ii) improve the material strength. For the first mean, using thicker lining, the stress concentration can be eased, but in a limit range. Moreover, it is not economical. Statistics show that the excavation may cost 40–60% of the total cost of a shaft construction. As deeper than 1000 m, a 10 mm decrease in lining thickness, the total cost will decrease by around 1% (reinforced concrete structure) and 0.25% (concrete structure). For the other mean, using high-strength concrete, the elastic ultimate bearing capacity will increase in a certain extent. But high strength concrete usually needs high strength cement and kinds of admixtures, and high-level construction technology, which will also lead to higher cost. Besides, as we all know, concrete materials with higher strength have more significant brittleness and very low residual strength. Thus, sudden failure will be more likely to occur in lining. Both theoretical and practical results illustrate that total cost will increase significantly and usually the failure cannot be avoided either by increasing thickness or by improving higher strength.

The usually used two means are neither economic nor practical. Herein, in order to conquer these problems and increase the ultimate bearing capacity of lining, an alternative way we proposed in this work is to introduce the idea, functionally graded materials (FGMs). The concept of FGM was firstly proposed in 1984 by Japanese materials scientists as a mean of preparing thermal barrier materials. Continuous changes in the composition, microstructure, porosity, etc. of material result in gradients in such properties as mechanical strength and thermal conductivity [2]. Compositional micro/macrostructure gradient can not only dismiss undesirable effects such as stress concentration [3] but also generate unique positive function [4]. Then, its applications have been expanded to also the components of chemical plants, solar energy generators, heat exchangers, nuclear reactors, and high efficiency combustion systems [5]. Some people also carried out some works on the functionally graded performances of construction materials [6, 7]. Shen et al. [8] developed a functionally graded material system with a spatially tailored fiber distribution to produce a four-layered, functionally graded fiber-reinforced cement composite. Dias et al. [9] discussed the use of statistical mixture designs to choose formulations and presented ideas for the production of functionally graded fiber cement components. In this work, in order to improve the stress distributions, inverse analysis is carried out firstly. We let the Young's modulus be a function of radius r, that is, E(r), not just a constant E as usual, and the Poisson's ratio μ be a constant. Then, we induce the expressions E(r) based on the preassumed stress distributions. In this way, we can get an ideal model, the functionally graded lining. But this ideal model is not practical for concrete material used in lining. Thus, we choose the multilayered lining as an alternative. We analyze the simplest one, a double-layered lining with hydrostatic pressure acting on the outer boundary, to assess the support performance of functionally graded lining. Finally, model tests are carried out to check the theoretical results.

2. Inverse analysis of functionally graded lining

There have been many analytical results on FG thick-walled hollow cylinder subjected to temperature loads or stresses. In these works, material parameters are preassumed to be a certain function about radius. Usually, the Poisson's ratio is set as a constant but the Young's modulus or shear modulus is defined as different types of functions about radius like linear variation [10, 11], power law variation [12–21], exponential variation [18, 22, 23], and other forms [24, 25]. All these abovementioned works mainly calculated the stress and displacement distributions in a FG thick-walled hollow cylinder with given Young's modulus E(r) or shear modulus G(r). Obviously, the stresses and displacement distributions are dependent on the given E(r) or G(r). This is a forward problem. Instead, in order to obtain a stress distribution that we want it to be, we preassume the stress distribution and let the Young's modulus E(r) be undetermined. The undetermined E(r) can be back-induced according to the preassumed stress distribution and loadings. This process is usually called inverse analysis [26]. The problem we discussed in this section is such a process. The FG lining is subjected to a hydrostatic pressure *p*.

2.1. Basic assumptions

If we want to give full play to the support material, the ideal state is that the whole lining enters into plastic yielding at the same time. We confine attention to the plane strain problem. Compressive stress is defined as positive in this chapter. Because of the axial symmetry of the problem, the tangential and radial stresses, σ_θ and σ_r, in the rock mass will, respectively, be the maximum and minimum principal stresses, that is, $\sigma_1 = \sigma_\theta$ and $\sigma_3 = \sigma_r$. And we assume the failure of the lining concrete is ruled by the Tresca criterion, $\sigma_1 - \sigma_3 = c$, where c is a constant reflecting the material strength. In this situation, the ideal stress state can be preassumed as $\sigma_\theta - \sigma_r = c$. It should be noted that the values of Young's modulus E(r) in different radius r may be different from each other, but for one point, it has the same value in any directions. It means that materials of the FG linings discussed here are isotropic. Besides, there is an implicit requirement behind "the whole lining enters into plastic yielding simultaneously": the FG lining should have a constant strength, although the Young's modulus may be different in the radial direction.

2.2. Basic equations and solution

The Young's modulus E(r) only depends on the radius r. So, the problem shown in **Figure 1** is a plane strain axisymmetric problem. For small strain problem, the strain-displacement relations, constitutive equations, and equilibrium equations can be given, respectively, as

$$\varepsilon_r = \frac{du}{dr} \tag{1}$$

$$\varepsilon_\theta = \frac{u}{r} \tag{2}$$

$$\varepsilon_r = \frac{1-\mu^2}{E(r)}\left(\sigma_r - \frac{\mu}{1-\mu}\sigma_\theta\right) \tag{3}$$

$$\varepsilon_\theta = \frac{1-\mu^2}{E(r)}\left(\sigma_\theta - \frac{\mu}{1-\mu}\sigma_r\right) \tag{4}$$

$$\frac{d\sigma_r}{dr} + \frac{\sigma_r - \sigma_\theta}{r} = 0 \tag{5}$$

where ε_θ and ε_r are the tangential and radial strains, respectively; u is the radial displacement; and μ is the Poisson's ratio.

Substituting $\sigma_\theta - \sigma_r = c$ into Eq. (5) and solving the partial differential equation gives the expression of the radial stress,

$$\sigma_r = A + c\ln r \tag{6}$$

where A and c are constants to be determined, the values of which can be determined by the stress boundary conditions on the inner boundary $r = R_{in}$, $\sigma_r = 0$, and outer boundary $r = R_{out}$, $\sigma_r = p$. We have $A = -p\ln R_{in}/\ln(R_{out}/R_{in})$ and $c = p/\ln(R_{out}/R_{in})$. And then substituting A and c into Eq. (6) and $\sigma_\theta = \sigma_r + c$, we get

$$\sigma_r = p\frac{\ln(r/R_{in})}{\ln(R_{out}/R_{in})} \tag{7}$$

$$\sigma_\theta = p\frac{1+\ln(r/R_{in})}{\ln(R_{out}/R_{in})} \tag{8}$$

Combining Eqs. (1)–(5), (7), and (8), the following ordinary differential equation can be obtained,

$$\frac{dE_*(r)}{dr}\left[r+\frac{1-2\mu}{1-\mu}r\ln\left(r/R_{\text{in}}\right)\right]=-2E_*(r) \tag{9}$$

where $E_*(r)=1/E(r)$. We can get the solution of Eq. (9) as

$$E(r)=B\left[\frac{1-2\mu}{1-\mu}\ln\left(r/R_{\text{in}}\right)+1\right]^{\frac{2(1-\mu)}{1-2\mu}} \tag{10}$$

where B is constant of integration. To differentiate, we name the linings with Young's modulus satisfying Eq. (10) as ideal FG lining. **Figure 2** shows the dimensionless Young's modulus as μ = 0.0, 0.25, and 0.5. Note that Eq. (10) is not available for incompressible materials, μ = 0.5. In this case, E(r) can be solved directly by Eq. (9) as

$$E(r)=Br^2 \tag{11}$$

which agrees with the result by Nie and Batra [15, 27].

It can be seen from **Figure 2** and Eqs. (10) and (11) that in order to achieve the preassumed stress distribution $\sigma_\theta - \sigma_r = c$, the Young's modulus $E(r)$ needs to be monotonously increasing with radius r and increases faster as the Poisson's ratio is larger.

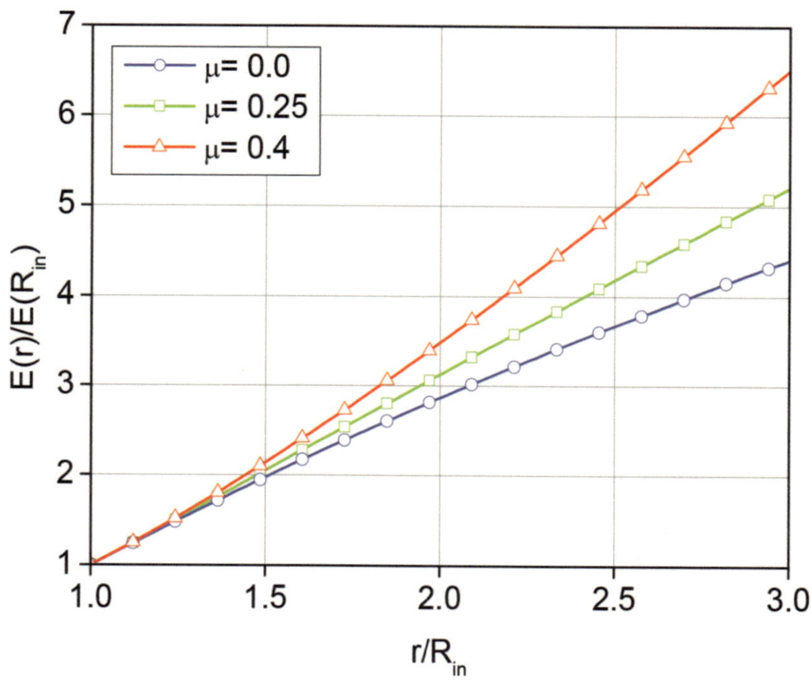

Figure 2. Dimensionless Young's modulus with respect to radius for different Poisson's ratios.

2.3. Ultimate bearing capacity of ideal FG lining

For the ideal FG lining, the stresses in any point can satisfy the equation $\sigma_\theta - \sigma_r = c$ as we preassumed, when $\sigma_\theta - \sigma_r = c = \sigma_{cFG}$ (σ_{cFG} is the UCS of the ideal FG lining material), the whole FG lining will enter into plastic state at the same time. So, for the ideal FG lining, the elastic ultimate bearing capacity p^e_{FG} is actually the same with the plastic ultimate bearing capacity p^p_{FG}. In this way, the ultimate bearing capacity of the ideal FG lining can be obtained by substituting Eqs. (7) and (8) into $\sigma_\theta - \sigma_r = \sigma_{cFG}$,

$$p^e_{FG} = p^p_{FG} = \sigma_{cFG}\ln\left(R_{out}/R_{in}\right) \tag{12}$$

3. Ultimate bearing capacities of double-layered lining

3.1. Model description and basic equations

According to the inverse analysis in Section 2, we know that the ideal FG lining material should mainly satisfy two requirements: (i) constant strength and (ii) continuously increasing Young's modulus in radial direction (see **Figure 3(B)**). Currently, most of the linings are constructed using concrete-based materials. But in view of the state of the art of concrete, it is scarcely possible to satisfy the second point, that is, continuously changing Young's modulus. So, in order to apply the FG idea in practical engineering, we use multilayered lining as an alternative (see **Figure 3(C)**). In this way, it is required that every layer of the multilayered lining should have almost equal strength and that each layer should have larger Young's modulus than its adjacent inner layer, that is, $E_i > E_{i-1}$. We herein, as an example, conduct the elastoplastic analysis of the simplest one, a double-layered lining subjected to hydrostatic in situ stress on the outer boundary (see **Figure 4**), calculating its elastic and plastic ultimate bearing capacities and comparing them with the corresponding parameters of the traditional single-layered lining and the ideal FG lining.

Similar to Section 2, the deformation pattern is restricted by the plane strain condition as well. Material behaviors of both layers are modeled by a finite strain elastoplastic flow theory based on the Tresca yield function. Perfectly elastoplastic behavior is assumed for the support material.

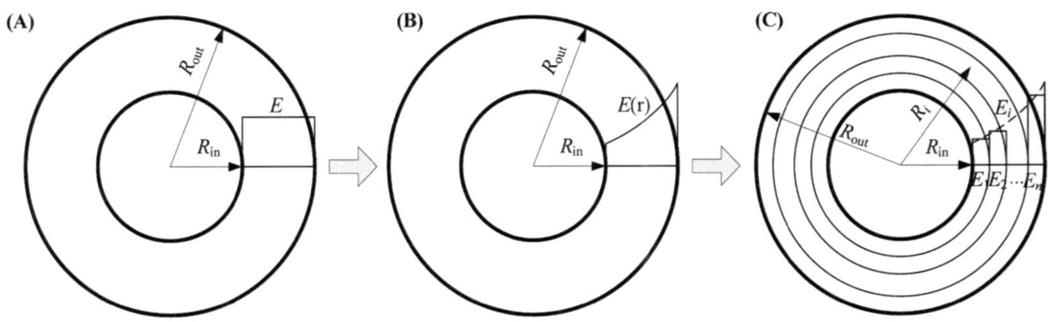

Figure 3. (A) Traditional single-layered lining; (B) ideal FG lining; and (C) multilayered lining.

$$\sigma_{\theta i} - \sigma_{ri} = \sigma_{ci} \quad (i = 1, 2) \tag{13}$$

The subscript i = 1 and 2 denote the inner and outer layers, respectively (similarly hereinafter). The stresses in both layers satisfy the equilibrium Eq. (5); as we use in this section, we just need to add the subscript i for each layer, the same with the strain-displacement relations, Eqs. (1) and (2).

E_i, μ_i, and σ_{ci} (i = 1 and 2) denote Young's modulus, Poisson's ratio, and uniaxial compressive strength, respectively (see **Figure 4**). The inner and outer radii of lining are R_{in} and R_{out}, respectively. Although the two layers are not bond with each other, the problem we discuss is axisymmetric, so the outer boundary of the inner layer and the inner boundary of the outer layer should be always together. The radius of the interboundary is R_1. The calculated mechanical model is shown in **Figure 4**.

As we know, for a thick-walled hollow cylinder under outer and/or inner hydrostatic pressures, the most significant stress concentration will occur along the inner boundary, so the plastic yielding will initiate formation of the inner boundary. It is the same for each layer of a multilayered lining. The initiation and extension path of plastic yielding in a multilayered lining depend on the values of the Young's modulus E_i, the strength σ_{ci}, and the radius R_i of each layer. For the double-layered lining, according to the initiation and extension mode of plastic yielding, there will exist four different cases of elastic and plastic zones. Both layers stay in elastic state as the loading is small (see **Figure 5(A)**). As the loading is relatively large, plastic

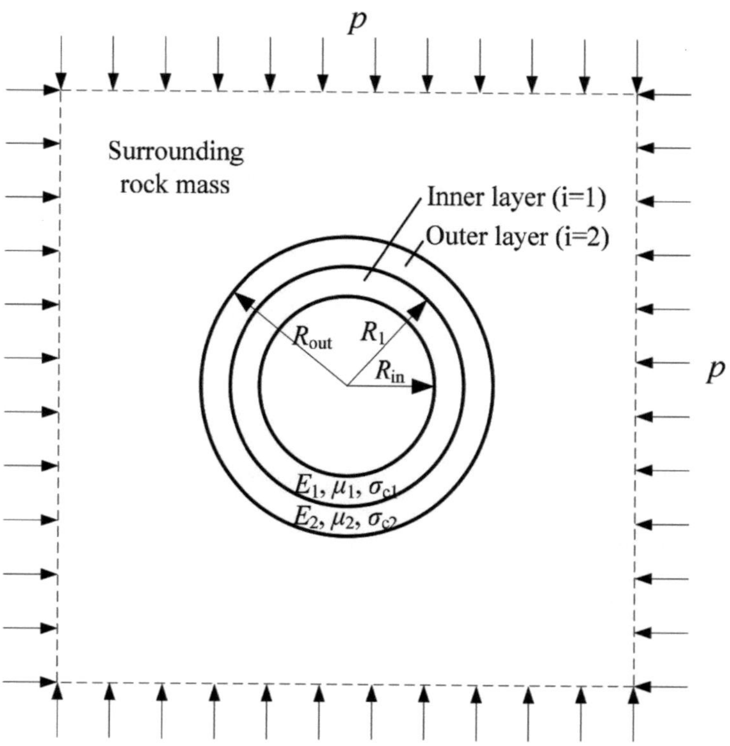

Figure 4. Double-layered thick-walled hollow lining under hydrostatic in situ stress.

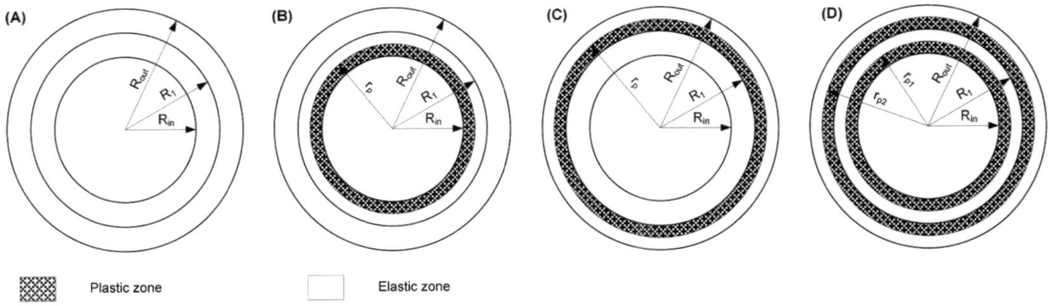

Figure 5. Four cases of double-layered lining under hydrostatic in situ stresses: (A) both layers in elastic state; (B) plastic yielding occurs only in inner layer, $R_{in} < r_p \leq R_1$; (C) plastic yielding occurs only in outer layer, $R_1 < r_p \leq R_{out}$; and (D) plastic yielding occurs in both layers, $R_{in} < r_{p1} \leq R_1$, $R_1 < r_{p2} \leq R_{out}$.

yielding may occur only in the inner layer (see **Figure 5(B)**) or only in the outer layer (see **Figure 5(C)**) or in both layers (see **Figure 5(D)**). r_p, and r_{pi} (i = 1 and 2) in **Figure 5** denote the radii of plastic zones. As mentioned above, the main aim of the elastoplastic analysis in this section is to calculate the elastic and plastic ultimate bearing capacity. So, we just give the analyzing processes, in detail, of Case A assessing the elastic ultimate bearing capacity and Case D assessing the plastic ultimate bearing capacity.

3.2. Stresses and displacement in elastic zone

For an axisymmetric plane strain problem, the radial and tangential stresses, σ_{ri} and $\sigma_{\theta i}$, and the radial displacement, u_{ri}, can, respectively, be given by the following expressions [28],

$$\sigma_{ri}^e = A_i - \frac{B_i}{r^2} \tag{14}$$

$$\sigma_{\theta i}^e = A_i + \frac{B_i}{r^2} \tag{15}$$

$$u_{ri}^e = \frac{1+\mu_i}{E_i} \frac{A_i r^2 (1-2\mu_i) + B_i}{r} \tag{16}$$

where A_i and B_i (i = 1 and 2) are constants to be determined. The superscript e represents elasticity.

3.3. Stresses and displacement in plastic zone

Substituting Eq. (13) into Eq. (5) gives the stresses in the plastic zone,

$$\sigma_{ri}^p = \sigma_{ci} \ln r + C_i \tag{17}$$

$$\sigma_{\theta i}^p = \sigma_{ci}(1 + \ln r) + C_i \tag{18}$$

where C_i (i = 1 and 2) are constants to be determined.

In the plastic zone, total radial and tangential strains, ε_r and ε_θ, can be decomposed into elastic and plastic parts as

$$\varepsilon_{ri} = \varepsilon_{ri}^e + \varepsilon_{ri}^p \tag{19}$$

$$\varepsilon_{\theta i} = \varepsilon_{\theta i}^e + \varepsilon_{\theta i}^p \tag{20}$$

Similarly, the superscript e and p represent elastic and plastic, respectively.

In order to determine the displacement field in the plastic region, a plastic flow rule is needed. By assuming that the elastic strains are relatively small in comparison to the plastic strains and that an associated flow rule is valid, the plastic parts of radial and tangential strains may be related for the plane strain condition

$$\varepsilon_{ri}^p + \varepsilon_{\theta i}^p = 0 \tag{21}$$

From Eqs. (1), (2), (19)–(21) the differential equation for the radial displacement can be expressed as

$$\frac{du_i}{dr} + \frac{u_i}{r} = f_i(r) \tag{22}$$

where $f_i(r) = \varepsilon_{ri}^e + \varepsilon_{\theta i}^e$.

By using the generalized Hooke law with consideration of the plane strain assumption, $\varepsilon_{\theta i} = 0$, the elastic strains can be given as

$$\varepsilon_{ri}^e = \frac{1}{E_i}\left(\sigma_{ri} - \mu_i\sigma_{\theta i} - \mu_i^2\sigma_{ri} - \mu_i^2\sigma_{\theta i}\right) \tag{23}$$

$$\varepsilon_{\theta i}^e = \frac{1}{E_i}\left(\sigma_{\theta i} - \mu_i\sigma_{ri} - \mu_i^2\sigma_{\theta i} - \mu_i^2\sigma_{ri}\right) \tag{24}$$

It should be noted that σ_{ri} and $\sigma_{\theta i}$ in Eqs. (23) and (24) are stresses in the plastic zone, so, they can be expressed by Eqs. (17) and (18), respectively. So, the function $f_i(r)$ is

$$f_i(r) = \frac{(1+\mu_i)(1-2\mu_i)}{E_i}\left[\sigma_{ci}(1+2\ln r) + 2C_i\right] = D_i \ln r + F_i \tag{25}$$

where $D_i = \frac{2(1+\mu_i)(1-2\mu_i)\sigma_{ci}}{E_i}$, $F_i = \frac{(1+\mu_i)(1-2\mu_i)(\sigma_{ci}+2C_i)}{E_i}$, C_i (i = 1 and 2) are constants to be determined. Substituting Eq. (25) into Eq. (22) and solving the differential equation, we can obtain the displacement in the plastic zone,

$$u_{ri}^p = \frac{r}{4}(2D_i \ln r - D_i + 2F_i) + \frac{G_i}{r} \tag{26}$$

where G_i (i = 1 and 2) are constants to be determined.

3.4. Elastic ultimate bearing capacity

The elastic ultimate bearing capacity is the load p, as plastic yielding initiates along the inner boundaries of either layer. It can be calculated by elastic analysis of Case A (see **Figure 5(A)**). The boundary and continuity conditions include at the inner boundary of the inner layer $r = R_{\text{in}}$, $\sigma_{r1}^{e} = 0$; at the interboundary $r = R_1$, $\sigma_{r1}^{e} = \sigma_{r2}^{e}$, and $u_1^e = u_2^e$; and at the outer boundary of the outer layer $r = R_{\text{out}}$, $\sigma_{r2}^{e} = p$. These conditions can, respectively, give the following equations,

$$A_1 - \frac{B_1}{R_{\text{in}}^2} = 0 \tag{27}$$

$$A_1 - \frac{B_1}{R_1^2} = A_2 - \frac{B_2}{R_1^2} \tag{28}$$

$$\frac{1+\mu_1}{E_1}\frac{A_1R_1^2(1-2\mu_1)+B_1}{R_1} = \frac{1+\mu_2}{E_2}\frac{A_2R_1^2(1-2\mu_2)+B_2}{R_1} \tag{29}$$

$$A_2 - \frac{B_2}{R_{\text{out}}^2} = p \tag{30}$$

We have four variables A_i and B_i (i = 1 and 2) and four equations Eqs. (27)–(30). When the plastic yielding initiates from the inner boundary of the inner layer, the stresses at $r = R_{\text{in}}$ should satisfy the Tresca criterion,

$$\left(\sigma_{\theta 1}^{e} - \sigma_{r1}^{e}\right)\Big|_{r=R_{\text{in}}} = \sigma_{c1} \tag{31}$$

Similarly, if the plastic yielding initiates from the inner boundary of the outer layer, we have

$$\left(\sigma_{\theta 2}^{e} - \sigma_{r2}^{e}\right)\Big|_{r=R_1} = \sigma_{c2} \tag{32}$$

From Eqs. (31) and (32), we can obtain two critical loads p_1^e and p_2^e, respectively. The ratio between the two critical loads can be given as

$$\frac{p_1^e}{p_2^e} = \frac{(\Delta_2E_1 + \Delta_3E_2)\sigma_{c1}}{\Delta_1E_1\sigma_{c2}} \tag{33}$$

where $\Delta_1 = 2(1-\mu_2^2)R_1^2$, $\Delta_2 = (\mu_2+1)(2\mu_2-1)(R_1 - R_{\text{in}})(R_1 + R_{\text{in}})$, and $\Delta_3 = (\mu_1+1)(-2R_1^2\mu_1 + R_{\text{in}}^2 + R_1^2)$. For simplicity sake, we take $\mu_1 = \mu_2 = 0.5$, considering the material as incompressible. In this situation, the ratio becomes

$$\frac{p_1^e}{p_2^e} = \frac{R_{\text{in}}^2E_2\sigma_{c1}}{R_1^2E_1\sigma_{c2}} = \lambda^2\frac{R_{\text{in}}^2}{R_1^2} \tag{34}$$

where $\lambda = \sqrt{(E_2/\sigma_{c2})/(E_1/\sigma_{c1})}$. According to the value of the ratio, we have the following three cases:

i. When $p_1^e/p_2^e < 1$ or $R_1/R_{\text{in}} > \lambda$, plastic yielding will initiate from the inner boundary of the inner layer and the elastic ultimate bearing capacity p^e can be given as

$$p^e = p_1^e = \frac{\sigma_{c1}}{2}\left[\frac{E_2 R_{\text{in}}^2}{E_1}\left(\frac{1}{R_1^2} - \frac{1}{R_{\text{out}}^2}\right) + 1 - \frac{R_{\text{in}}^2}{R_1^2}\right] \tag{35}$$

ii. When $p_1^e/p_2^e = 1$ or $R_1/R_{\text{in}} = \lambda$, plastic yielding will initiate from both inner boundaries of the inner and outer layers simultaneously and the elastic ultimate bearing capacity p^e can be given as

$$p^e = p_1^e = p_2^e = \frac{\sigma_{c1}}{2}\left[\frac{\sigma_{c2}}{\sigma_{c1}}\left(1 - \frac{E_1}{E_2}\right) + 1 - \frac{E_2}{E_1}\frac{R_{\text{in}}^2}{R_{\text{out}}^2}\right] \tag{36}$$

iii. When $p_1^e/p_2^e > 1$ or $R_1/R_{\text{in}} < \lambda$, plastic yielding will initiate from the inner boundary of the outer layer and the elastic ultimate bearing capacity p^e can be given as

$$p^e = p_2^e = \frac{\sigma_{c2}}{2}\left[\frac{E_1 R_1^2}{E_2}\left(\frac{1}{R_{\text{in}}^2} - \frac{1}{R_1^2}\right) + 1 - \frac{R_1^2}{R_{\text{out}}^2}\right] \tag{37}$$

3.5. Plastic ultimate bearing capacity

As the whole lining turns into plastic yielding state, the corresponding load is the ultimate bearing capacity, which can be calculated by analyzing Case D (see **Figure 5(D)**). In the same way as the elastic analysis, the elastoplastic analysis can be carried out by using the boundary and continuity conditions, which include at the inner boundary of the inner layer $r = R_{\text{in}}$, $\sigma_{r1}^p = 0$; at the elastic-plastic interface of the inner layer $r = r_{p1}$, $\sigma_{r1}^p = \sigma_{r1}^e$, and $u_1^p = u_1^e$; at the interboundary $r = R_1$, $\sigma_{r1}^e = \sigma_{r2}^p$, and $u_1^e = u_2^p$; at the elastic-plastic interface of the outer layer $r = r_{p2}$, $\sigma_{r2}^p = \sigma_{r2}^e$, and $u_2^p = u_2^e$; and at the outer boundary of the outer layer $r = R_{\text{out}}$, $\sigma_{r2}^e = p$. These conditions can, respectively, give the following equations,

$$\sigma_{c1}\ln R_{\text{in}} + C_1 = 0 \tag{38}$$

$$\sigma_{c1}\ln r_{p1} + C_1 = A_1 - \frac{B_1}{r_{p1}^2} \tag{39}$$

$$\frac{r_{p1}}{4}\left(2D_1\ln r_{p1} - D_1 + 2F_1\right) + \frac{G_1}{r_{p1}} = \frac{1+\mu_1}{E_1}\frac{A_1 r_{p1}^2(1-2\mu_1) + B_1}{r_{p1}} \tag{40}$$

$$A_1 - \frac{B_1}{R_1^2} = \sigma_{c2}\ln R_1 + C_2 \tag{41}$$

$$\frac{1+\mu_1}{E_1}\frac{A_1 R_1^2(1-2\mu_1) + B_1}{R_1} = \frac{R_1}{4}\left(2D_2\ln R_1 - D_2 + 2F_2\right) + \frac{G_2}{R_1} \tag{42}$$

$$A_2 - \frac{B_2}{r_{p2}^2} = \sigma_{c2} \ln r_{p2} + C_2 \tag{43}$$

$$\frac{r_{p2}}{4}(2D_2 \ln r_{p2} - D_2 + 2F_2) + \frac{G_2}{r_{p2}} = \frac{1+\mu_2}{E_2} \frac{A_2 r_{p2}^2 (1 - 2\mu_2) + B_2}{r_{p2}} \tag{44}$$

$$A_2 - \frac{B_2}{R_{\mathrm{out}}^2} = p \tag{45}$$

We have 12 variables A_i, B_i, C_i, D_i, F_i, and G_i ($i = 1$ and 2), but just eight equations (Eqs. 38–45), while the four variables D_i and F_i ($i = 1$ and 2) can be determined using the material parameters and values of other variables. So, all variables can be calculated and the stresses and displacements in both layers can be obtained. Besides, in order to determine the relationship between the load, p, and the radii of plastic zones, r_{p1} and r_{p2}, we need two more equations. According to the extension path of plastic yielding, we know that the stresses along the both inner boundaries of two elastic zones should satisfy the Tresca criterion as well, which gives

$$\frac{2B_1}{r_{p1}^2} = \sigma_{c1} \tag{46}$$

$$\frac{2B_2}{r_{p2}^2} = \sigma_{c2} \tag{47}$$

If we take the Poisson's ratio $\mu_1 = \mu_2 = 0.5$, the relationship between the load and the two radii of the plastic zones can be obtained as

$$p = \frac{\sigma_{c1}}{2}\left[1 + \ln\left(\frac{r_{p1}^2}{R_{\mathrm{in}}^2}\right) - \frac{r_{p1}^2}{R_1^2}\right] + \frac{\sigma_{c2}}{2}\left[1 + \ln\left(\frac{r_{p2}^2}{R_1^2}\right) - \frac{r_{p2}^2}{R_{\mathrm{out}}^2}\right] \tag{48}$$

$$\frac{r_{p2}^2}{r_{p1}^2} = \frac{\sigma_{c1} E_2}{\sigma_{c2} E_1} \tag{49}$$

By letting $r_{p1} = R_1$ and $r_{p2} = R_{\mathrm{out}}$ of Eq. (48), we can get the plastic ultimate bearing capacity p^P as

$$p^p = \sigma_{c1} \ln\left(\frac{R_1}{R_{\mathrm{in}}}\right) + \sigma_{c2} \ln\left(\frac{R_{\mathrm{out}}}{R_1}\right) \tag{50}$$

It can be seen from Eq. (49) that for the given material parameters, the ratio between the two radii of the outer and inner plastic zones keeps equal the coefficient λ we defined in Section 3.4, that is, $\lambda = r_{p2}/r_{p1}$. According to the ratio, we have the following three different modes of plastic yielding:

i. When $R_{\mathrm{out}}/R_1 > \lambda$, the inner layer will first finish the plastic yielding and the corresponding critical load at this state is

$$p = \frac{\sigma_{c1}}{2} \ln\left(\frac{R_1^2}{R_{\mathrm{in}}^2}\right) + \frac{\sigma_{c2}}{2}\left[1 + \ln\left(\lambda^2\right) - \lambda^2 \frac{R_1^2}{R_{\mathrm{out}}^2}\right] \tag{51}$$

ii. When $R_{out}/R_1 = \lambda$, the two layers will finish the plastic yielding simultaneously and the corresponding critical load is the plastic ultimate bearing capacity (see Eq. (50)).

iii. When $R_{out}/R_1 < \lambda$, the outer layer will first finish the plastic yielding and the corresponding critical load at this state is

$$p = \frac{\sigma_{c1}}{2}\left[1 + \ln\left(\frac{R_{out}^2}{\lambda^2 R_{in}^2}\right) - \frac{1}{\lambda^2}\frac{R_{out}^2}{R_1^2}\right] + \frac{\sigma_{c2}}{2}\ln\left(\frac{R_{out}^2}{R_1^2}\right) \tag{52}$$

3.6. Elastic and plastic ultimate bearing capacities of traditional single-layered lining (TSL)

In order to assess the support performance of a FG lining, we will compare the elastic ultimate bearing capacity of the FG lining with that of the traditional single-layered lining with constant Young's modulus and Poisson's ratio. The tangential and radial stresses σ_θ and σ_r are also the maximum and minimum principle stresses, respectively. They can be given by the classic Lame solution, which can be referred in most classic elastic mechanics books [29]. As we mentioned earlier, the plastic zone for this case will initiate from the inner boundary of the lining $r = R_{in}$. So, as the stresses along the inner boundary satisfy the Tresca criterion, that is, $\sigma_\theta - \sigma_r = \sigma_c$ (σ_c is the uniaxial compressive strength (UCS) of the traditional lining material), the loading acting on the outer boundary is just the elastic ultimate bearing capacity of the traditional lining p_{TSL}^e,

$$p_{TSL}^e = \sigma_c\left(1 - R_{in}^2/R_{out}^2\right)/2 \tag{53}$$

As the outer boundary turns into plastic state, the loading is the plastic ultimate bearing capacity,

$$p_{TSL}^p = \sigma_c \ln\left(R_{out}/R_{in}\right) \tag{54}$$

3.7. Analysis and discussions

In order to investigate the support characteristics of different kind of linings, numerical examples are illustrated in this section. Three different types of linings, that is, traditional single-layered lining (TSL), ideal FG lining (IFGL), and double-layered FG lining (DFGL), are considered. The parameters are listed in **Table 1**. In order to facilitate comparative analysis, the values of dimensions, R_{in}, R_1, and R_{out}, the Young's modulus, E, and the UCS, σ_c, are taken based on the model test shown in the following section.

3.7.1. Elastic and plastic ultimate bearing capacity

Figure 6 shows the support characteristic curves (SCCs) of the three different types of linings and the ground response curves (GRCs) of a very good quality quartzite under different in situ stresses given by Hoek and Brown [1], assuming the convergence that has finished as the lining is installed, u_i = 0.3 mm. The displacements of the several given coordinate values in **Figure 6** are the radial displacement of the outer boundary of lining under support pressure, smaller by 0.3 mm than the tunnel deformation on the excavation boundary. It can be easily

	TSL	IFGL	DFGL
Radii, R (m)	$R_{in} = 0.3$, $R_{out} = 0.5$	$R_{in} = 0.3$, $R_{out} = 0.5$	$R_{in} = 0.3$, $R_1 = 0.4$, $R_{out} = 0.5$
Young's modulus, E (GPa)	41	Eq. (11), $E(R_{in}) = 20$	$E_1 = 25$, $E_2 = 41$
Poisson's ratio, μ	0.5	0.5	0.5
UCS, σ_c (MPa)	50	50	$\sigma_{c1} = \sigma_{c2} = 50$

Table 1. Parameters of numerical examples.

seen from **Figure 6** that, as the in situ stress $p_0 = 27$ MPa, all three lining structures can work in elastic state. As $p_0 = 54$ MPa, the IFDL and DFGL structures can still work in elastic state, with approximate 8.4 and 2 MPa elastic bearing capacity unused, respectively. But the TSL cannot work in elastic state in such an in situ stress. Under higher in situ stress, $p_0 = 81$ MPa, all three lining structures will turn into plastic state.

Figure 7 shows the support pressure versus radii of plastic zones. Presenting the relationships in this way helps to visualize the extension path of plastic yielding. The points labeled by same letter in **Figures 6** and **7** indicate the same moment as the support pressure increases. For the TSL, plastic yielding extends radially outward, initiating from its inner boundary (see point A in **Figures 6** and **7**, p = 16.00 MPa) and finishing at the outer boundary (point B, p = 25.54 MPa). For the IFGL, as we mentioned above, the elastic and the plastic ultimate bearing capacities are the same (point C, p = 25.54 MPa). But for the DFGL, the extension mode of plastic yielding

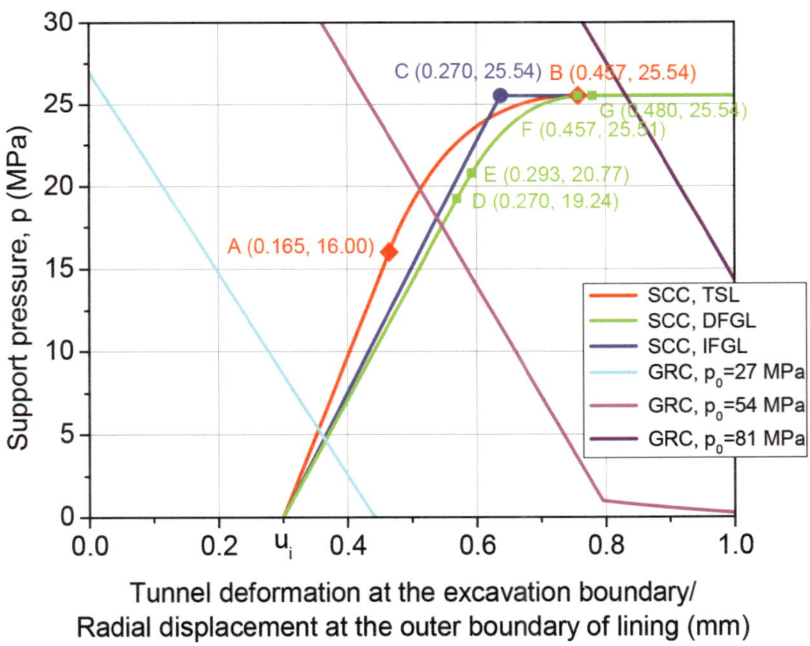

Figure 6. Support characteristic curves of three different types of linings and ground response curves under different in situ stresses.

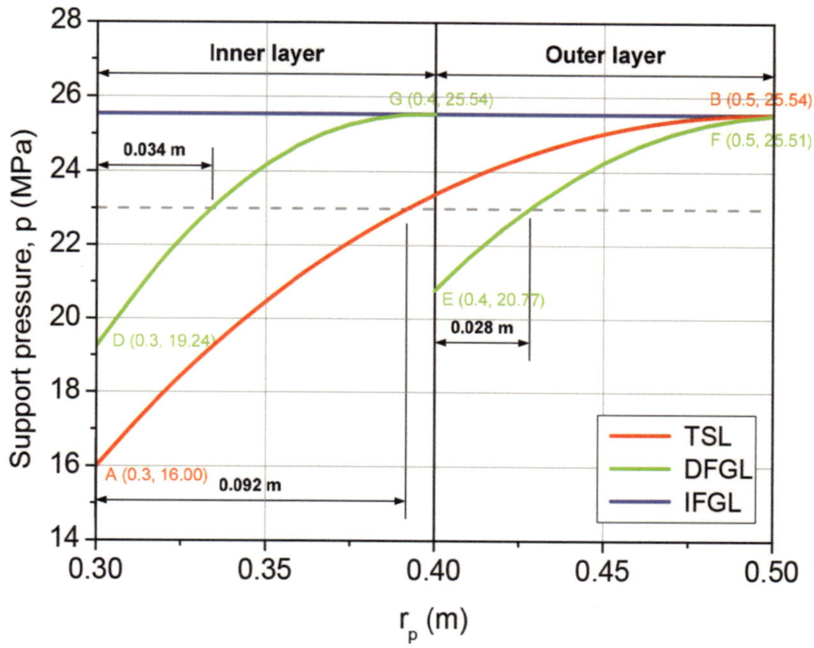

Figure 7. Extension path of plastic yielding.

becomes more complex, depending on the Young's modulus and Poisson's ratio of the materials of both layers. For the incompressible material, that is, $\mu = 0.5$, the extension path can be determined by Eqs. (35)–(37), (50), and (51). Taking the DFGL given in **Table 1** as example, the extension of plastic yielding can be divided into four steps: (i) plastic yielding initiates from the inner boundary of the inner layer (point D, p = 19.24 MPa) and extends radially (see **Figure 5(B)**); (ii) along the inner boundary of the outer layer, plastic yielding begins to occur (point E, p = 20.77 MPa); in this stage, plastic and elastic zones exist alternately in both layers; (iii) the outer layer firstly turns into plastic state totally (point F, p = 25.51 MPa); and (iv) the inner layer turns into plastic state totally (point G, p = 25.54 MPa).

As can be seen from the SCCs in **Figures 6** and **7**, the elastic ultimate bearing capacity of the DFGL is 19.24 MPa, an increase of 20.3% over that of the TSL, 16.00 MPa, and the IFGL can reach to 25.54 MPa, an increase of 59.6%, and, it should also be noted that the increase in the elastic ultimate bearing capacity is proportional to the relative thickness of lining, t/R_{in}. In the numerical examples, $t/R_{in} = 0.667$, which may be higher than most used in real engineering. We will discuss later the effect of t/R_{in} on the elastic ultimate bearing capacity. Unlike the elastic ultimate bearing capacity, the plastic ultimate bearing capacities of all three linings are the same, 25.54 MPa, which is due to the same UCS.

One more thing that needs to be noted is that the behavior model for the lining concrete we used is perfectly elastoplastic (see **Figure 8(a)**), but, as known to all, the concrete material should be strain-softening (see **Figure 8(b)**), which means that the strength will decrease after the concrete turns into plastic yielding state. For the strain-softening TSL, the strength decrease

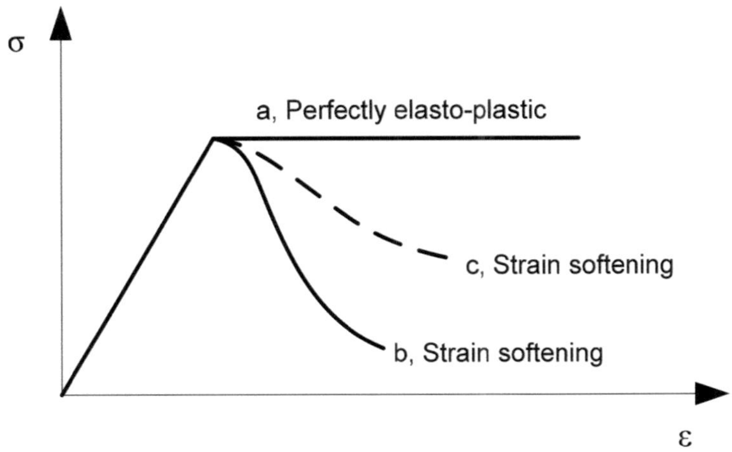

Figure 8. Material behavior models.

will affect its bearing capacity significantly, may be much smaller than that of the perfectly elastoplastic TSL, because the plastic yielding initiates from the inner boundary and extends outward. But for the strain-softening DFGL, the postfailure strength of the plastic zone in the outer layer can be improved in some degree (see **Figure 8(c)**) because of the support action of the elastic zone in the inner layer (see **Figures 5(D)** and **7**). In this way, the support performance of lining can be improved.

3.7.2. Optimized double-layered FG lining

If we want to make the most of a multilayered lining material, the optimum case should be that plastic yielding initiates and finishes in each layer simultaneously, while, for a uniform thick multilayered FG lining like the DFGL in **Table 1**, the optimum case cannot happen. In order to achieve the optimum case for a double-layered lining constructed with incompressible material, according to the above elastoplastic analysis, we have

$$\frac{R_1}{R_{in}} = \frac{R_{out}}{R_1} = \lambda \tag{55}$$

It can be seen from Eq. (55) that for a circular tunnel supported by a double-layered FG lining, if it requires a net space with radius R_{in}, the thickness δ can be chosen as $\delta = (\lambda^2 - 1)R_{in}$ and the radius of the interboundary R_1 should be λR_{in}. Herein, the effect of the convergence on the required net space is ignored.

3.7.3. Effect of thickness on elastic ultimate bearing capacity

As we can see, from the numerical examples, the elastic ultimate bearing capacity can be improved significantly by using the FG linings, either IFGL or DFGL. And, the improvement degree, as mentioned above, is related to the relative thickness as well. So, it is necessary to discuss the effect of lining thickness on the elastic ultimate bearing capacity. The relationships between the dimensionless elastic ultimate bearing capacity and the relative thickness, t/R_{in}, are to be given as follows. For the TSL, the relationship can be given according to Eq. (53),

$$\frac{p^e_{TSL}}{\sigma_c} = \frac{1}{2}\left[1 - \frac{1}{(1 + t/R_{\text{in}})^2}\right] \tag{56}$$

For the IFGL, it can be deduced according to Eq. (12),

$$\frac{p^e_{IFGL}}{\sigma_c} = \frac{p^p_{IFGL}}{\sigma_c} = \ln(1 + t/R_{\text{in}}) \tag{57}$$

For the ODFGL, it can be deduced according to Eqs. (36) and (55),

$$\frac{p^e_{ODFGL}}{\sigma_c} = \frac{t/R_{\text{in}}}{1 + t/R_{\text{in}}} \tag{58}$$

For an uniform thick DFGL, that is, $R_1 = (R_{in} + R_{out})/2$, its elastic ultimate bearing capacity is also related to the values of the Young's modulus. If $R_1/R_{\text{in}} > \lambda$, according to Eq. (35), we have

$$\frac{p^e_{DFGL}}{\sigma_c} = \frac{1}{2}\left\{\frac{E_2}{E_1}\left[\frac{1}{(1 + t/2R_{\text{in}})^2} - \frac{1}{(1 + t/R_{\text{in}})^2}\right] + 1 - \frac{1}{(1 + t/2R_{\text{in}})^2}\right\} \tag{59}$$

if $R_1/R_{\text{in}} < \lambda$, according to Eq. (37), we have

$$\frac{p^e_{DFGL}}{\sigma_c} = \frac{1}{2}\left\{\frac{E_1}{E_2}\left[(1 + t/2R_{\text{in}})^2 - 1\right] + 1 - \left[\frac{1}{2} + \frac{1}{2(1 + t/R_{\text{in}})}\right]^2\right\} \tag{60}$$

It should be noted that Eqs. (56)–(60) are obtained on the assumption that the UCS of each lining material is constant. **Figure 9** illustrates the curves between the dimensionless elastic ultimate bearing capacity with respect to the relative thickness. The curve of the DFGL in **Figure 9** is plotted based on the values of Young's modulus list in **Table 1**, E_1 = 25 GPa and E_2 = 41 GPa. The increases on elastic ultimate bearing capacity of different types of linings with respect to relative lining thickness are listed in **Table 2**. As can be seen from **Figure 9** and **Table 2** that, if we use the IFGL and ODFGL, the increases will become more significant, especially the IFGL. But the opposite situation happens for the DFGL. The elastic ultimate bearing capacity is improved more significantly as the lining is thinner.

3.7.4. Stresses and radial displacement distributions

In order to compare the stress distributions in different types of linings, we take the value of support pressure p = 15 and 23 MPa as examples and the other parameters are the same as in **Table 1**. **Figures 10** and **11** illustrate the radial and tangential stress distributions in the radial direction. The main difference between the three linings is the tangential stress σ_θ and the radial stress distributions are almost the same. As p = 15 MPa (see **Figure 10**), all the three linings are in elastic state. As we know, the tangential stress concentration is the main reason of plastic yielding. Usually, the largest tangential stress concentration occurs along the inner boundary of the lining and the stress there is two dimensional, so the inner boundary of the lining is the most plastic-yielding-prone zone. In this zone, as can be seen from **Figure 10** that the tangential stress concentration of the TSL $\sigma_\theta/p = 3.125$, the largest tangential stress concentration of the TSL, while the tangential stress concentration factor σ_θ/p along the inner boundary of the DFGL and that of the IFGL are 2.599 and 1.958, decreases of 16.8 and 37.3% compared to the TSL, respectively. And the largest tangential stress concentration of the DFGL

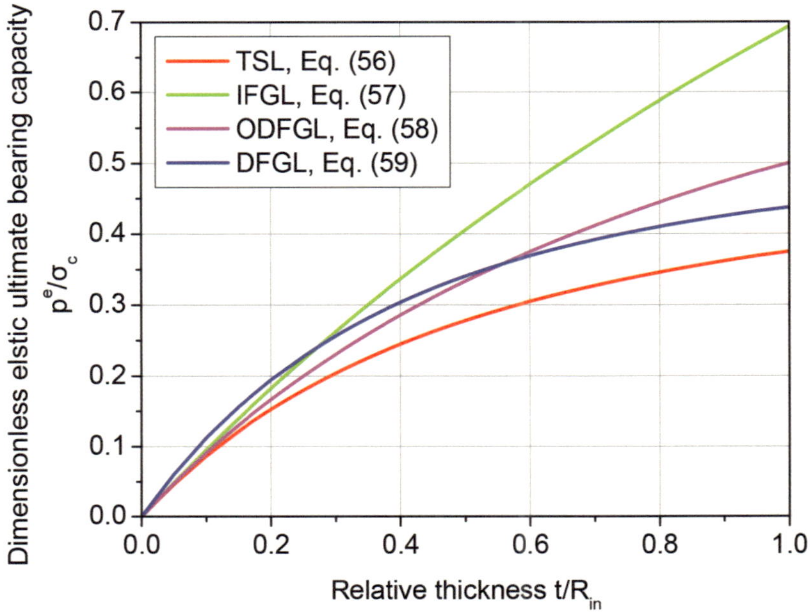

Figure 9. Dimensionless elastic ultimate bearing capacity with respect to relative thickness.

t/R_{in}	p_{TSL^e}/σ_c	p^e_{IFGL}/σ_c	Increment (%)	p^e_{ODFGL}/σ_c	Increment (%)	p^e_{DFGL}/σ_c	Increment (%)
0.05	0.046	0.049	4.96	0.048	2.44	0.061	30.83
0.10	0.087	0.095	9.83	0.091	4.76	0.113	29.72
0.20	0.153	0.182	19.34	0.167	9.09	0.195	27.65
0.30	0.204	0.262	28.52	0.231	13.04	0.257	25.77
0.40	0.245	0.336	37.39	0.286	16.67	0.304	24.07
0.50	0.278	0.405	45.97	0.333	20.00	0.340	22.53
0.60	0.305	0.470	54.26	0.375	23.08	0.369	21.12
0.70	0.327	0.531	62.28	0.412	25.93	0.392	19.83
0.80	0.346	0.588	70.04	0.444	28.57	0.410	18.66
0.90	0.361	0.642	77.55	0.474	31.03	0.425	17.58
1.00	0.375	0.693	84.84	0.500	33.33	0.437	16.59

Note: All the increases are calculated comparing with the TSL.

Table 2. Increases on elastic ultimate bearing capacity.

http://dx.doi.org/10.5772/intechopen.81435

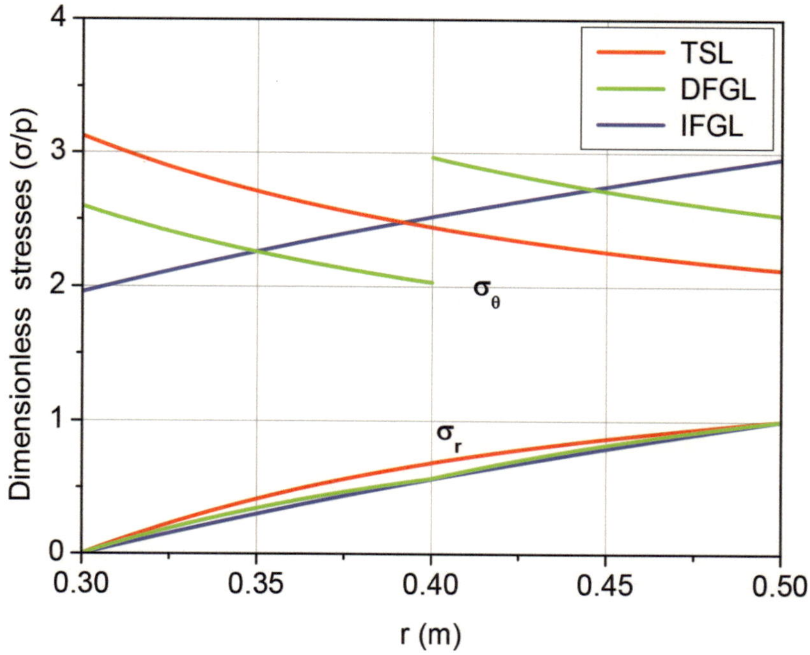

Figure 10. Stress distributions in the radial direction (p = 15 MPa).

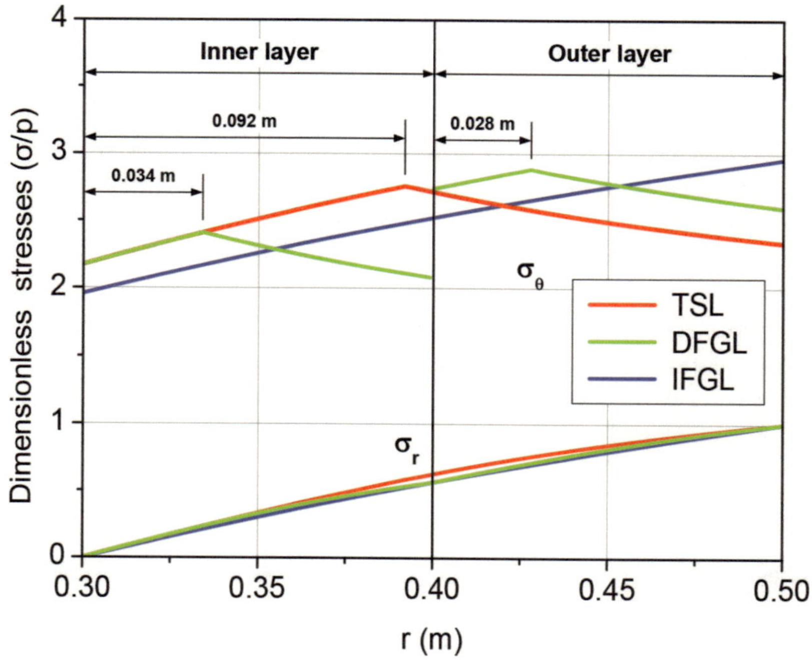

Figure 11. Stress distributions in the radial direction (p = 23 MPa).

and that of the IFGL, $\sigma_\theta/p = 2.966$ and 2.958, occur along the inner boundary of the outer layer and the outer boundary of the lining, respectively, where both are in three dimensional stress. So, the stress distributions in both the DFGL and IFGL are more reasonable than that in the TSL. As p = 23 MPa (see **Figure 11**), the given support pressure is larger than the elastic ultimate bearing capacities of the TSL and DFGL, so parts of the two linings turn into plastic state. But the IFGL is still in elastic state. The thickness of plastic zone in the TSL is about 0.092 m, 46% of the total thickness of the lining, while, for the DFGL, the thickness of plastic zone in both layers is about 0.062 m total, 31% of the total thickness of the lining. So, according to the stress distributions and the plastic area, we can come to that both the DFGL and IFGL are superior to the TSL.

4. Model tests

4.1. Mixture of model material

In order to testify the superiorities of the FG linings over the TSL, we carried out a model test to compare the support characteristics. Due to the difficulties of constructing a multilayered concrete lining, the simplest multilayered lining, a double-layered lining structure, is chosen. As mentioned earlier, in order to obtain a FG concrete lining, we need a FG concrete mixture (FGC), producing concrete with same strength but different Young's modulus.

The main factors influencing the strength and/or the Young's modulus contain the aggregate properties (like the quality [30], the type and maximum aggregate size [31], the grade [32],the Young's modulus [33], etc.), the water/binder ratio [34], the chemical admixtures (like the air-void mixture [35], shrinkage-reducing mixture [36], retarders and accelerators [37], superplasticizer [38], etc). Besides, the steel or polyfibers [39], usually used in practice, also have influence on the concrete strength and Young's modulus. In order to obtain the FGC, we conducted orthogonal test and a large number of single factor tests, investigating the influences of almost all of the abovementioned factors on the Young's modulus and compressive strength of concrete. Experimental data were analyzed statistically. Test results of multivariate analysis of variance (MANOVA) with 95% confidence level ($\alpha = 0.05$) show that the quantity and modulus of coarse aggregate are two significant factors influencing the Young's modulus, but the compressive strength is slightly influenced by the two factors. The poly fiber also has such an influence just not that significant. So, the quantity and modulus of coarse aggregate the poly fiber are chosen as the major factors to prepare the FGC.

4.1.1. Raw materials

The mixture test and the model tests later in this chapter are based on Portland Cement 42.5R as the cementing material. The physical parameters of cement are listed in **Table 3**. The coarse aggregate is crushed limestone with the size between 5 and 20 mm. **Figure 12** shows the size

Type of cement	P.O 42.5R
Specific surface area (m^2/kg)	330
Initial setting time (min)	≥60
Final setting time (h)	6
Compressive strength (3d) (MPa)	≥24.0
Compressive strength (28d) (MPa)	≥48.0
Flexural strength (3d) (MPa)	≥4.5
Flexural strength (28d) (MPa)	≥7.0

Table 3. Parameters of cement.

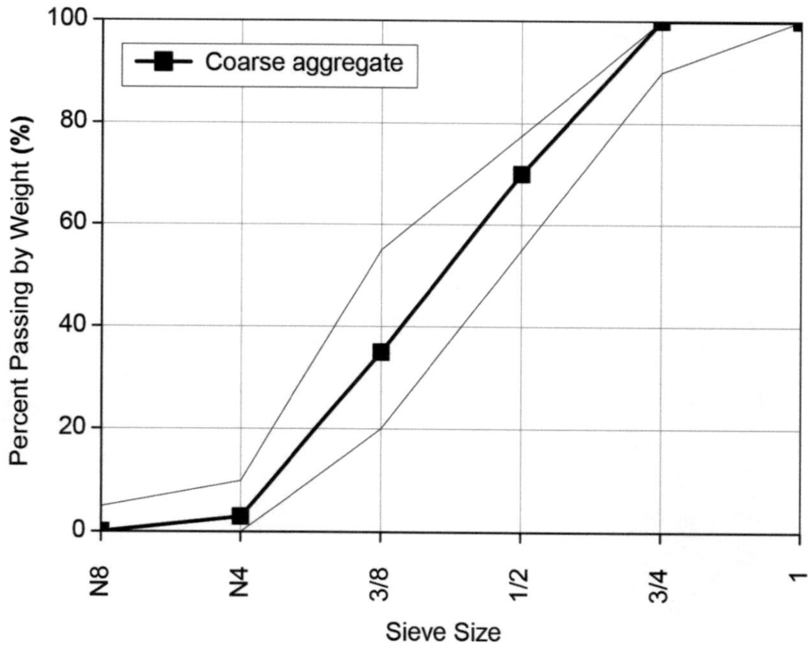

Figure 12. Grading of coarse aggregates.

distribution of the coarse aggregate, including the corresponding ASTM C 33 limits. It can be seen that the coarse aggregate is within the ASTM C 33 coarse aggregate grading band [40]. The fine aggregate was river sand with a fineness modulus of 3.2 (see **Figure 13**). Polypropylene fibers, GRPF-12 mm (**Figure 14**), are added in concrete mixes at different volumetric fractions. Specifications are listed in **Table 4**. The superplasticizer (JM-PCA(V)) is a complex additive based on carboxylic grafted polymeric, produced by Nanjing Subote New Materials Co., Ltd., China.

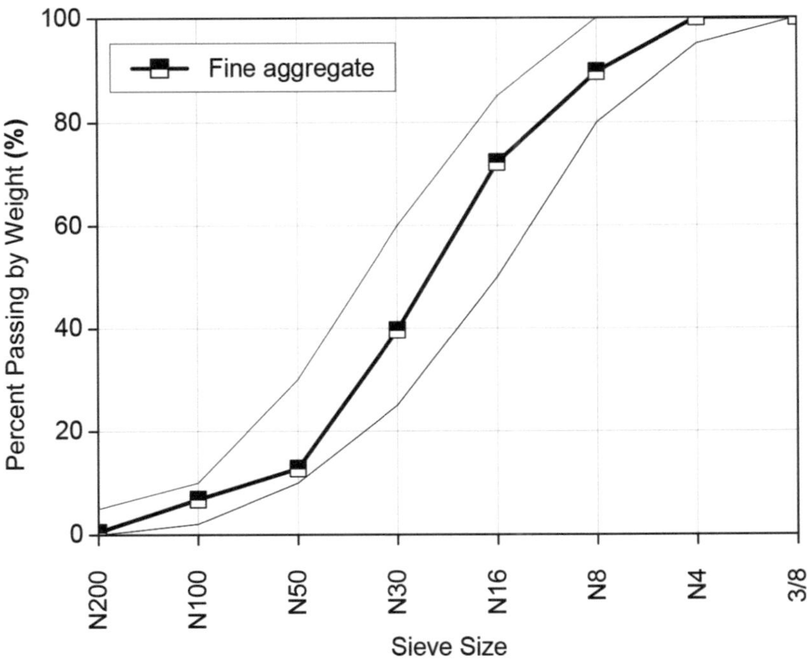

Figure 13. Size distribution for fine aggregate with fineness modulus = 3.2.

Figure 14. Polypropylene fiber.

Material composition	100% polypropylene
Density (g/cm^3)	0.91
Equivalent diameter (μm)	27.69
Fracture strength (MPa)	641
Initial modulus (GPa)	8.5
Elongation at break (%)	30
Length (mm)	12
Melting point (°C)	>160
Material shape	Monofilament
Chemical properties	Acid-proof, alkali-proof, nontoxic

Table 4. Properties of polypropylene fiber.

No.	Water (kg/m^3)	Cement (kg/m^3)	Sand (kg/m^3)	Aggregate (kg/m^3)	Water reducer (kg/m^3)	Polypropylene fiber (kg/m^3)	Properties			
							Compressive strength, σ_c (Mpa)	Young's modulus, E (Gpa)	Poisson's ratio	Slum (cm)
1	156	390	585	1325	8.775	8	54.6	40.2	0.228	8.5
2	187	467	702	1040	10.508	—	51.3	41.0	0.235	17.0
3	218	546	819	795	12.285	4	45.9	37.7	0.235	12.0
4	255	686	1029	—	17.493	4	50.2	26.9	0.234	6.0
5	255	686	1029	—	17.493	8	49.1	24.8	0.233	6.0
6	255	686	1029	—	17.493	12	46.0	24.1	0.23	4.0

Table 5. Proportions and mechanical properties of FGCs (kg/m^3).

4.1.2. FGC mixture

Based on the results of tests, six concrete mixtures and the parameters are given in **Table 5**.

4.2. Model test program

4.2.1. Lining structures and dimensions

As mentioned earlier, in the model tests, we conduct two lining structures, that is, the traditional single-layered lining (**Figure 15(I)**) and the double-layered FG lining (**Figure 15(II)**), assessing the support characteristics. Because the two layers of the DFGL need to be poured separately, the interface between them is unavoidable. In order to reduce the interface effect to the results, the TSL is also poured twice like the DFGL. The pouring process is to be given later.

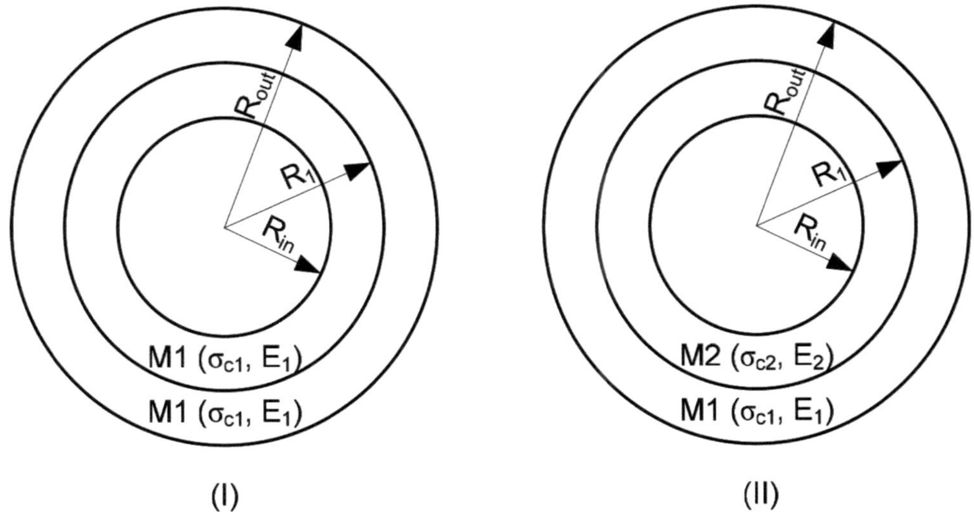

Figure 15. Two types of lining structures: (I) TSL and (II) DFGL.

Considering the working space of the test platform, the dimensions are chosen as $R_{in} = 300$ mm, $R_1 = 400$ mm, and $R_{out} = 500$ mm. It should be noted that the problem we discussed in our theoretical models is a plain strain problem; so in order to get linings that can simulate the state of a plain strain, the linings should be long enough, comparing with the dimensions of the cross section. The length in our test is 2300 mm (**Figure 16**).

4.2.2. *Material properties*

Combining the theoretical results and the functionally graded concrete (FGC), two types of concrete, M1 and M2, are poured using the two mixtures, No. 2 and 5 in **Table 5**, respectively. The mechanical parameters of the lining concretes are listed in **Table 6**. The parameters are

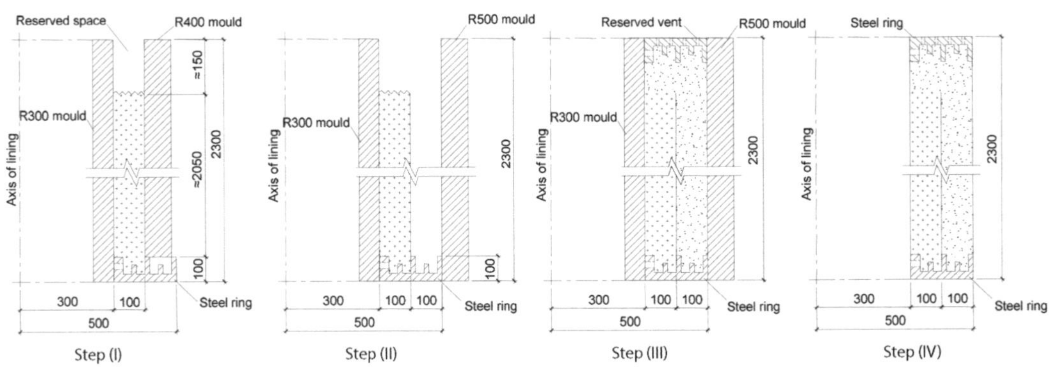

Figure 16. Pouring process of DFGL (mm).

Lining types		UCS, σ_c (MPa)	Young's modulus, E (GPa)	Poisson's ratio, μ	Slump (cm)
Single-layered lining	Inner layer	51.260	40.687	0.235	19
	Outer layer	50.120	41.047	0.236	17
Two-layered FGL	Inner layer	49.634	25.391	0.235	16
	Outer layer	48.583	41.199	0.233	7

Table 6. Mechanical properties of lining concretes.

obtained by testing the specimens poured using the real lining concrete mixtures, not the results in **Table 5**. The specimens and the linings are all cured in water (20 ± 2°C) for 28 days.

4.2.3. Casting of the lining

We firstly pour the inner layer and then the outer layer. The process in detail is as follows:

i. Choose 400- and 300-mm-radius steel cylinders (R400 and R300) as the outer and inner moulds of the lining inner layer, respectively, and fix them on the bottom steel ring, making sure they are located with the same axis (**Figure 16(I)**).

ii. Wipe release agent and pour concrete of the inner layer (**Figure 16(I)**). In order to fix the top steel ring and ensure the contact between the concrete and the top steel ring, a certain space should be reserved and the surface should be rough (**Figure 16(I)**).

iii. 24 hours later, tear down the outer mold.

iv. Choose 500-mm-radius steel cylinder (R500) as the outer mold of the lining outer layer, wipe release agent, and fix it around the bottom steel ring (**Figure 16(II)**).

v. Pour the concrete of the outer layer (**Figure 16(III)**) and fix the top steel ring.

vi. Tear down R300 and R500 (**Figure 16(IV)**).

vii. 48 hours later, move lining into the special curing box.

4.2.4. Monitoring scheme

As mentioned earlier, both ends of linings are poured in steel rings, we can say both ends are fixed in the horizontal cross section, which, of course, will lead to difference between the theoretical model and the test linings. So, three monitoring cross sections are all in the middle part of the linings. The monitoring parameters contain the radial displacements on the inner surface (all displacement sensors are fixed on a vertical steel cylinder with magnetic stands, the axial displacements at both ends, and the tangential and axial strains on the inner and outer surfaces). Due to the limited working space, displacements only in the upper two cross sections are monitored. **Figure 17** shows the layout of monitoring points.

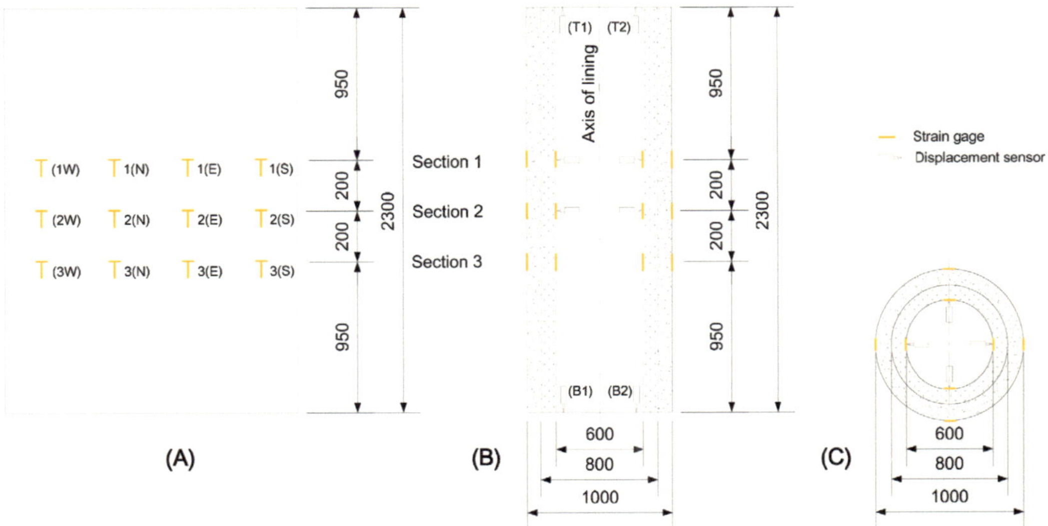

Figure 17. Layout of monitoring points (mm): (A) unfolded section; (B) vertical section; and (C) horizontal section.

4.3. Results and discussions

As mentioned earlier, the load process is accomplished in three steps. In this section, we just analyze the third step, that is, confining pressure step. In addition, due to the complexity and the long loading time, only part of displacement sensors and strain gages work regularly, so, in our analysis, only effective data are illustrated. For the test results, the sign convention is defined positive for tension strain and outward displacement, and negative for compressive strain and inward displacement. All data, related to displacements, strains, and pressures, are collected with interval of 5 seconds.

Figure 18 illustrates the curves of the confining pressure versus loading time. **Figure 19** illustrates the convergences along inner boundaries of linings versus support pressure, containing model test and theoretical results. As can be seen from **Figure 19**, the confining pressures of both the TSL and DFGL drop suddenly after reaching the maximum values. For comparison, the results of theory and model test are listed in **Table 7**. It can be noted that the maximum confining pressure that the DFGL can bear in model test is approximately 20.44 MPa, 24.56% greater than that of the TSL, 16.41 MPa. The two maximum confining pressures are close to the theoretical elastic ultimate bearing capacities of the DFGL and TSL (19.24 and 16.00 MPa, corresponding to points D and A in **Figure 19**), respectively.

4.3.1. Effect of Poisson's ratio on ultimate bearing capacity

It should be noted that the two theoretical elastic ultimate bearing capacities of the DFGL and TSL are obtained by assuming the lining concrete as incompressible, that is, $\mu = 0.5$. According to the theoretical analysis, the plastic ultimate bearing capacities are just affected by the material strengths, having nothing to do with the Poisson's ratio. The only one elastic ultimate

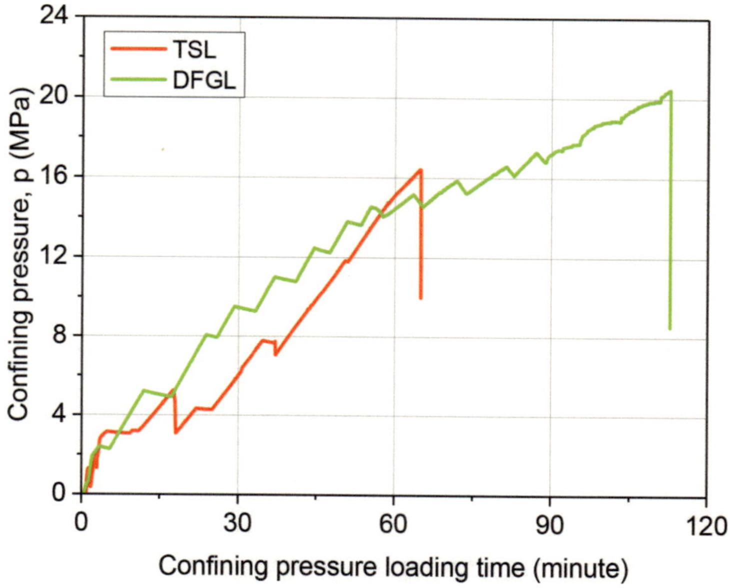

Figure 18. Loading scheme.

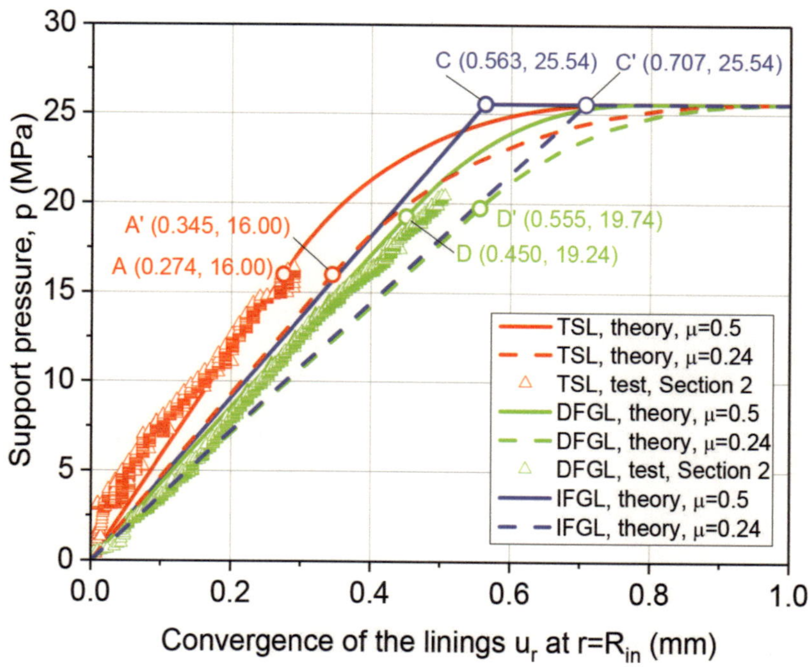

Figure 19. Convergences along inner boundaries of linings versus support pressure.

Lining types	Elastic ultimate bearing capacities ($\mu = 0.5$), p_e (MPa)	Plastic ultimate bearing capacities, p_p (MPa)	Maximum confining pressure in model tests, p_{max} (MPa)
TSL	16.00	25.54	16.41
DFGL	19.24 (19.74[1])	25.54	20.44
IFGL	25.54	25.54	

[1]The value for $\mu = 0.24$.

Table 7. Ultimate bearing capacities of linings.

bearing capacity related to the Poisson's ratio is that of the DFGL (see **Table 7**). Let $\mu = 0.24$ (see **Table 6**), we can obtain the elastic ultimate bearing capacities, $p^e_{DFGL} = 19.74$ MPa (see point D' in **Figures 18** and **19**), which is closer to the maximum confining pressure from the model tests. So, the maximum confining pressures of the TSL and DFGL should be their elastic ultimate bearing capacities, which can also be confirmed by **Figure 18**.

4.3.2. *Effects of the Poisson's ratio on radial displacement*

From **Figure 19**, in the elastic stage, the Poisson's ratio affects the radial displacement of all the three lining structures in a same rule, that is, for a given confining pressure, the radial displacement for $\mu = 0.24$ is greater than that for $\mu = 0.5$. The Poisson's ratios of the concrete in the model tests are about 0.24. So, the p-u_r curves of TSL and DFGL from model tests should be closer to those theoretical ones as $\mu = 0.24$ than as $\mu = 0.5$. The opposite happens, that is, the convergences by model tests are smaller than the analytical values, being closer to the

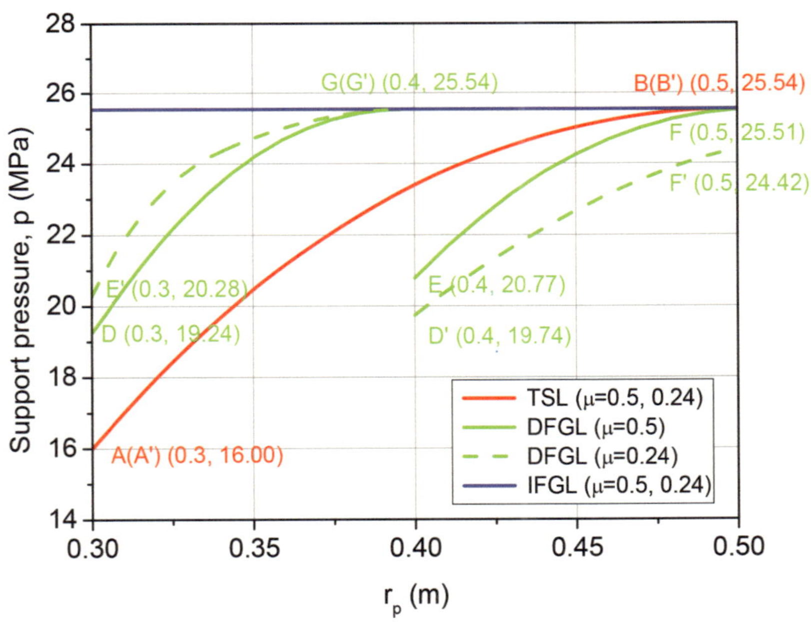

Figure 20. Extension path of plastic yielding.

analytical results as $\mu = 0.5$. This situation may be resulted from the restrictions of steel rings at both ends of linings in our model tests (**Figure 16**).

4.3.3. Effects of the Poisson's ratio on plastic zone

Figure 20 illustrates the extension paths of plastic yielding of all the three lining structures as $\mu = 0.24$ and 0.5. The value of the Poisson's ratio only affects the plastic yielding path of the DFGL, but not that of the IFGL and TSL, which is because the Poisson's ratio is related to the displacement and the displacement continuous conditions are used in the elastoplastic analysis of the DFGL, but not in the IFGL and TSL. For $\mu = 0.24$, the plastic yielding will (i) initiate from the inner boundary of the outer layer (point D′ in **Figures 19** and **20**, where p = 19.74 MPa), not from the inner layer as $\mu = 0.5$; (ii) the inner boundary of the inner layer turns into plastic yielding at point E′, where p = 20.28 MPa; (iii) the outer layer totally turns into plastic state at point F′, where p = 24.42 MPa; and followed by (iv) the inner layer turns into plastic state at point G′, where p = 25.54 MPa.

5. Conclusions

The purpose of the work is to show a new way to improve the elastic ultimate bearing capacity of a circular single-layered concrete lining. We proposed the concept of functionally graded concrete lining, which is based on the inverse analysis of assuming the concrete failure ruled by the Tresca criterion, and the theoretical analysis can verify the effectiveness of the method. But the requirements of the theoretical ideal functionally graded lining, that is, the constant compressive strength and continuously increasing Young's modulus in radially outward direction, can hardly be achieved for concrete material. So, we choose the multilayered lining as the alternative.

Elastoplastic analysis is carried out, and the elastic and plastic ultimate bearing capacities are obtained. The results show that a double-layered functionally graded lining has greater elastic ultimate bearing capacity but the same plastic ultimate bearing capacity comparing with the traditional single-layered lining. The distributions of stresses in ideal functionally graded lining and double-layered functionally graded lining are more reasonable, maximum tangential stress concentration decreasing and no longer in the inner boundary of lining. Even if in the plastic stage, the areas of plastic zone in functionally graded linings are smaller than that in single-layered lining. And the opposite happens for the radial displacement, which is resulted from the lower Young's modulus of the inner layer. Besides, the Poisson's ratio has effect on the radial displacement, elastic ultimate bearing capacity, and extension path of plastic yielding.

Finally, model tests are conducted in order for verifying the theoretical results. A traditional single-layered lining and a double-layered functionally graded lining are poured and loaded in axial and confining pressures. The maximum confining pressures that the linings can bear in test are only slightly greater than the corresponding theoretical elastic ultimate bearing capacity. Combined with the curves of pressure versus radial displacement, the maximum pressures can be regarded as the respective elastic bearing capacity. In this way, the theoretical results are verified by the model tests and the functionally graded lining would make a great deal of practical sense.

Acknowledgements

This research work was supported by the Natural Science Foundation of China (Grant No. 51704117, 11572126). Their financial support is gratefully acknowledged.

Conflict of interest

The authors declare that they have no conflict of interest.

Author details

Ning Zhang[1,2]*, Aizhong Lu[1] and Xuguang Chen[2]

*Address all correspondence to: zning1125@ncepu.edu.cn

1 Institute of Hydroelectric and Geotechnical Engineering, North China Electric Power University, Beijing, China

2 Shandong Provincial Key Laboratory of Marine Environment and Geological Engineering, Ocean University of China, Qingdao, China

References

[1] Hoek E, Brown ET. Underground Excavations in Rock. London: The Institution of Mining and Metallurgy; 1980

[2] Koizumi M. FGM activities in Japan. Composites Part B: Engineering. 1997;**28**:1-4

[3] Mirsalehi M, Azhari M, Amoushahi H. Stability of thin FGM microplate subjected to mechanical and thermal loading based on the modified couple stress theory and spline finite strip method. Aerospace Science and Technology. 2015;**47**:356-366

[4] Gasik MM. Functionally graded materials: Bulk processing techniques. International Journal of Materials and Product Technology. 2010;**39**:20-29

[5] Jha DK, Kant T, Singh RK. A critical review of recent research on functionally graded plates. Composite Structures. 2013;**96**:833-849

[6] Yang JJ, Hai R, Dong YL. Effect of the component gradient distribution of the strength of cement-based composite materials. Journal of the Chinese Ceramic Society. 2002;**30**:804-807

[7] Hai R, Wu KR, Yang JJ. Effect of reactive aggregate gradient distribution in cement-based functionally gradient materials on its mechanical property. Journal of Tongji Unlversity (Natural Science). 2006;**34**:786-789

[8] Shen B, Hubler M, Paulino GH, Struble LJ. Functionally-graded fiber-reinforced cement composite: Processing, microstructure, and properties. Cement and Concrete Composites. 2008;**30**:663-673

[9] Dias CMR, Savastano H Jr, John VM. Exploring the potential of functionally graded materials concept for the development of fiber cement. Construction and Building Materials. 2010;**24**:140-146

[10] Shi Z, Zhang T, Xiang H. Exact solutions of heterogeneous elastic hollow cylinders. Composite Structures. 2007;**79**:140-147

[11] Xiang H, Shi Z, Zhang T. Elastic analyses of heterogeneous hollow cylinders. Mechanics Research Communications. 2006;**33**:681-691

[12] Jabbari M, Sohrabpour S, Eslami MR. General solution for mechanical and thermal stresses in a functionally graded hollow cylinder due to nonaxisymmetric steady-state loads. Journal of Applied Mechanics. 2003;**70**:111-118

[13] Eslami MR, Babaei MH, Poultangari R. Thermal and mechanical stresses in a functionally graded thick sphere. International Journal of Pressure Vessels and Piping. 2005;**82**:522-527

[14] Batra RC, Iaccarino GL. Exact solutions for radial deformations of a functionally graded isotropic and incompressible second-order elastic cylinder. International Journal of Non-Linear Mechanics. 2008;**43**:383-398

[15] Nie GJ, Batra RC. Material tailoring and analysis of functionally graded isotropic and incompressible linear elastic hollow cylinders. Composite Structures;**92**:265-274

[16] Nie GJ, Batra RC. Static deformations of functionally graded polar-orthotropic cylinders with elliptical inner and circular outer surfaces. Composites Science and Technology. 2010; **70**:450-457

[17] Tutuncu N, Temel B. A novel approach to stress analysis of pressurized FGM cylinders, disks and spheres. Composite Structures. 2009;**91**:385-390

[18] Theotokoglou EE, Stampouloglou IH. The radially nonhomogeneous elastic axisymmentric problem. International Journal of Solids and Structures. 2008;**45**:6535-6552

[19] Horgan CO, Chan AM. The pressurized hollow cylinder or disk problem for functionally graded isotropic linearly elastic materials. Journal of Elasticity. 1999;**55**:43-59

[20] Dai HL, Fu YM, Dong ZM. Exact solutions for functionally graded pressure vessels in a uniform magnetic field. International Journal of Solids and Structures. 2006;**43**:5570-5580

[21] Batra RC, Bahrami A. Inflation and eversion of functionally graded non-linear elastic incompressible circular cylinders. International Journal of Non-Linear Mechanics. 2009; **44**:311-323

[22] Tutuncu N. Stresses in thick-walled FGM cylinders with exponentially-varying properties. Engineering Structures. 2007;**29**:2032-2035

[23] Marin L, Lesnic D. The method of fundamental solutions for nonlinear functionally graded materials. International Journal of Solids and Structures. 2007;**44**:6878-6890

[24] Shokrolahi-Zadeh B, Shodja HM. Spectral equivalent inclusion method: Anisotropic cylindrical multi-inhomogeneities. Journal of the Mechanics and Physics of Solids. 2008;**56**: 3565-3575

[25] Chen YZ, Lin XY. An alternative numerical solution of thick-walled cylinders and spheres made of functionally graded materials. Computational Materials Science. 2010;**48**:640-647

[26] Lu AZ, Jiang BS. Inverse Problems in Rock Mechanics. Beijing: Coal Industry Press; 1998

[27] Batra RC. Optimal design of functionally graded incompressible linear elastic cylinders and spheres. AIAA Journal. 2008;**46**:2050-2057

[28] Xu ZL. A Concise Course in Elasticity. 3rd ed. Beijing: Higher Education Press; 2002

[29] Sp T, Jn G. Theory of Elasticity. 3rd ed. New York: McGraw-Hill Book Company; 1970

[30] Beshr H, Almusallam AA, Maslehuddin M. Effect of coarse aggregate quality on the mechanical properties of high strength concrete. Construction and Building Materials. 2003;**17**:97-103

[31] Grabiec AM, Zawal D, Szulc J. Influence of type and maximum aggregate size on some properties of high-strength concrete made of pozzolana cement in respect of binder and carbon dioxide intensity indexes. Construction and Building Materials. 2015;**98**:17-24

[32] Meddah MS, Zitouni S, Belâabes S. Effect of content and particle size distribution of coarse aggregate on the compressive strength of concrete. Construction and Building Materials. 2010;**24**:505-512

[33] Zhou FP, Lydon FD, Barr BIG. Effect of coarse aggregate on elastic modulus and compressive strength of high performance concrete. Cement and Concrete Research. 1995;**25**:177-186

[34] Chi JM, Huang R, Yang CC, Chang JJ. Effect of aggregate properties on the strength and stiffness of lightweight concrete. Cement and Concrete Composites. 2003;**25**:197-205

[35] Nambiar EKK, Ramamurthy K. Air-void characterisation of foam concrete. Cement and Concrete Research. 2007;**37**:221-230

[36] Yoo D-Y, Banthia N, Yoon Y-S. Effectiveness of shrinkage-reducing admixture in reducing autogenous shrinkage stress of ultra-high-performance fiber-reinforced concrete. Cement and Concrete Composites. 2015;**64**:27-36

[37] Claisse PA. Chapter 24—Admixtures for concrete. In: Claisse PA, editor. Civil Engineering Materials. Boston: Butterworth-Heinemann; 2016. pp. 251-258

[38] Gołaszewski J. Influence of cement properties on new generation superplasticizers performance. Construction and Building Materials. 2012;**35**:586-596

[39] Afroughsabet V, Ozbakkaloglu T. Mechanical and durability properties of high-strength concrete containing steel and polypropylene fibers. Construction and Building Materials. 2015;**94**:73-82

[40] ASTM. C 33 Standard Specification for Concrete Aggregates. Philadelphia, PA: American Society for Testing and Materials; 2003

Chapter 8

Cumulative Tensile Damage and Consolidation Effects on Fracture Properties of Sandstone

Martin Šperl and Miloš Drdácký

Additional information is available at the end of the chapter

http://dx.doi.org/10.5772/intechopen.81434

Abstract

The presence of cracks in many historical objects indicates the action of external forces accompanied by internal strain gradients. This is usually a repetitive process, and damage cumulation may occur. A study of these effects requires a suitable methodology for testing historical stone that has been subjected to repeated tension strains. The chapter presents the results of a pilot experimental assessment of changes in the mechanical characteristics of sandstone due to accumulation of damage. The Young modulus and the Poisson number were investigated, using a verified methodology for testing stone in simple tension and in cyclic simple tension/compression loading. The results show that the first tension load displacement can be approximated very satisfactorily by a power function, and the optical digital image correlation (DIC) method again demonstrated its capacity and suitability for measuring the complex deformation field on porous surfaces and on naturally well-structured surfaces. The chapter further presents a methodology for investigating fracture phenomena in sandstone treated for consolidation. It shows the preparation of test specimens with a cyclic loading generated crack, control of the test specimen preparation, and verification by means of X-ray micro-CT and DIC techniques. The chapter illustrates an influence of various consolidation agents on the toughness of cracked specimens.

Keywords: damage cumulation, sandstone, historical monuments, simple tension, cyclic loading, stone consolidation

1. Introduction

Visible cracks are present on most historical buildings, structures, and objects of art made of porous brittle or quasi-brittle materials. Stone monuments form an important category of these structures.

IntechOpen

© 2018 The Author(s). Licensee IntechOpen. This chapter is distributed under the terms of the Creative Commons Attribution License (http://creativecommons.org/licenses/by/3.0), which permits unrestricted use, distribution, and reproduction in any medium, provided the original work is properly cited.

In principle, visible cracks may develop due to external forces acting under various time situations, for example, incidental shocks or gradually increasing loads, or due to internal strain gradients, which mostly generate cumulative damage, such as fatigue phenomena.

Cumulative tensile damage may occur in stone monuments due to repeated environmental uneven volumetric change. Stone objects are frequently subjected to repeated changes in temperature from solar irradiation, when the heat is transferred unevenly inside the material. The accompanying different thermal dilations may cause dangerous stresses that lead to cumulative crack propagation, typically initiated in the interior defects that are usually present. In addition, a combination of temperature and moisture dilation effects may worsen the situation, especially in sandstone or in specific fragile stones, e.g., in the so-called "opuka" stone (Cretaceous marly stone) used in the Czech Republic. In the case of these materials, stone elements in the interior environment are also threatened by environmental fatigue. Let us present a typical example that has been described in greater detail in Drdácký [1].

The second author of this paper investigated the case of a severe failure of the sandstone tracery of the triforium of St. Vitus Cathedral, where repeated uneven humidity loading (wetting and drying) and temperature loading (heating and cooling) were the cause of the collapse. A detailed survey discovered that this type of damage was present or had been repaired on most of the tracery arches of the triforium, where further failures subsequently occurred. The extent of the dangerous stresses was verified using simple thermal loading modeling and computations. It was proved that the state of stress in the damaged places reached or surpassed the material strength levels (**Figure 1**) [2].

The triforium tracery elements are chiseled out of a single piece of sandstone. They are connected with the lower window beam of the nave or the transept. The triforium structure is situated on the border between the exterior and the interior, and its ceiling plates are exposed to the exterior environment forming the roof structure. The temperature of the air fluctuates in the exterior and the interior of the cathedral. The stone elements of the triforium include parts of different proportions that have been composed into a single unit. Due to rapid changes in the ambient temperature, situations can arise when the temperature of the massive parts is different from the temperature of the subtler elements. This situation generates a stress inside

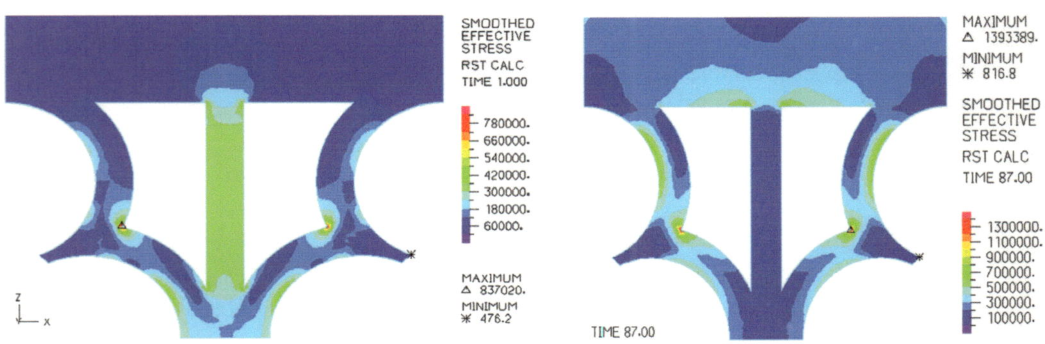

Figure 1. Distribution of the effective stress in the slice for coordinate y = 1.37; 3D model of own weight (left) and 3D model of warming up (right) [1, 2]. (Image by P. Beran).

the structure. The same effect is caused by the change in moisture content that always accompanies temperature changes.

Similar damage has been observed on other historic buildings, for example, on the church of Notre Dame du Sablon in Brussels (see **Figure 2**) where several failed original elements have been replaced.

Adverse effects of this type have led to a call for studies of the behavior of similar stone structures under extreme climate effects. These studies require greater knowledge than what is available of the fracture characteristics of the historical stone. Numerous works on the fracture behavior of rock have been published, mostly based on tests on cylindrical stone specimens under cyclic compression according to the ISRM standards (e.g., Ning et al. [3], Bagde and Petroš [4], Backers [5]). Quantitative toughness data are calculated, or qualitative data are assessed on the basis of these tests. A classical work by Attewell and Farmer [6] and also a comprehensive work by Jian-Qing et al. [7], which review fatigue damage variables and discuss a broad range of NDT measurement methods, in particular ultrasonic and acoustic emission methods, are worth mentioning. However, this type of testing methodology is not suitable for studies of the effects of conservation treatment on the fatigue behavior of stone. No pure tensile fatigue tests on stone have been found in the literature, only three-point or four-point bending tests (e.g., Cardani and Meda [8]). There is a lack of data on the behavior of stone under repeated uniaxial tensile loads, and until now no suitable methodology for testing stone toughness before and after conservation treatment has been suggested and accepted.

Figure 2. Repaired tracery with damaged elements in Notre Dame du Sablon (photo by M. Drdácký).

This paper presents the results of an experimental pilot assessment of changes in the mechanical properties of sandstone from Božanov (a typical material used on the Charles Bridge in medieval times) resulting from accumulated damage. The Young modulus and the Poisson number were observed. In addition, a methodology for testing stone in tension and in cyclic tension/compression loading is suggested and verified.

2. Experiments

Božanov stone is a grayish beige gross grain strong arkose sandstone without marked layering (**Figure 3**). The material used in the tests was extracted from the 11th arch of the Charles Bridge parapet wall, which had to be replaced during recent repairs. It has a rather deeply located detachment crack parallel to the surface, probably due to some previous surface treatment which had locked moisture inside the stone. The material can be characterized as a quasi-brittle inelastic silicate composite.

Test specimens $50 \times 50 \times 200\ mm^3$ in dimensions were fixed into rectangular steel tubes with axially welded flat steel hangers, which served to fix the set into the hydraulic grips of the loading frame (**Figure 4**). Two-component Sikadur-31 CF RAP resin was used for gluing. The fixtures had to be prepared with great precision in order to ensure perfect alignment and perpendicular arrangement for tension and combined loading. The specimens in prismatic form were cut with negligible geometrical imperfections in a precise prismatic form.

The specimens were loaded into an Instron 1343 electrohydraulic testing frame with force capacity of 100 kN (**Figure 5**) using an Instron 100 kN load cell, a FastTrack 8800 controller, two Instron extensometers type 2620-602 (accuracy class 1, with maximum possible error of 0.5% of the read value, measurement base l_0 = 50 mm), and WaveMatrix measurement software. The extensometers were placed on opposite sides of the specimen, and their read values were arithmetically averaged. The contacts on the surface of the stone were provided with glued thin aluminum sheets to protect the extensometer wedges. The deformations were also

Figure 3. Macro photo of the Božanov sandstone (arkose) structure (photo J. Frankl).

Figure 4. Test compression/tension specimens (photo by M. Šperl).

Figure 5. Arrangement of pure tension tests on the stone specimens (photo by M. Šperl).

recorded and evaluated using a contactless telecentric digital camera and the digital image correlation (DIC) method. This method utilizes a sequence of acquired images that represents a specimen surface deformation process. In this sequence, DIC observes the displacements of small rectangular regions (templates) on the sample with a distinguishable distribution of gray-scale intensities [9, 10]. The displacements obtained from this method are utilized for calculating the strains. In the case of stone, the random surface pattern, a prerequisite for the method, is formed by the distinct natural texture of the surface of the specimen.

The DIC algorithm used in this work has two main steps. First, an integer value of the pixel displacement is evaluated using a normalized cross-correlation function for a central pixel of a square image template in the reference image. The second step in the matching process is known as the Lucas-Kanade algorithm [11]. This step takes into account the reference template's own deformation. The method searches for an affine transformation that projects the template onto the deformed image using minimization of the sum-squared differences between the template and the deformed image. The LK algorithm is an iterative nonlinear optimization method. The integer values of the pixel displacement estimated in the first step were passed to this iterative process as initial guess values. The longitudinal deformation was measured using two rows of points. Each point represents the center of the rectangular template, the displacement of which was measured by DIC. Multiple points were used, and their displacements were finally averaged for noise reduction. Subsequently, the longitudinal engineering strain was calculated using the measured displacements as

$$\varepsilon_{eng} = \frac{l - l_0}{l_0} \tag{1}$$

Knowing the measured applied force and the dimensions of the cross section of the specimen, the engineering stress can finally be evaluated.

Twelve experiments were carried out—four on the specimens from **Figure 4** and eight on prismatic beams cut out of the broken parts after the tension tests. The tests therefore involved (i) compression, (ii) very-low-cycle fatigue, (iii) tension, (iv) fatigue (an alternating nonsymmetrical cycle)—all on the specimens from **Figure 4**—and (v) three-point bending tests on small size beams with a 20×20 mm^2 cross section.

The effect of consolidation on the fracture behavior was studied on prismatic beams of the same cross section as described above provided in its center with a notch and a generated crack. The crack generation procedure is described in detail by Drdácký et al. [12]. Specimens with similar length of crack were consolidated applying the most typical agents: elastified Steinfestiger 300 (30% concentration of the active substance), Paraloid B-72 (2% concentration), and Funcosil 100 (10% concentration of the active substance). After maturing they were tested in standard static three-point bending.

3. Test results

The compression force was applied in three stages: −1.5, −26, and −37 MPa. Deformation was measured during the tests in order to study the change in Young's modulus and the Poisson number with relation to applied load (**Figures 6** and **7**). Hysteresis and irreversible deformation were observed, and the Poisson number reached a value of 0.23. The diagram shows that the tested sandstone increased in stiffness with an increasing number of compression cycles at the beginning of the tests. The change in the slope of the interpolated linear approximations of the load-displacement diagrams reflects this fact. The modulus of elasticity rose from an initial value of 11,360 MPa to a value of 19,237 MPa for the third cycle of loading, i.e., a 70% increase. This differs from the results of other tests on sandstone as presented in the cited literature. At the same time, hysteresis occurs with irreversible deformations. During the first loading cycle to −1.5 MPa, no irreversible deformation was observed. It seems that irreversible deformation occurs after the material has been subjected to some limit compression load. The Czech standard for tests on natural stone prescribes that the modulus of elasticity is to be evaluated after three loading cycles.

During the tension tests, the velocity of the crosshead movement was intentionally kept very low at a value of 5 μm/min. Due to this low loading velocity, it was possible to evaluate the overall energy absorbed during crack origin and propagation upon the initiation of a crack at a

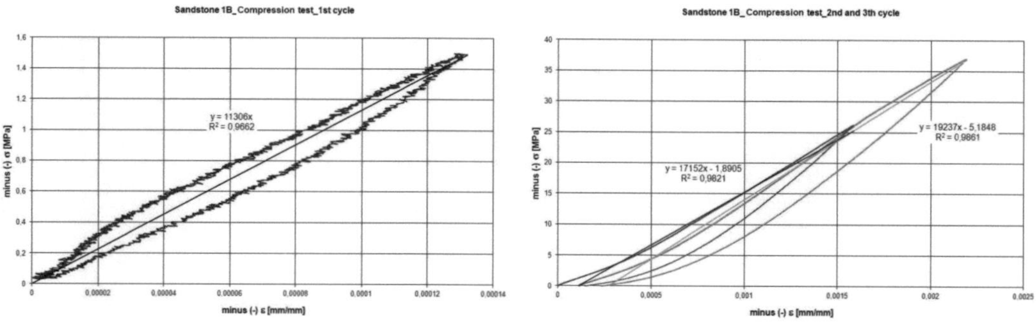

Figure 6 First compression loading cycle (left) and second and third loading cycles (right), with corresponding linear approximations.

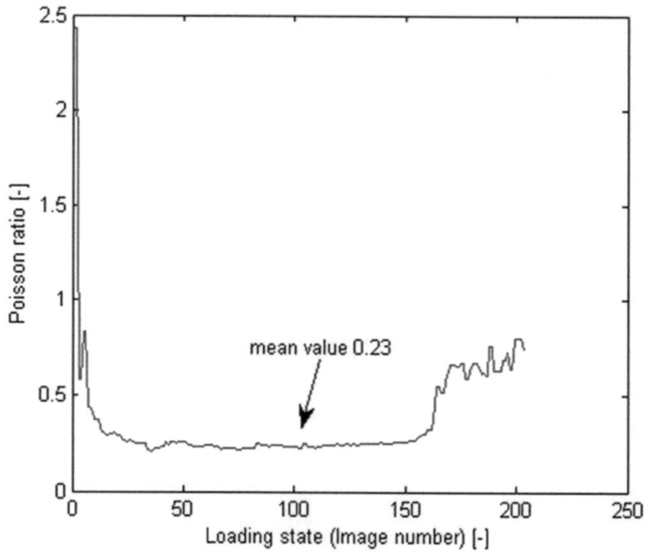

Figure 7 Poisson ratio measured using the DIC method.

defect present within the specimen (**Figure 8**). The load-displacement graph shows the overall energy absorbed during the test (0.4 J) and during crack growth (0.16 J). This corresponds to the behavior of highly brittle material. The strength decreased (to 1.43 MPa) due to the defect in the material. The course of the stress and deformation up to fracture can be described well by means of a suitable power function (see **Figure 9**), which presents a detailed diagram.

It follows from **Figure 9** that a suitable power function appropriately describes the material behavior up to rupture. The tangent to the curve gives the modulus of elasticity value:

$$\sigma = \boldsymbol{k} \cdot \varepsilon^{n} \text{ ....by derivation after } \Sigma \text{ this yields: } E = \frac{d\sigma}{d\varepsilon} = k.n.\varepsilon^{n-1} \qquad (2)$$

The power function seems to be ideal for describing the deformation behavior of this sandstone.

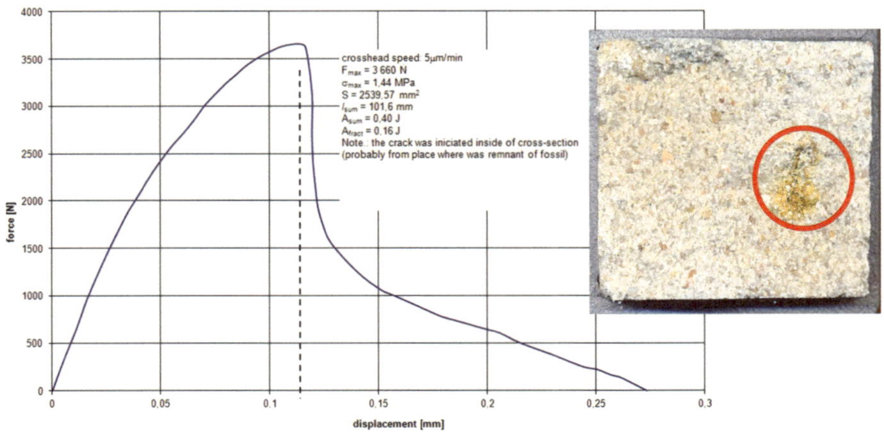

Figure 8. Load-crosshead displacement diagram for the tension test with a calculation of the absorbed energies and an indication of the probable place with an initial defect in the cross section of the specimen.

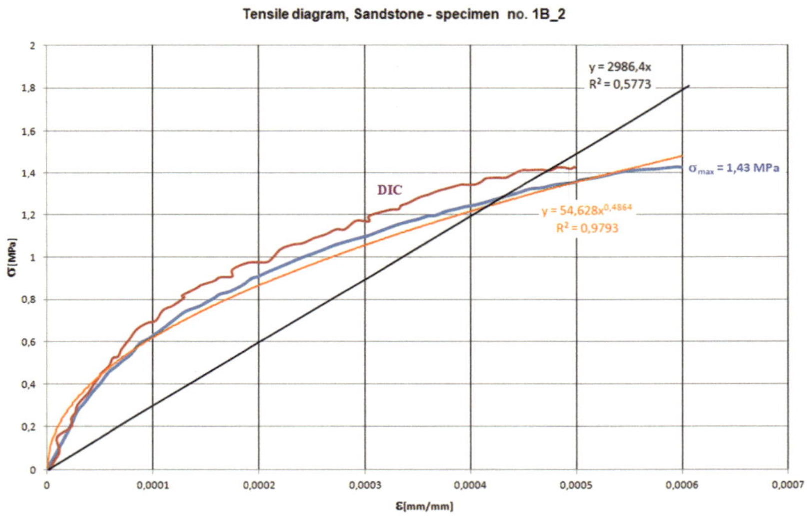

Figure 9. Tension diagram of the tested Božanov sandstone, comparing the conservative measurement with the DIC measurement and presenting a suitable power function approximation.

The DIC record exhibits a slight difference in comparison with the conservative measurement approach. This may be caused by a small change in the geometric relations between the specimen and the camera objective during loading. However, the agreement is still very good.

A sinusoidal altering nonsymmetrical loading cycle with stress limits of +1.5 and −2MPa was applied during the very-low-fatigue test. The mean value of the stress was −0.25 MPa, the stress double amplitude was 3.5 MPa, and the frequency of loading was 0.2 Hz (**Figure 10**).

http://dx.doi.org/10.5772/intechopen.81434

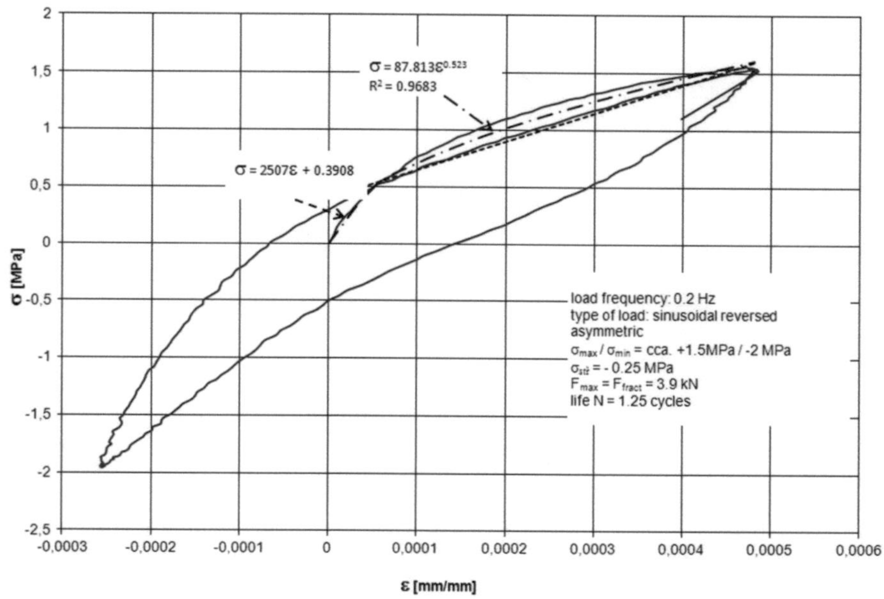

Figure 10. Low-cycle fatigue diagram.

The maximum tension stress ($\sigma_{max} \approx 1.5$ MPa) approached the limit stress (strength) in tension of the tested material (the static strength was 1.43 MPa on a weakened profile; see above). The resulting number of cycles was only 1.25; the fracture occurred while the maximum tension load was being attained in the second cycle.

Figure 10 also indicates that a power function can be used to approximate the first loading course. The parameters of the constant however are different from that of the previous case. This could be a result of the presence of an interior defect in the first case or merely the wide distribution of the material's characteristics. The change in the fatigue loop's tension branch is also noteworthy. In comparison to the first cycle, it is significantly straighter and more horizontal. It is even possible to approximate it with a linear function. It signifies considerable damage to the material resulting from the first loading cycle, corresponding to the stress applied. The hysteresis exhibited by this specimen is quite high. The compression component is steeper than the tension component [14].

The nonsymmetrical altering sinusoidal loading cycle was also used in the fatigue tests, with stress limits of +0.265 and −0.354 MPa. There were two modes of cycling velocity used—for the main loading sequence a frequency of 0.25 and 0.01 Hz at certain stages for groups of three cycles measuring deformations, based on which the second cycle was evaluated and the E modulus calculated. Such detailed measurements were attained up to 2000 cycles. The specimen was then loaded to the point of failure. The asymmetry characteristics of the cycle were kept the same to allow for comparability of results (R = −1.33 and stress span $\Delta\sigma = 0.619$ MPa). The overall lifespan reached 224,908 loading cycles. Typical results are presented in **Figure 11**.

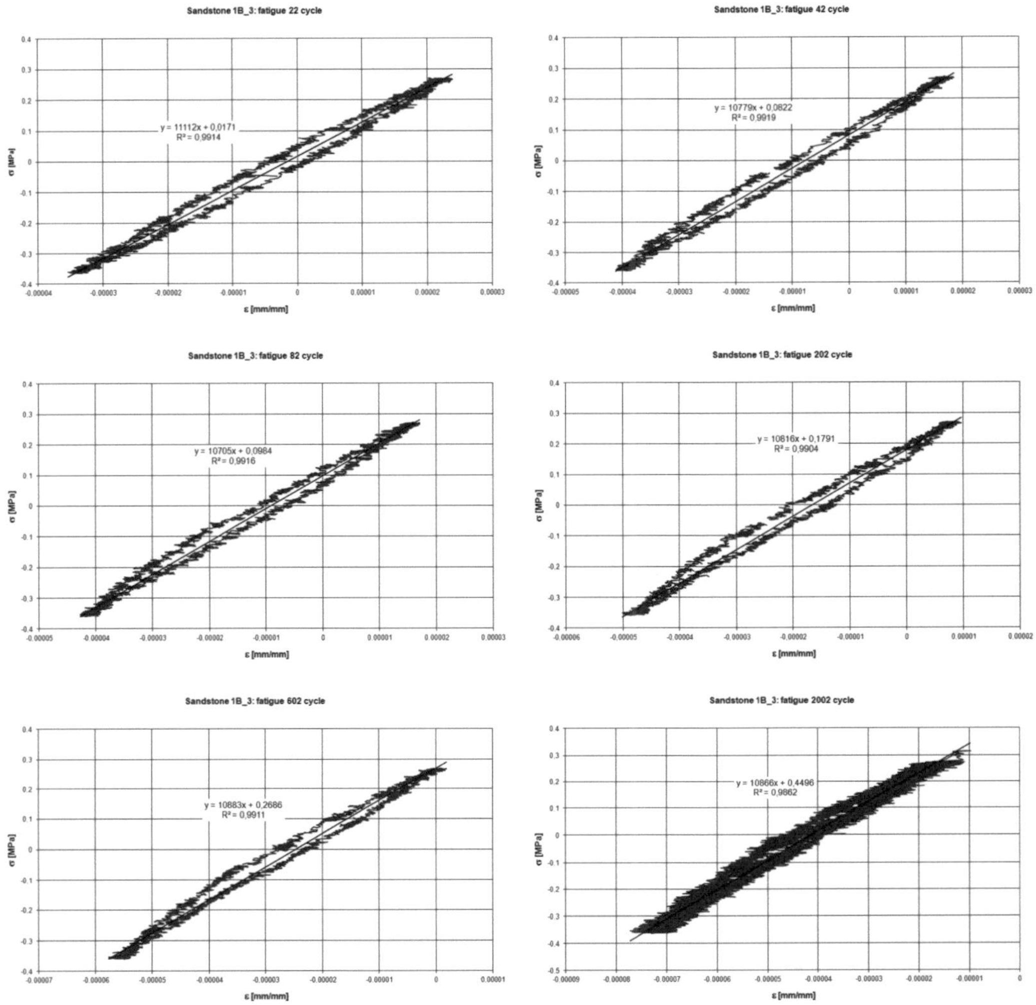

Figure 11. Stress-strain ratios after a given number of cycles: 22 (top left), 42 (top right), 82 (center left), 202 (center right), 602 (bottom left), and 2002 (bottom right).

The modulus of elasticity E changed during the cycling. Its value dropped from the initial figure of 11,214–10,705 MPa after 82 loading cycles (a decrease of about 5%). Then the value stabilized around 10,865 MPa after 2002 cycles (**Figure 12**). The change was influenced by fatigue damage in the material.

In **Figure 11**, the observed shift of the measured data toward negative strain values with the increasing number of cycles is caused by thermal dilation of the specimens. The average change in temperature in the testing hall was about 4°C. The temperature change dilations were checked by computations, taking into account the coefficient of dilation of sandstone equal to $\alpha = 10 \times 10^{-6}\ K^{-1}$ and the equation $\Delta l = \alpha \cdot \Delta T \cdot l$. **Figure 13** presents examples of the temperature drift.

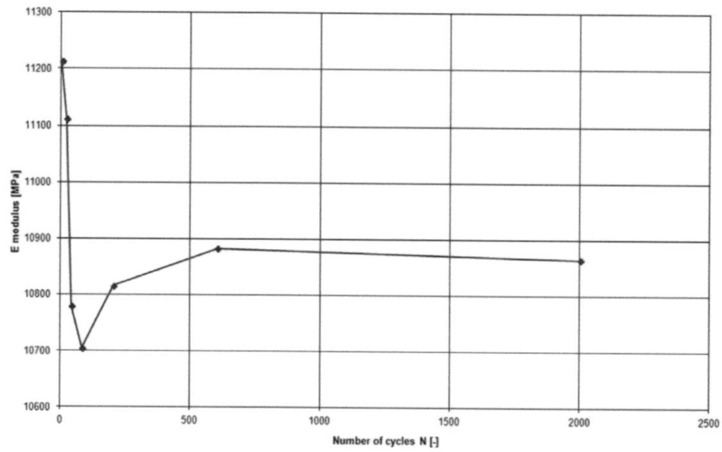

Figure 12. Change in the modulus of elasticity during cycling.

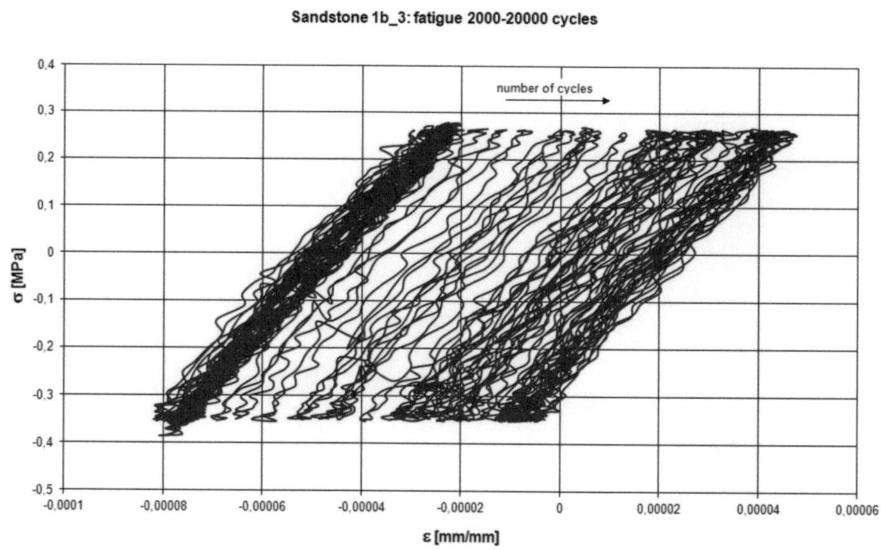

Figure 13. Temperature caused a drift in specimen deformation due to dilation (2000–20,000 cycles).

Figure 14 presents a very rough representation of the probable life estimate of the tested sandstone between two applied stress levels. Of course, many more tests will be required before the true Wöhler curve can be constructed.

After the uniaxial and fatigue tests, small beams with a cross section of 20×20 mm^2 were prepared from the broken specimens and were subjected to three-point bending loading (**Figure 15**). The measured flexural strength reached much higher values (about 4.5 MPa) than the strength in the tension tests, which was influenced by a rather low span to height ratio (about 3.75).

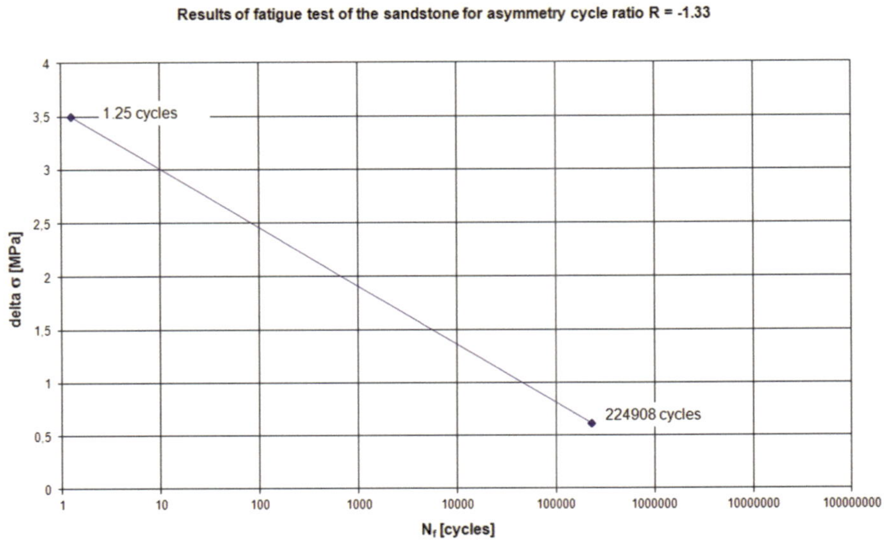

Figure 14. Fatigue test results for Božanov sandstone for cycle asymmetry of R = −1.33.

Figure 15. Three-point bending test arrangement (photo O. Vála).

However, the moduli of elasticity were lower than those measured during the previous tests (though still higher than those measured by the first author on thin plates during tests on the characteristics of the Charles Bridge sandstone—Čechová et al. [13]). It is clear that both the load-displacement diagram and the change in the modulus of elasticity can be well approximated by a suitable mathematical model (**Figure 16**).

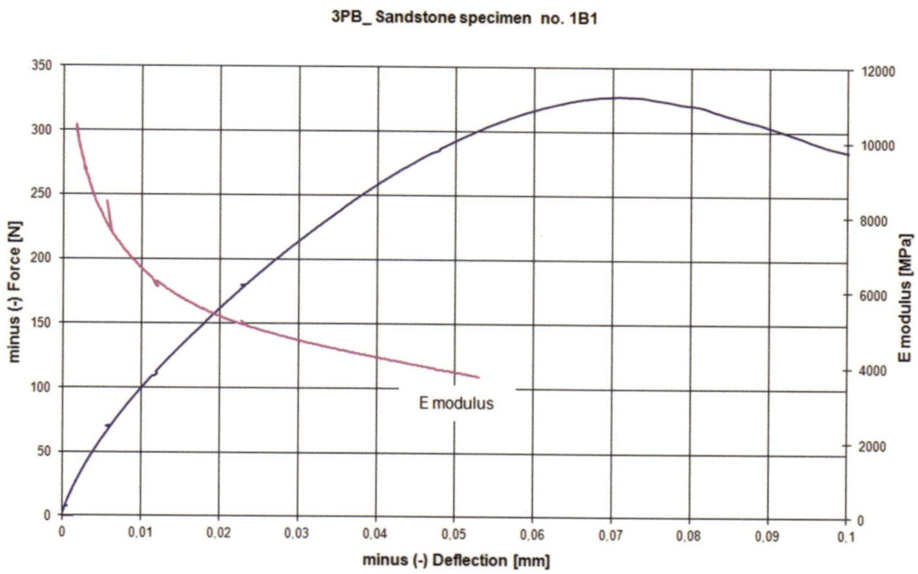

Figure 16. Loading force-deflection diagram during three-point bending testing and calculated modulus of elasticity.

In the study on consolidation effects influencing crack propagation and toughness, three types of specimens were tested—notched beams without cracks, notch beams with cracks generated by cycling, and notched beams with cracks after consolidation [14]. They were loaded in three-point bending configuration.

The results show that consolidation increased the load-carrying capacity of the cracked sandstone specimens. The reference specimen with the initiation notch attained an average bending strength of 5.05 MPa, the notched specimen with the cycled crack one of 4.56 MPa, and the consolidated specimen one of about 6 MPa.

4. Conclusions

Although it is difficult to generalize the results of the pilot tests described here, they provide important findings for further planning of stone fatigue tests.

The methodology of tension tests using specimens inserted and glued in tubular fixtures is functional and has been verified. However, it is very demanding from the point of view of specimen preparation, and it requires very skilled staff. It can be recommended mainly for special small-series tests. For research requiring tests on a large numbers of specimens, it is necessary to develop some other techniques. However, the pilot tests were successfully completed, and the energies for crack initiation and propagation in the given sandstone have been determined.

The optical DIC method again proved its capacity and its suitability for measurements of a complex deformation field on porous and naturally well-structured surfaces.

The modulus of elasticity may be a good parameter for assessing damage accumulation, though it is difficult to measure on real structures (see **Figure 11**). In the next series of tests, the correlation between the modulus of elasticity and NDT techniques will be studied, especially ultrasonic measurements for possible application in situ.

It follows from **Figure 13** that the first tension load displacement can be approximated very satisfactorily by means of a power function.

The tests with consolidated specimens proved the positive effects of stone impregnation on strength and toughness. All notched and cracked specimens after consolidation with ethyl silicates KSE 100 and KSE 300 and Paraloid B-72 attained values above full profile strength which was tested on identical beams. The KSE 300 consolidation agent attained the best result in the range of 145% as far as strength is concerned and 161% in the toughness value (K_{IC}).

The toughness of untreated notched and cracked specimens reached almost the same values (0.33:0.35 MPa√m), which mean that with such a heterogeneous material, one cannot expect stress concentration factors similar to those of metals.

Acknowledgements

The authors acknowledge the kind support of the Czech Science Foundation Project GAČR P105/12/G059 and the kind collaboration with Dr. Ivan Jandejsek who made the DIC measurements and evaluated the data.

Author details

Martin Šperl* and Miloš Drdácký

*Address all correspondence to: sperl@itam.cas.cz

Institute of Theoretical and Applied Mechanics of the Academy of Sciences of the Czech Republic, Prague, The Czech Republic

References

[1] Drdácký MF. Impact of climate change on building structures. In: Lefèvre RA, Sabbioni C, editors. Climate Change and Cultural Heritage. Scienze e materiali del patrimonio culturale, 10. Bari: EDIPUGLIA, s. r. l.; 2010. pp. 139-153

[2] Beran P, Drdácký M. Influence of temperature changes on stresses in the Triforium tracery of St Vitus' Cathedral in Prague. In: Oñate E, Papadrakakis M, Schrefler B, editors. Proceedings Computational Methods for Coupled Problems in Science and Engineering II Coupled Problems. Barcelona: CIMNE; 2007. pp. 433-436

[3] Ning L, Ping Z, Yunsheng C, Gunter S. Fatigue properties of cracked, saturated and frozen sandstone samples under cyclic loading. International Journal of Rock Mechanics and Mining Sciences. 2003;**40**:145-150

[4] Bagde MN, Petroš MN. Fatigue properties of intact sandstone samples subjected to dynamic uniaxial cyclical loading. International Journal of Rock Mechanics and Mining Sciences. 2005;**42**:237-250

[5] Backers T. Fracture Toughness Determination and Micromechanics of Rock Under Mode I and Mode II Loading. PhD Thesis. Universität Potsdam 2004. Scientific Technical Report STR 05/05, GeoForschungsZentrum Potsdam; 2004. p. 138

[6] Attewell PB, Farmer IW. Fatigue behaviour of rock. International Journal of Rock Mechanics and Mining Sciences and Geomechanics Abstracts. 1973;**10**(1):1-9

[7] Jian-Qing X, De-Xin D, Fu-Liang J, Gen X. Fatigue damage variable and evolution of rock subjected to cyclic loading. International Journal of Rock Mechanics and Mining Science. 2010;**47**:461-468. DOI: 10.1016/j.ijrmms.2009.11.003

[8] Cardani G, Meda A. Marble behaviour under monotonic and cyclic loading in tension. Construction and Building Materials. 2004;**18**:419-424

[9] Lin Q, Labuz JF. Fracture of sandstone characterized by digital image correlation. International Journal of Rock Mechanics and Mining Sciences. 2013;**60**:235-245

[10] Peters WH, Ranson WF. Digital imaging techniques in experimental stress analysis. Optical Engineering. 1982;**21**:427-431

[11] Lucas BD, Kanade T. An iterative image registration technique with an application to stereo vision. In: Proceedings of Imaging Understanding Workshop; 1981. pp. 121-130

[12] Drdácký M, Šperl M, Jandejsek I. Consolidation effects on sandstone toughness. In: Hughes J, Howind T, editors. Proceedings of the 13th Int. Congress on the Deterioration and Conservation of Stone "SCIENCE and ART: A Future for Stone". Vol. 2. Paisley: University of West Scotland; 2016. pp. 687-694

[13] Čechová E, Drdácký M, Frankeová D, Lesák J, Slížková Z, Valach J, et al. Technology Guidelines for Restoration of the XIth Arch of the Charles Bridge in Prague. Research Report of ITAM AS CR. Prague: ITAM; 2010. p. 139

[14] Šperl M, Drdácký M. Non-standard experimental tests of sandstone and its pre-cracking for fracture testing. In: IOP Conference Series: Materials Science and Engineering 379; 2018. DOI: 10.1088/1757-899X/379/1/012029

Key Performance Criteria Influencing the Selection of Construction Methods Used for the Fabrication of Building Components in the Middle East

Alireza Moghayedi and Abimbola Windapo

Additional information is available at the end of the chapter

http://dx.doi.org/ 10.5772/intechopen.81673

Abstract

There is a lack of an efficient systematic approach to the selection of appropriate construction methods for building projects. Not only various innovative methods are now available, but also established methods may often be adapted inappropriately, without recourse to the necessary scientific foundation of their efficiency. The result is that there is a low level of performance on building projects. This study examines how key performance criteria were used in the selection of construction methods on projects. The study employed an extant review of the literature, cross-section survey of construction managers of building projects and experts interview in the Middle East to identify and evaluate the influencing of the key performance criteria on selecting construction methods for building projects. It emerged from the Pearson Correlation Coefficient and Analytical Hierarchy Process analysis that key performance criteria consisting of time, quality, and cost have strong positive significant roles in the selection of construction methods used on building projects and that these selection criteria differed depending on the building components. The study concludes that the likelihood of a construction method being selected for use on projects in the Middle East depends on its ability to shorten the duration, improving the quality and reduce the cost of projects.

Keywords: analytical hierarchy process (AHP), building project, construction method, Middle East, Pearson correlation coefficient, performance criteria

1. Introduction

In the construction of buildings, there are common problems and challenges of a low level of productivity and efficiency. According to Wambeke et al. [1], 58% of building construction

IntechOpen

© 2018 The Author(s). Licensee IntechOpen. This chapter is distributed under the terms of the Creative Commons Attribution License (http://creativecommons.org/licenses/by/3.0), which permits unrestricted use, distribution, and reproduction in any medium, provided the original work is properly cited.

projects exceed the scheduled time, and 15 out of 20 projects exceed their original approved budgets. Kamali et al. [2] and Ren et al. [3] found that different construction methods influence project performance in various ways and impact on the productivity of construction projects; deficient methods decrease the productivity of projects. Furthermore, Forbes and Ahmed [4] posited that the choice of construction method significantly impacts on the cost, time, and quality of buildings, and adopting inappropriate methods increases the cost and duration of projects, as well as decreasing the quality and lifespan of buildings. Currently, the construction industry has been revolutionized and is experiencing changes, with the rapid growth of technology and the introduction of new building materials and modern construction methods [5]. Furthermore, the new generation of building regulations has been enacted to increase the efficiency and improve the quality of buildings and infrastructure [6]. As a result, construction managers, as decision makers, have to choose appropriate construction methods from those available. Therefore, to achieve construction project performance objectives, there is a need for adequate information and knowledge to help construction managers to make good choices of construction methods.

This research examines the selection of construction methods and materials on building projects in the Middle East by using the Pearson correlation coefficient (PCC) and the analytical hierarchy process (AHP) to determine the weight and influence of each key criterion. The PCC and AHP are also helpful to consider as the total weight of different construction methods used in the fabrication of selected building components. The AHP is an effective mathematical method used in solving multicriteria decision-making problems [7]. It has been applied to many decision-making problems related to construction management. However, AHP is unable to handle the inherent subjectivity and ambiguity associated with the mapping of an individual's perception to an exact number [8]. In this condition, the Pearson correlation coefficient (PCC) analysis is applied in verifying the AHP weight by measuring the relationship between the key criteria used in the selection and the level of use of each construction method.

The focus of the study is on the construction of buildings because these are the most common types of construction projects that make use of a wide variety of methods and components. With adequate knowledge and comprehensive data, the most suitable construction method, complementing the objectives, and condition of the project, can be selected.

2. Literature review

2.1. Overview of construction methods and key performance criteria

The construction method is a technical procedure to transform construction resources (materials, workforce, and equipment) into constructed products [9]. According to Haidar [10], the construction methods adopted affect the work activities and the work sequence. Construction planning and management techniques are without value if construction methodologies are not selected appropriately and if those selected are not optimal [11]. Each construction method has different specifications and aspects. Information such as cost, time, quality, ease of construction, and availability of method and skill is used by construction managers in the selection and use of appropriate construction methods on construction projects [12].

According to Monghasemi et al. [13], shorter time, lower cost, and higher quality are the primary project objectives which should be considered in the selection of construction methods. However, among these three primary factors, cost and quality are single dimension elements; this means that choosing the method with less cost, or with higher quality, will not guarantee or improve the other influencing factors. Time is a multidimensional element; choosing the method with a shorter construction time will reduce the labor cost and the error caused by labor [13, 14].

Ferrada and Serpell [15] posited that it is important to consider other parameters, such as the availability of materials in the market, the supply of a skilled workforce, ease of transportation, and ease of implementation; these all impact on the project objectives. The above suggests that selecting suitable construction methods depends on understanding the outcome of each method in relation to the final project outcome. Therefore, in this study, the impact and effect of each primary construction project performance criterion (cost, time, and quality), combined with ease of construction and the availability of method and skill, focus on six common components of buildings: the foundation, the frame of the structure, the roof, the wall, the flooring, and the façade. This effect applies to all buildings, regardless of their size and type.

The factor called "ease of construction" concerns the use of less intensive labor on construction sites to reduce the duration of construction and concomitant labor cost; however, the material cost of this method may be higher than those of more established construction methods [16].

2.2. Overview of the use of analytical hierarchy process and Pearson correlation coefficient in multicriteria decision analysis in selecting construction methods and materials

The analytic hierarchy process (AHP) is a structured technique for organizing and analyzing complex decisions, based on mathematics and psychology. It is a multicriteria decision-making approach [17]. In construction management, many decisions are made based on various criteria such as time, cost, and quality; thus, these decisions could be made by construction experts by assigning weights to the different criteria. It is important to determine the structure of the problem and explicitly evaluate the many criteria. The AHP method has been used in various areas of construction management. For instance, Sangiorgio et al. [18] measured the relative project performance among a set of criteria; Ng [19] evaluated environmental benefit of building designs using a weighting AHP; Hossaini et al. [20] assessed the lifecycle sustainability of a six-story wood frame and concrete frame buildings (respectively) in Vancouver; Wong and Li [21] analyzed the selection of intelligent building systems in Hong Kong; Pan [22] used a multi-criteria decision model to select the most suitable bridge construction method in China; Zayed et al. [23] evaluated the highway construction projects risks in China;

Shapira and Goldenberg [24] created an equipment selection model for construction projects; Al-Harbi [25] evaluated the problem of contractor prequalification; and Skibniewski and Chao [26] measured the technical and economic factors of the selection process of a tower crane in construction projects.

However, there is limited research undertaken involving the use of multicriteria models as a basis for the selection of construction methods using the key performance criteria. Reza et al. [27] used the AHP as a tool to assess the sustainability of floor systems in the city of Tehran. In a similar approach, Nadoushani et al. [28] utilized AHP to select façade systems for a building, based on criteria of sustainability. Hosseini et al. [29] used a simple scoring aggregation procedure, combined with the AHP method, to select the unsurpassed types of exterior walls to reconstruct in earthquake areas; Akadiri et al. [30] applied a multicriteria evaluation model for the selection of sustainable roofing materials for building projects in the United Kingdom. In all these studies, the AHP method was used for selecting only one component of the building, reflecting its particular characteristics. Even though AHP has been used for tackling multicriteria decision-making problems, it is required to validate the weight value of the criteria because of the discrete scale used in AHP, which cannot handle the uncertainty and ambiguity present in deciding the priorities of different attributes [31]. Therefore, the correlation coefficient analysis is commonly used in construction research to validate the research results and to measure the relationship between two random variables [32].

The Pearson correlation coefficient (PCC) has been used to determine and validate the relationship between different variables in construction projects, such as the relationship between total project cost and total material cost; between uncertainty factors and risk contingency value; between intelligence attributes of the Integrated Building Management Systems (IBMS) and the operational benefits; and between productivity and safety performance [21, 33–35].

3. Research objectives

The primary purpose of this study is to evaluate the key performance criteria influencing the selection of construction methods, which are currently being used on building projects in the Middle East and whether the selection criteria differ according to building components. According to Ferrada and Serpell [15], selecting appropriate construction methods based on their performance objectives is a contemporary topic in construction management that is progressing and expanding gradually. This technique involves evaluating, classifying, and suggesting the most appropriate construction method that best fits the project conditions [36]. There is limited research that examines the selection of construction methods for fabrication of different building components; most of the existing AHP models [27, 28, 30, 37] employ only one building component, such as the wall, the roof, or the floor. Therefore, this research examines the key performance criteria (time, cost, quality, ease of construction, and availability of method and skill) affecting the selection of 28 construction methods used in the fabrication of 6 building components, by employing multicriteria models as a basis for matching construction methods to performance criteria.

The novelty of this research is in the selection of construction methods for six different building components based on the five performance criteria, which are not only very complex issues due to involving a multi-criteria decision, but also some of these criteria may also influence some additional problems. Therefore, this research presents an AHP approach to determine the weight of the five key performance criteria in the selection of construction methods, for six main building components, and validates the key criteria weight of the construction method, through a PCC statistical analysis.

This approach is straightforward and provides the total weight value of the different performance criteria, which impacts on the selection of construction methods. It also provides a holistic overview of the most important criteria for construction managers and engineers, who are the decision makers on the building project.

4. Research method

The study made use of a sequential mixed-method research approach in evaluating the selection of construction methods currently adopted in the Middle East and in determining the contribution and relationship between the key performance criteria and the level of use of identified construction methods, differentiated by building components. First of all, the research identifies common construction methods that are currently employed on projects in the Middle East. To do this, the information on construction methods used was gathered and classified based on data obtained from construction industry experts. Secondly, the influence and significance of each performance criterion in selecting construction methods used on the building projects were obtained from the data collected from questionnaires. The questionnaires were completed by construction managers on 200 building projects in five countries, namely Iran, United Arab Emirates, Turkey, Egypt, and Qatar. The construction managers had different work experience and levels of education, as classified in **Tables 1** and **2**. Each questionnaire consisted of two parts:

Part 1 sought demographic information related to the construction manager.

Part 2 sought information concerning the construction methods and criteria used in the selection of each construction method used on the project.

The study also sought to know the perceptions of the construction managers regarding the performance in terms of cost, time, quality, ease of construction, and availability of the method/skill of the identified construction methods, on a scale of 1–3, where 3 = High, 2 = Moderate, and 1 = Low. If all respondents scored the performance of the construction method as 3, it was recorded as High (see **Table 4**); also when 2 < 3, it was recorded as Moderate performance, while scores of less than 2 were classified as Low.

The data were also analyzed using descriptive and inferential statistics, consisting of the analytical hierarchy process (AHP) pair-wise comparison and the Pearson correlation coefficient (PCC) analysis. The AHP weight of each performance criterion and a total weight of available construction methods in the different building components were determined using Expert

Choice Version 11.5. To validate the AHP weight values and determine the extent of the relationship between the criteria used in the selection of the construction methods and the level of use of the construction methods on building projects, the Pearson correlation coefficient (PCC) analysis was conducted using IBM SPSS Statistics version 23. The most commonly used correlation coefficients in construction research are Spearman's rank correlation coefficient and the Pearson correlation coefficient [38]. The Pearson correlation coefficient is a measure of the strength of the linear dependence between random variables, while Spearman's rank correlation coefficient is a nonparametric measure of statistical dependence between two variables and is an indication of correlation between ranks of the values of random numbers instead of correlation between values [39]. Therefore, in this research, the Pearson correlation coefficient is preferred over the Spearman's rank correlation coefficient, since Pearson is the correlation between variates, while Spearman is the correlation between the ranks of the variates.

Tables 1 and **2** indicate that 81% of the respondents have more than 5 years of work experience in the construction industry and that 77% of respondents hold a university degree. The levels of education and work experience the respondent has in the construction sector are of relevance to the study because the higher the education level and work experience of the respondent the better the credibility and reliability of the information provided via questionnaires which focus on their knowledge of construction management and methods.

Reliability in quantitative research indicates that the scores received from the respondents are consistent and stable over time; reliability is often assessed through reliability coefficients [40]. In order to check that the collected data and scores were reliable, a statistical analysis was made of the reliability and internal consistency of the data. These data had been provided by the 200 respondents on the key project performance criteria, which influenced the selection of construction methods for projects. To do this, the Cronbach's alpha coefficient of each criterion and the internal consistency ratio of the overall influencing criteria were calculated. The values of Cronbach's alpha that are commonly used to determine the internal reliability, consistency, and co-variation among variables related to the measurement of each construct often range from 0 to 1 [41]. The results of the test showed that dependency among the five identified criteria was equal to 0.838, which indicates a high reliability and internal consistency of data collected across the five criteria. Also, Cronbach's alpha coefficients of each

Years of experience	Individual percentage					Overall percentage (%)
	Iran (%)	UAE (%)	Turkey (%)	Egypt (%)	Qatar (%)	
<5	18	16	20	23	17	19
6–10	34	38	32	33	36	35
11–15	28	24	25	23	29	26
16–20	9	13	14	11	10	11
>20	11	9	9	10	8	9

Source: Researcher's field survey.

Table 1. Respondents' work experience.

Level of education	Individual Percentage					Overall percentage (%)
	Iran (%)	UAE (%)	Turkey (%)	Egypt (%)	Qatar (%)	
Vocational degree	19	24	25	26	23	23
Bachelor degree	54	55	57	62	60	58
Postgraduate degree	27	21	18	12	17	19

Source: Researcher's field survey.

Table 2. Respondents' education level.

	Cronbach's alpha
Cost	0.829
Time	0.783
Quality	0.762
Ease of construction	0.768
Availability of method and skill	0.868

Table 3. Cronbach's alpha coefficient of each influencing criterion.

criterion were greater than 0.7, which indicate sufficient reliability and internal consistency of the data collected for each criterion, as illustrated in **Table 3**.

5. Data presentation and analysis

This research evaluates five criteria (cost, time, quality, ease of construction, and availability of method/skill), which influence the selection of construction methods on building projects. **Table 4** summarizes and evaluates the types of construction methods used, distributed by the number of projects, and the perceived performance attributes of each construction method.

Table 4 indicates the characteristics of each construction method, in terms of technology, three primary factors (cost, time, and quality) and two combinational influential factors (ease of construction and availability of method and skill), which influence the selection of construction methods on building projects. Each method was evaluated based on the technology utilized in two categories of the conventional method and modern innovative method. Furthermore, the cost (material cost and labor cost) [42], time (construction duration), quality, ease of construction, and availability of method and skill of each method are evaluated in three classes, namely low, moderate, and high.

It can be seen from **Table 4** that the common construction methods used in the foundation, structure, roofing, wall, façade, and flooring of the projects studied are steel formwork, concrete framed structure, steel decking, Leca blocks, composite façade, and laminate flooring, respectively. However, **Table 4** also shows that the construction methods with a perceived

Method	Number of projects	Technology	Cost	Time	Quality	Ease of construction	Availability of method and skill	Overall performance
Foundation								
Steel formwork	106	Innovative	High (85)	High (100)	High (80)	High (50)	High (20)	High
Brick formwork	54	Conventional	Low (50)	Low (40)	Low (8)	Low (45)	High (25)	Low
Wood formwork	22	Conventional	Low (5)	Moderate (5)	Low (2)	Moderate (20)	High (15)	Moderate
Cement hollow block (CHB) formwork	18	Conventional	Moderate (15)	Low (18)	Moderate (10)	Low (5)	High (5)	Moderate
Structure								
Concrete frame structure	88	Conventional	Low (85)	Low (65)	Moderate (80)	Low (25)	High (30)	Moderate
Bolted steel frame structure	56	Conventional	Moderate (45)	Moderate (55)	Moderate (40)	Moderate (50)	Moderate (15)	Moderate
Welded steel frame structure	45	Conventional	Moderate (35)	Moderate (44)	Moderate (30)	Low (30)	High (20)	Moderate
Light steel frame structure (LSF)	11	Innovative	High (5)	High (10)	High (6)	High (11)	Low (0)	High
Roofing								
Steel decking	76	Innovative	High (65)	High (75)	High (70)	High (30)	High (25)	High
Reinforced concrete slab (one/two way)	33	Conventional	High (0)	Low (0)	Low (10)	Low (5)	High (30)	Low
Polystyrene inter-Joist	29	Conventional	Low (20)	High (25)	Moderate (10)	High (10)	High (5)	High
Hollow core	26	Conventional	Moderate (15)	Moderate (25)	Moderate (20)	Moderate (10)	High (5)	Moderate
Double tee	22	Innovative	Low (20)	High (15)	High (15)	High (10)	Moderate (5)	Moderate
Cobiax	14	Innovative	Moderate (10)	Moderate (12)	Moderate (10)	Moderate (5)	Moderate (0)	Moderate

http://dx.doi.org/ 10.5772/intechopen.81673

Method	Number of projects	Technology	Cost	Time	Quality	Ease of construction	Availability of method and skill	Overall performance
Wall								
Leca block	90	Innovative	Moderate (75)	High (75)	High (85)	High (45)	Moderate (35)	High
Autoclaved aerated concrete (AAC) block	58	Innovative	Moderate (45)	Moderate (50)	High (55)	High (35)	Moderate (20)	High
Clay hollow blocks	30	Conventional	Low (30)	Low (15)	Low (10)	Low (25)	High (10)	Moderate
Silica block	16	Innovative	High (10)	High (10)	High (15)	High (9)	Moderate (0)	High
Cement hollow blocks	6	Conventional	Moderate (0)	Low (0)	Moderate (0)	Low (5)	High (6)	Moderate
Façade								
Composite façade	81	Conventional	Moderate (80)	Moderate (76)	Moderate (60)	Moderate (40)	High (10)	Moderate
Stone façade	56	Conventional	Moderate (20)	Low (15)	Moderate (55)	Low (5)	High (30)	Moderate
Steel façade	24	Innovative	High (5)	High (22)	High (20)	High (10)	High (0)	Moderate
Brick façade	21	Conventional	Low (15)	Low (10)	Moderate (10)	Low (20)	High (15)	Moderate
Glass façade	18	Innovative	High (5)	High (17)	High (15)	High (5)	High (5)	Moderate
Flooring								
Laminate flooring	78	Innovative	Low (75)	High (78)	High (60)	High (65)	High (20)	High
Ceramic flooring	64	Conventional	Moderate (30)	Moderate (50)	Moderate (60)	Moderate (30)	High (25)	Moderate
Stone flooring	42	Conventional	High (10)	Low (20)	Moderate (40)	Low (15)	Moderate (25)	Moderate
Parquet flooring	16	Conventional	High (0)	Low (10)	High (15)	Moderate (15)	High (0)	Moderate

Table 4. Distribution of construction methods by perceived performance attributes.

high overall performance are steel formwork, light steel frame structure (LSF), steel decking, polystyrene inter-joist, Leca block, autoclaved aerated concrete (AAC) block, silica block, and laminate flooring. By comparison, the low performing construction methods are brick formwork and reinforced concrete slab (one/two way).

Figures 1–5 show the distribution of the five key performance criteria used by the construction managers in the selection of each construction method.

It can be seen from **Figures 1–5** that from a ranking perspective, the performance criteria with the most influence on the selection of the construction methods studied are time, quality, cost, ease of construction, and availability of method and skill, respectively.

5.1. Influence of identified criteria on selection of construction methods

The study sought to know from the construction managers, the key criteria responsible for the choice of construction methods used on their projects. The influence and rank of each criterion in the selection of construction methods used in the fabrication of different components of the examined building projects are analyzed and presented in **Table 5**.

Table 5 shows that from a ranking perspective, in the foundation, structures, and roofing, the primary criterion influencing the choice of the construction method employed is time. However, in the other components consisting of the wall, flooring, and façade, quality is the key criterion influencing the selection of the construction methods used. **Table 5** also shows the overall level of influence and rank of the criteria used in the selection of construction

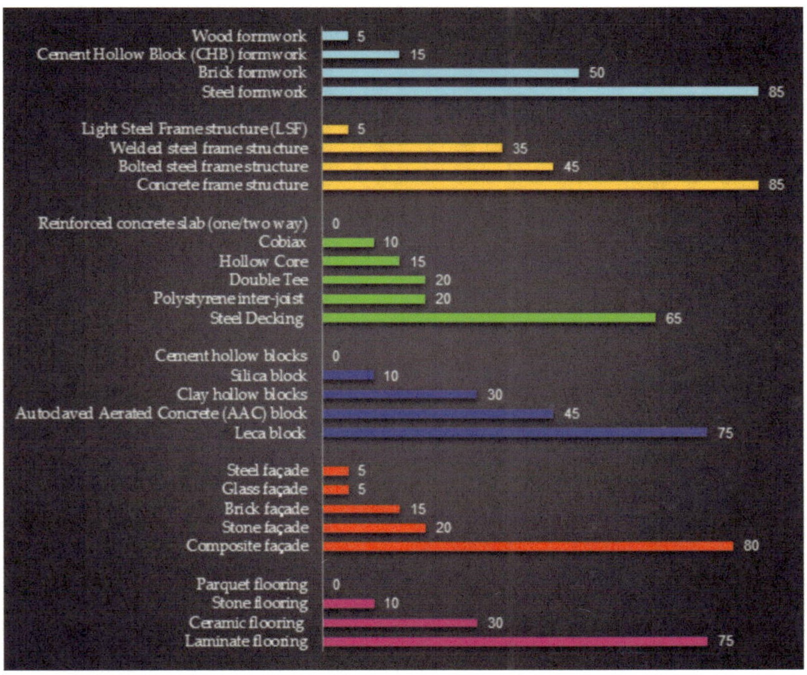

Figure 1. Distribution of cost as key performance in the selection of construction methods by construction managers.

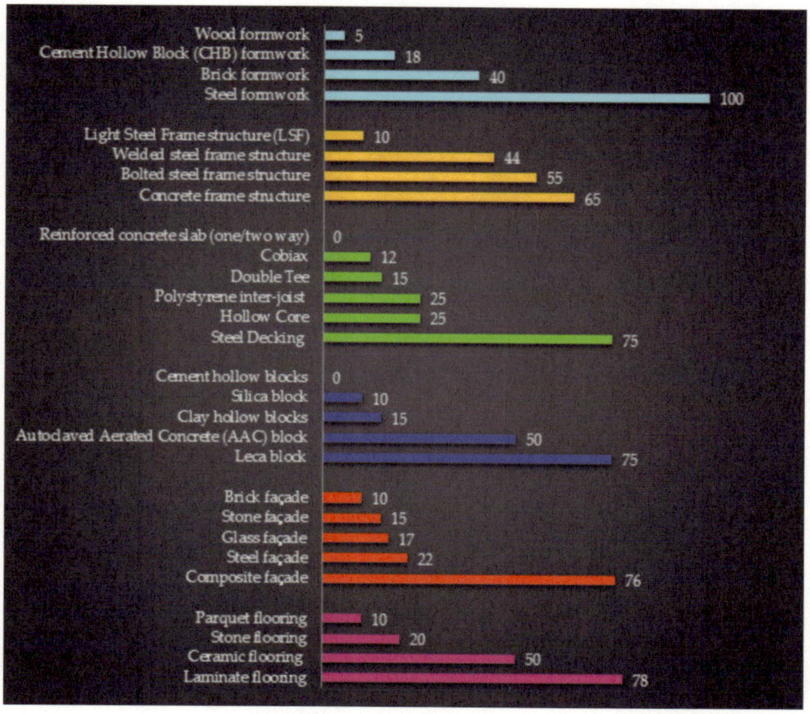

Figure 2. Distribution of time as key performance in the selection of construction methods by construction managers.

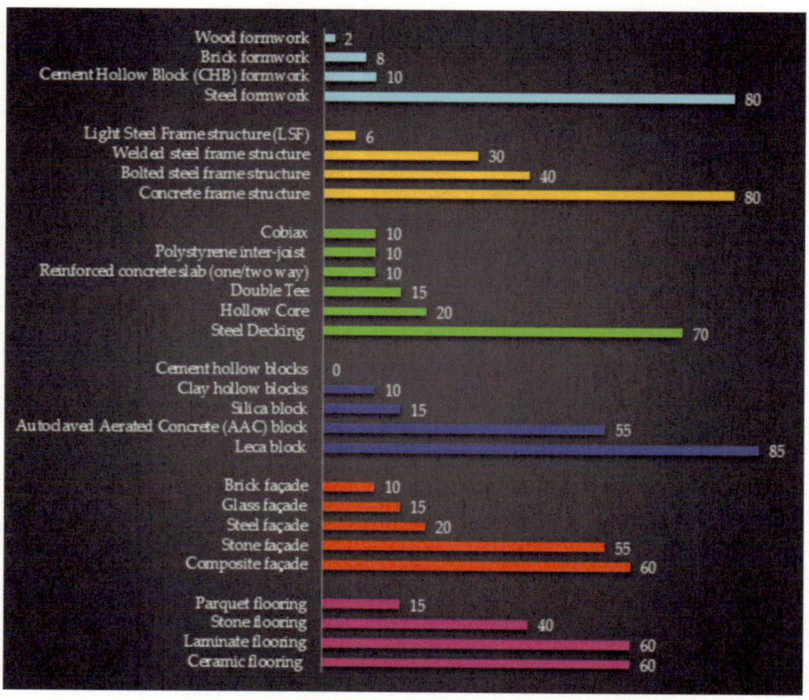

Figure 3. Distribution of quality as key performance in the selection of construction methods by construction managers.

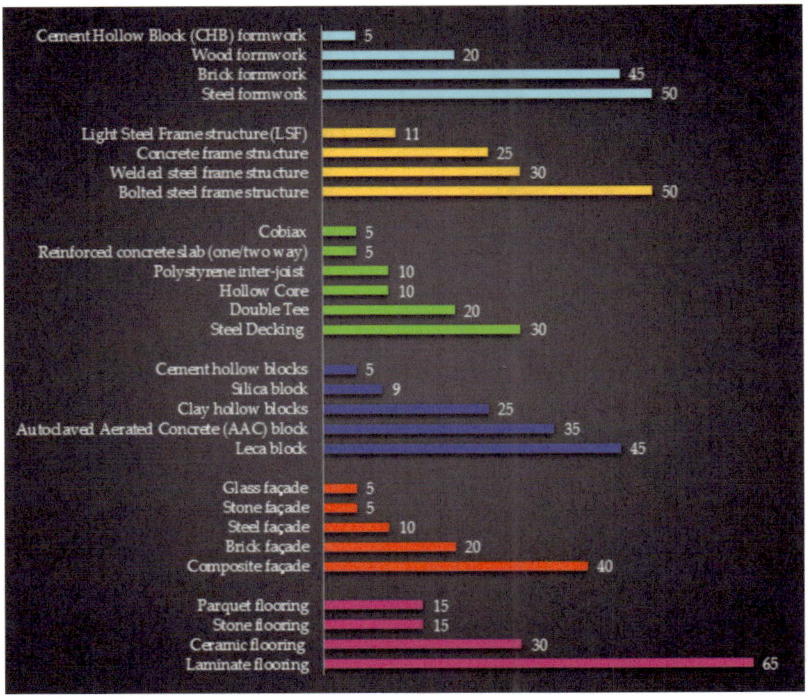

Figure 4. Distribution of ease of construction as key performance in the selection of construction methods by construction managers.

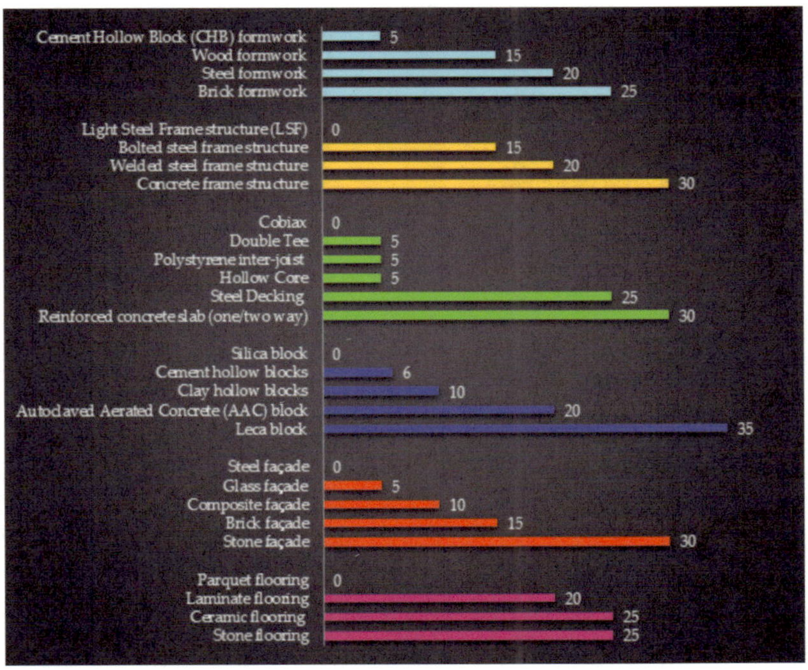

Figure 5. Distribution of availability of method and skill as key performance in the selection of construction methods by construction managers.

	Foundation	Rank	Structure	Rank	Roofing	Rank	Wall	Rank	Façade	Rank	Floor	Rank	Overall	Rank
Cost	27.0	1	25.5	1	26.8	1	22.6	3	24.6	2	24.8	3	25.2	1
Time	25.7	2	25.0	2	22.9	3	24.1	2	17.9	4	22.1	1	23.0	3
Quality	16.6	4	22.9	3	23.8	2	24.8	1	27.2	1	28.3	1	23.9	2
Ease of construction	19.9	3	17.0	4	14.1	4	17.9	4	19.4	2	14.2	4	17.2	4
Availability of method and skill	10.8	5	9.5	5	12.4	5	10.7	5	10.9	5	10.6	5	10.8	5

Table 5. Level of influence and rank of the criteria used in the selection of construction methods, distributed by building projects.

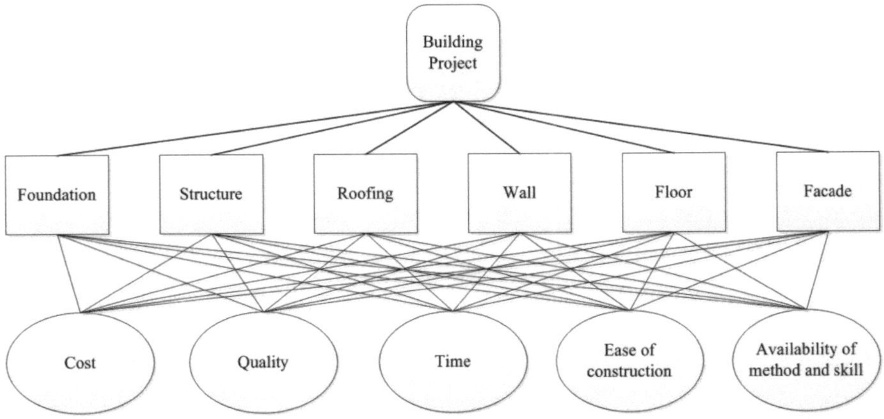

Figure 6. AHP model of the criteria influencing the selection of construction methods within building projects.

methods based on the data collected. **Table 5** indicates that time (25.2%) is the primary criterion that is most frequently used by construction managers in the selection of construction methods on building projects in the Middle East, followed by quality (23.9%), cost (23.0%), and ease of construction (17.2%), while the least used criterion is the availability of method and skill (10.8%).

5.2. Prioritizing the criteria used in the selection of construction methods used within building projects

The AHP model shown in **Figure 6** was developed to evaluate the importance of the different criteria used in the selection of construction methods (within the building components of the projects). Pairwise comparisons included cost, time, quality, ease of construction, and availability of method and skills, as shown in **Figure 6.**

The responses of construction experts, obtained in the pairwise comparisons, were used as inputs to determine the ranking of the selection criteria by each respondent. **Table 6** shows the weight and rank of the criteria used in the selection of construction methods within each building component studied, along with the inconsistency ratio in each component of construction, within the building project, from the AHP paired comparison matrix.

Table 6 shows that the inconsistency ratios of each component are less than 0.10 [43], which indicates the sufficient consistency in each matrix; therefore, all generated eigenvectors are considered. As shown in **Table 6**, it can be deduced that the construction method selection criteria weighting is in the following order at each stage:

1. Foundation time with weight value of 44.54% is more important than the other factors (cost (29.72%), quality (14.69%), ease of construction (7.31%), and availability of method and skill (3.74%)) in selecting foundation methods;

2. Structure cost with a weight value of 41.27% is the most important factor, followed by quality (24.19%), time (20.22%), availability of method and skill (10.39%), and ease of construction (3.92%), respectively;

3. Roofing quality with a weight value of 39.61% is the most important factor, followed by time and cost with weight values of 23.9 and 22.42%, ease of construction (8.21%), and availability of method and skills (5.86%);
4. Wall time and cost with weight values of 39.08 and 34.03% respectively, are the most important factors in selecting the wall methods, followed by quality (13.40%), ease of construction (7.25%), and availability of method and skill (6.25%);
5. Façade quality with a weight value of 42.45% has a significant role in the selection of the façade construction method, followed by cost (27.28%), time (17.72%), ease of construction (8.36%), and availability of method and skill (4.19%); and
6. Flooring quality with a weight value of 37.61% and time with a weight value of 32.1% have a more important influence on the selection of a flooring method, followed by cost (14.21%), ease of construction (9.84%), and availability of method and skill (6.24%).

From a ranking perspective, the expert construction managers rated the key performance criteria (weight value) according to their importance in the selection of construction methods on building projects.

Table 7 presents the overall weight value and rank of each criterion used in the selection of construction methods, based on the AHP paired comparison matrix.

As shown in **Table 7**, the inconsistency ratio is less than 0.10, and time (29.60%) is the most important criterion used in the selection of construction methods in building projects, followed by quality (28.66%), cost (28.16%), and ease of construction (7.48%), and the least important is the availability of method and skill (6.11%). It is therefore anticipated that construction methods will be selected based on the criterion of time before other considerations.

5.3. Relationship between the level of use of the construction methods and selection criteria

To validate the calculated AHP weight value of each criterion in the previous section, the association between the criteria used in the selection of the construction methods and the level of use of the construction methods on building projects is determined, using the Pearson correlation coefficient (PCC). The results of the correlation coefficients are shown in **Table 8**.

Table 8 shows that there are positive significant relationships between the criteria used in the selection of construction methods identified by the respondents, and the level of use of the construction methods, in six different building components. This indicates that as the value of the criterion used in the selection of a construction method increases, the influence of using that construction method is likely to increase and vice-versa. Touran [44] proposed following convenient system to quantify the correlation coefficients: weak 0.15; moderate 0.45; and strong 0.80. For instance, time (0.983) and cost (0.978) have a strong positive significant relationship on the selection of foundation methods, which means that an increase in the value of time and cost of a construction method (shorter duration and cheaper product) will result in an increase in the possibility of using the method and vice-versa. **Table 6** shows that quality is the most significant criterion when selecting construction methods for roofing and that all criteria are

	Foundation	Rank	Structure	Rank	Roofing	Rank	Wall	Rank	Façade	Rank	Floor	Rank
Cost	29.72	2	41.27	1	22.42	3	34.03	2	27.28	2	14.21	3
Time	44.54	1	20.22	3	23.90	2	29.08	1	17.72	3	32.10	2
Quality	14.69	4	24.19	2	39.61	1	13.40	3	42.45	1	37.61	1
Ease of construction	7.31	3	3.92	5	8.21	4	7.25	4	8.36	4	9.84	4
Availability of method and skill	3.74	5	10.39	4	5.86	5	6.25	5	4.19	5	6.24	5
Inconsistency ratio	0.06		0.07		0.02		0.02		0.04		0.04	

Table 6. Weight value and rank of the criteria used in the selection of construction methods within each component of building projects.

http://dx.doi.org/ 10.5772/intechopen.81673

	Weight value (%)	Inconsistency ratio	Rank
Cost	29.60		1
Time	28.66		2
Quality	28.16	0.04	3
Ease of construction	7.48		4
Availability of method and skill	6.11		5

Table 7. Weight and rank of criteria used in the selection of construction methods on building projects.

	Foundation	Structure	Roofing	Wall	Façade	Floor
Cost	0.978*	0.994*	0.875*	0.992**	0.902*	0.903
Time	0.983*	0.961*	0.879*	0.993**	0.814	0.953*
Quality	0.921	0.986*	0.940**	0.978**	0.960**	0.974*
Ease of construction	0.873	0.436	0.779	0.970**	0.658	0.840
Availability of method and skill	0.614	0.949	0.687	0.965**	0.390	0.734

*Correlation is significant at the 0.05 level (two-tailed).
**Correlation is significant at the 0.01 level (two-tailed).

Table 8. Correlation between criteria used in the selection and level of use of the construction methods within different building components.

significant when considering construction methods for the wall components. Also, **Table 8** presents the overall association between each criterion used in the selection of construction methods and the level of use of the construction method within the sample population.

As shown in **Table 8**, time (0.933) is the most associated criterion used in the selection of construction methods within the building projects studied followed by quality (0.932) and cost (0.931). These three factors have a strong positive significant relationship with the level of use of construction methods within the building projects assessed, while the ease of construction (0.771) and availability of method and skill (0.723) have a moderate positive significant relationship with the construction methods used within the building projects.

6. Discussion of findings

The main objectives of this study were to assess the criteria influencing the selection of construction methods used in building projects and whether selection criteria differed according to building components. The collected data from expert surveys and 200 building projects were analyzed using AHP and Pearson correlation coefficient to determine the rank and weight value of each selection criterion, as well as the relationship between the criteria and

Stage	Rank	Level of influence		AHP		Correlation coefficient	
		Factor	%	Factor	Weight value	Factor	Dependency
Foundation	1st	Time	27.0	Time	44.54	Time	0.983*
	2nd	Cost	25.7	Cost	29.72	Cost	0.978*
Structure	1st	Time	25.5	Cost	41.27	Cost	0.994*
	2nd	Cost	25	Quality	24.19	Quality	0.986*
Roofing	1st	Time	26.8	Quality	39.61	Quality	0.940**
	2nd	Quality	23.8	Time	23.9	Time	0.879*
Wall	1st	Quality	24.8	Time	39.08	Time	0.993**
	2nd	Cost	24.1	Cost	34.03	Cost	0.992**
Façade	1st	Quality	28.3	Quality	27.28	Quality	0.960**
	2nd	Cost	22.1	Cost	42.45	Cost	0.902*
Flooring	1st	Quality	27.2	Quality	37.61	Quality	0.974*
	2nd	Time	24.6	Time	32.1	Time	0.953*

*Correlation is significant at the 0.05 level (two-tailed).

**Correlation is significant at the 0.01 level (two-tailed).

Table 9. 1st and 2nd criteria used in the selection of construction methods in each stage of building projects.

the level of use of the construction methods within different components of building projects. Triangulating the results of the three analysis tests, as shown in **Table 9**, reveals that the AHP weight values and Pearson correlation coefficients correspond to each other in all the different stages of the building projects considered. However, the discrepancy between results in some stages (level of influence of criteria in selecting the construction method to AHP weight values and Pearson correlation coefficients) demonstrated that some of the project managers selected improper and inappropriate construction methods that were neither complement project objectives nor optimal methods. For instance, the construction method employed by project managers studied at the roofing stage was not the optimal method, due to a mismatch in the ranking of quality and time factors. In addition, construction methods selected for use in the structure and wall stages were improper and inappropriate.

The total weight value of different construction methods was achieved using perceived performance attributes of construction methods and AHP weight value of criteria in six building components. The total weight value assists decision makers in identifying the best suitable construction methods with the highest total weight value in each building component. To facilitate the comparison between construction methods within each component of building projects, the cluster bar chart of the total value of six building components was plotted in **Figure 7**.

Figure 7 shows that the best construction methods with the highest total weight value at each stage of building projects are steel framework in foundation stage with total weight value of 300, light steel frame structure (LSF) in structure stage with total weight value of 289.61, steel decking in roofing stage with total weight value of 300, Leca block in wall stage with total

http://dx.doi.org/ 10.5772/intechopen.81673

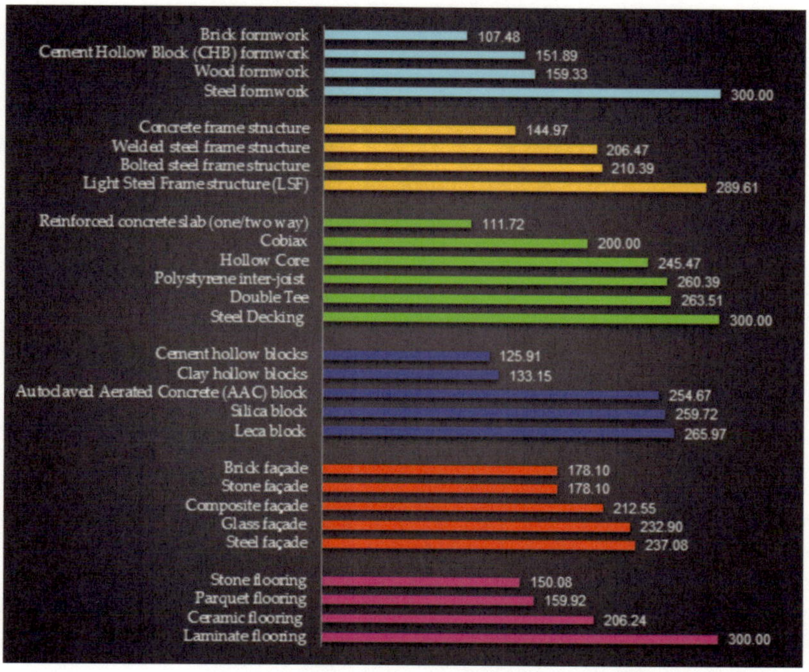

Figure 7. Total weight value of construction methods within each component of building projects.

weight value of 265.97, steel façade in the façade stage with total weight value of 237.08, and laminate flooring in the flooring stage with total weight value of 300.

The comparison between the distribution of construction methods and the total weight value of construction methods reveals that majority of the responding construction managers were not able to determine and select the best construction methods for different building components. It can be deduced that only 53% of construction managers have implemented the best construction method in the foundation stage, 5.5% in the structure stage, 38% in the roofing stage, 45% in the wall stage, 12% in the façade stage, and 39% in the flooring stage.

Table 10 shows the data analysis summary obtained from **Tables 5, 6**, and **8**, on the level of use of the criteria in the selection of construction methods, AHP weight value, and PCC relationship of each selection criterion to methods used in the overall stage of the building projects examined.

Cross-comparison of the PCC dependency results between criteria used in the selection and level of use of the construction methods (**Table 8**), with the AHP weight value of the criteria in the selection of each component of building, and the overall AHP weight value (**Tables 6** and **7**), validates the determined AHP weight values (See **Table 10**).

As demonstrated in **Table 10**, it emerged from the study that the results of three analysis tests established that the most important factor in the selection of construction methods is time, or duration of construction, with the highest level of use in the selection of construction methods, a weight value of 29.60% and a strong positive significant dependency to construction methods

	Level of influence	Rank	AHP weight value	Rank	PCC dependency	Rank
Time	25.2	1	29.6	1	0.933**	1
Quality	23.9	2	28.66	2	0.932**	2
Cost	23.0	3	28.16	3	0.931**	3
Ease of construction	17.2	4	7.48	4	0.771**	4
Availability of method and skill	10.8	5	6.11	5	0.723**	5

**Correlation is significant at the 0.01 level (two-tailed).

Table 10. Data analysis summary.

used. It can be inferred from this finding that adopting a process that has a shorter duration has two impacts on the project; first, it reduces the duration of the project, thereby decreasing the cost of the project by reducing the cost of labor equivalent to amended time [13, 14].

The second most important factor seen from **Table 10** is that quality highly influences the selection of construction methods, with a weight value of 28.66% and a strong positive significant dependency to construction method. It can be deduced that quality is the second priority of construction managers when selecting building materials and construction methods. Conversely, the new generation of building regulations and codes tends to increase the quality of buildings [6]; therefore, choosing poor quality methods and materials could lead to additional cost and time, due to failure to meet these building regulations, or owners' expectations.

Furthermore, the study found that the third factor influencing the selection of construction methods is cost, with a weight value of 28.16% and also a strong positive significant dependency to construction method (see **Table 10**). Unlike time, the cost is a single dimension element, and choosing the method or material with less cost will not guarantee or improve the other influencing factors [13]. For instance, cement hollow block (CHB) is cheaper than Leca block, but the construction time is longer and the implementation process is more difficult because the method uses more labor in the construction process and construction of the CHB wall would increase the overall cost of the project due to higher labor costs [42].

The fourth important factor influencing the selection of construction methods is the Ease of construction, with a weight value of 7.48% and moderate positive significant dependency to construction methods used. Finally, the factor with the least priority is the availability of methods and skills, with a weight value of 6.11% and also a moderate positive significant dependency on construction methods used. The data obtained provide evidence of the adequate supply of all methods and expertise in the construction market of the Middle East.

The findings of the study are consistent with those of previous studies by Noorzai et al. [45] and Monghasemi et al. [13], who acknowledge that the most important effective criterion for the selection of construction methods in building projects is time, followed by quality and cost, among other factors; Lam et al. [46] indicates that in order to enhance productivity and efficiency of building projects, a shorter time and higher quality methods, such as precast or prefabricated systems, should be employed in the construction stage of projects.

The above analysis evaluated the importance, significance, and priority of each criterion that influences the selection of construction methods in selected components of building projects and highlighted that the selection criterion differs in line with the different building components. This knowledge helps decision makers like construction managers and engineers to identify the importance (weight value) of each key performance criterion (time, cost, quality, ease of construction, and availability of method and skill) and best available construction method (total weight value) in different stages of building projects. They are then enabled to make proper decisions and select appropriate methods accordingly.

Selecting appropriate construction methods based on these priorities plays a significant role in achieving the project objectives and enhancing productivity and successful project delivery [47]. The other advantage of this knowledge is that building material manufacturers could use this guideline as a tool in the production of innovative technologies and modern construction methods for each component, according to the priority of the criterion used in its selection, so as to improve the productivity, efficiency, and quality of building projects.

7. Conclusion

The lack of an efficient systematic approach to the selection of appropriate construction methods for building projects necessitates a critical examination of the various construction methods currently available to identify the criteria used in their selection and their weight and to enable the optimum performance, success, and sustainability of building projects. Therefore, this study examines how key performance criteria were used in the selection of construction methods on projects and whether these selection criteria differed across different building components toward making recommendations for improving the selection of appropriate construction methods on projects. To achieve this objective, the study employed an AHP (structured mathematical) model and PCC (statistical) analysis to aggregate and compare relative weight values of different construction methods in six major building components, based on their performance criteria in construction projects completed in the Middle East.

The study found that 47% of construction managers had employed improper and inappropriate methods in the foundation stage, 94.5% in the structure stage, 62% in the roofing stage, 55% in the wall stage, 88% in the façade stage, and 61% in the flooring stage despite the availability of new equipment and facilities, innovative technologies, and modern construction methods in the Middle East.

The results of the AHP model and PCC analysis revealed that the most important criteria for the selection of construction methods and also the best component for each stage of a building project were steel framework in the foundation, light steel frame structure (LSF) in the structure, steel decking also in the roofing, Leca block in the wall, steel façade in the façade, and laminate in the flooring.

The study established the positive significant role and impact of time and quality performance criteria in the selection of construction methods. Based on the findings, it was concluded that the capacity of a method of construction to shorten the time span for the process (29.60%) and facilitate a higher quality of building (28.66%) rendered it more likely to be used on building projects.

To achieve project objectives, enhance productivity, and successful project delivery, building components that are suitably aligned to these criteria should be selected by project decision makers, while construction materials producers need to align their products to fit these criteria.

Finding the most suitable and sustainable construction methods for building components will accomplish a sustainable building, as is required in the Middle East and other developing countries. The detailed analysis and results of prioritizing the affecting factors from this research should be used as a platform and benchmark for future studies. Also, this platform should be utilized for evaluating the level of efficiency of building processes and as a guideline for improving the effectiveness of building processes by selecting optimum construction methods that are aligned to project objectives and targets.

It is possible to extend the straightforward and flexible model developed here to other construction methods and ultimately to the whole building project. The model criteria can also be modified, depending on the scope or focus of the study. Not only it is the proposed systematic approach applicable to different construction projects, but also it is useful globally to improve standards of construction, thereby benefiting communities and ensuring safer, sturdier, and more sustainable buildings.

Author details

Alireza Moghayedi* and Abimbola Windapo

*Address all correspondence to: mghali001@myuct.ac.za

Department of Construction Economics and Management, University of Cape Town, Cape Town, South Africa

References

[1] Wambeke BW, Hsiang SM, Liu M. Causes of variation in construction project task starting times and duration. Journal of Construction Engineering and Management. 2011;**137**(9):663-677

[2] Kamali M, Hewage K, Milani AS. Life cycle sustainability performance assessment framework for residential modular buildings: Aggregated sustainability indices. Building and Environment. 2018;**138**:21-41

[3] Ren Z, Shen G, Xue X. Failure caused by inappropriate construction methods: An expensive lesson. Journal of Management in Engineering. 2011;**29**(1):25-34

[4] Forbes LH, Ahmed SM. Modern Construction: Lean Project Delivery and Integrated Practices. United States of America: Crc Press; 2010

[5] Harris F, McCaffer R. Modern Construction Management. United Kingdom: John Wiley & Sons; 2013

[6] Imrie R. The role of the building regulations in achieving housing quality. Environment and Planning. B, Planning & Design. 2004;**31**(3):419-437

[7] Saaty RW. The analytic hierarchy process—What it is and how it is used. Mathematical Modelling. 1987;**9**(3-5):161-176

[8] Jaskowski P, Biruk S, Bucon R. Assessing contractor selection criteria weights with fuzzy AHP method application in group decision environment. Automation in Construction. 2010;**19**(2):120-126

[9] Ferrada X, Serpell A, editors. Construction Methods Selection: Lessons Learned from Chile. W102-Special Track 18th CIB World Building Congress May 2010 Salford. United Kingdom: CIB; 2010

[10] Haidar AD. Construction Program Management–Decision Making and Optimization Techniques. Switzerland: Springer; 2015

[11] Kandil A, El-Rayes K, El-Anwar O. Optimization research: Enhancing the robustness of large-scale multiobjective optimization in construction. Journal of Construction Engineering and Management. 2009;**136**(1):17-25

[12] Udaipurwala A, Russell AD. Computer-assisted construction methods knowledge management and selection. Canadian Journal of Civil Engineering. 2002;**29**(3):499-516

[13] Monghasemi S, Nikoo MR, Fasaee MAK, Adamowski J. A novel multi criteria decision making model for optimizing time–cost–quality trade-off problems in construction projects. Expert Systems with Applications. 2015;**42**(6):3089-3104

[14] Moghayedi A. Improving critical path method (CPM) by applying safety factor to manage delays. Scientia Iranica Transaction A, Civil Engineering. 2016;**23**(3):815

[15] Ferrada X, Serpell A. Selection of construction methods for construction projects: A knowledge problem. Journal of Construction Engineering and Management. 2014;**140**(4): B4014002

[16] Rahman MM. Barriers of implementing modern methods of construction. Journal of Management in Engineering. 2013;**30**(1):69-77

[17] Dalalah D, Al-Oqla F, Hayajneh M. Application of the analytic hierarchy process (AHP) in multi-criteria analysis of the selection of cranes. Jordan Journal of Mechanical and Industrial Engineering. 2010;**4**(5):567-578

[18] Sangiorgio V, Uva G, Fatiguso F. Optimized AHP to overcome limits in weight calculation: Building performance application. Journal of Construction Engineering and Management. 2017;**144**(2):04017101

[19] Ng C. An evidential reasoning-based AHP approach for the selection of environmentally-friendly designs. Environmental Impact Assessment Review. 2016;**61**:1-7

[20] Hossaini N, Reza B, Akhtar S, Sadiq R, Hewage K. AHP based life cycle sustainability assessment (LCSA) framework: A case study of six storey wood frame and concrete frame buildings in Vancouver. Journal of Environmental Planning and Management. 2015;**58**(7):1217-1241

[21] Wong JK, Li H. Development of intelligence analytic models for integrated building management systems (IBMS) in intelligent buildings. Intelligent Buildings International. 2009;**1**(1):5-22

[22] Pan N-F. Fuzzy AHP approach for selecting the suitable bridge construction method. Automation in Construction. 2008;**17**(8):958-965

[23] Zayed T, Amer M, Pan J. Assessing risk and uncertainty inherent in Chinese highway projects using AHP. International Journal of Project Management. 2008;**26**(4):408-419

[24] Shapira A, Goldenberg M. AHP-based equipment selection model for construction projects. Journal of Construction Engineering and Management. 2005;**131**(12):1263-1273

[25] Al-Harbi KMA-S. Application of the AHP in project management. International Journal of Project Management. 2001;**19**(1):19-27

[26] Skibniewski MJ, Chao L-C. Evaluation of advanced construction technology with AHP method. Journal of Construction Engineering and Management. 1992;**118**(3):577-593

[27] Reza B, Sadiq R, Hewage K. Sustainability assessment of flooring systems in the city of Tehran: An AHP-based life cycle analysis. Construction and Building Materials. 2011;**25**(4):2053-2066

[28] Nadoushani ZSM, Akbarnezhad A, Jornet JF, Xiao J. Multi-criteria selection of façade systems based on sustainability criteria. Building and Environment. 2017;**121**:67-78

[29] Hosseini ST, Lale Arefi S, Bitarafan M, Abazarlou S, Zavadskas EK. Evaluation types of exterior walls to reconstruct Iran earthquake areas (Ahar Heris Varzeqan) by using AHP and fuzzy methods. International Journal of Strategic Property Management. 2016;**20**(3):328-340

[30] Akadiri PO, Olomolaiye PO, Chinyio EA. Multi-criteria evaluation model for the selection of sustainable materials for building projects. Automation in Construction. 2013;**30**:113-125

[31] Medineckiene M, Zavadskas E, Turskis Z. Dwelling selection by applying fuzzy game theory. Archives of Civil and Mechanical Engineering. 2011;**11**(3):681-697

[32] Ang AH-S, Tang WH. Probability Concepts in Engineering: Emphasis on Applications in Civil & Environmental Engineering. New York: Wiley; 2007

[33] El Khalek HA, Aziz RF, Kamel HM. Risk and Uncertainty Assessment Model in International Construction Projects Using Fuzzy Logic; International Journal of Innovative Research in Engineering & Management. 2016;**3**(2):134-149

[34] Lee YB, Hwang J, editors. A Study on Correlation between Elements of AHP for Government R&D Programs. Proceedings of the International Symposium on the Analytic Hierarchy Process. Italy: ISAHP; 2011

[35] Sawacha E, Naoum S, Fong D. Factors affecting safety performance on construction sites. International Journal of Project Management. 1999;**17**(5):309-315

[36] Haidar AD. Techniques for Intelligent Decision Support Systems. Construction Program Management–Decision Making and Optimization Techniques. Switzerland; Springer; 2016. pp. 159-183

[37] Hosseini SA, de la Fuente A, Pons O. Multicriteria decision-making method for sustainable site location of post-disaster temporary housing in urban areas. Journal of Construction Engineering and Management. 2016;**142**(9):04016036

[38] Bakhshi P, Touran A. A method for calculating cost correlation among construction projects in a portfolio. International Journal of Architecture, Engineering and Construction; **1, 2012**(3):134-141

[39] Devore JL. Probability and Statistics for Engineering and the Sciences: Cengage Learning. Richard Stratton, United States of America; 2011

[40] Golafshani N. Understanding reliability and validity in qualitative research. The Qualitative Report. 2003;**8**(4):597-606

[41] Heale R, Twycross A. Validity and reliability in quantitative studies. Evidence-Based Nursing. Ebnurs. 2015;**18**(3):66-67

[42] Windapo A, Moghayedi A, Oliphant D, Adediran A. Exploring the components of cost on construction projects. Journal of Construction Business and Management. 2018;**2**(2):42-50

[43] Saaty TL. Decision Making for Leaders: The Analytic Hierarchy Process for Decisions in a Complex World. RWS Publications. United States of America; 1990

[44] Touran A. Probabilistic cost estimating with subjective correlations. Journal of Construction Engineering and Management. 1993;**119**(1):58-71

[45] Noorzai E, Hosseini A, Gharouni Jafari K, Aghaeipoor M. Providing a model to select an optimum multifamily housing method in Iran. Journal of Architectural Engineering. 2016;**23**(2):04016019

[46] Lam PT, Chan AP, Wong FK, Wong FW. Constructability rankings of construction systems based on the analytical hierarchy process. Journal of Architectural Engineering. 2007;**13**(1):36-43

[47] Kamali M, Hewage K. Development of performance criteria for sustainability evaluation of modular versus conventional construction methods. Journal of Cleaner Production. 2017;**142**:3592-3606

Chapter 10

Flyash as a Resource Material in Construction Industry: A Clean Approach to Environment Management

Mohammad Nadeem Akhtar and Nazia Tarannum

Additional information is available at the end of the chapter

http://dx.doi.org/10.5772/intechopen.82078

Abstract

The maximum amount of electricity is produced by most of the thermal power plants by burning coal at their operating facilities. Due to this activity, various types of secondary materials are generated. Any material resulting from coal-combustion processes may be called as a coal-combustion product (CCP). Among different CCPs reported worldwide by coal-burning power plants, flyash is the most common one. As per the characterization report, flyash is considered as a powdery material being collected by dust collectors installed in the thermal power plants with the use of coal as fuel. There are different problems related to flyash like requirement of large area of land for disposal and toxicity caused by flyash which leach to groundwater. The study has established flyash as air and water pollution source. It is considered as waste that may act as a resource material in construction industry, thereby acting as a resource for waste and environment management. Till a decade back, flyash was treated as waste material worldwide, but now it is developed as an environment savior.

Keywords: thermal power plants, coal-combustion product, flyash, waste material

1. Introduction

A naturally cementitious material that is obtained as a by-product after coal-combustion is flyash. In order to reduce pollution, flyash is extracted from the precipitators installed in smokestacks of coal-burning power plants. In near future, it is expected that a number of thermal power stations will increase with the increasing demand of power and coal. Flyash is spherical in shape and solidify in suspension form in exhaust gases. The composition of flyash constitutes silica (SiO_2), alumina (Al_2O_3), and iron oxide (Fe_2O_3). As per the need and use of flyash, their physical and chemical requirement may vary. The specific requisites for the use of flyash

IntechOpen

© 2018 The Author(s). Licensee IntechOpen. This chapter is distributed under the terms of the Creative Commons Attribution License (http://creativecommons.org/licenses/by/3.0), which permits unrestricted use, distribution, and reproduction in any medium, provided the original work is properly cited.

in concrete or soil stabilization are discussed in [1, 2]. According to Ref. [1], flyash is broadly classified into two classes, viz. Class F and Class C. The major difference between these two classes is the percentage of calcium, silica, alumina, and iron content. The chemical content of burned coal properties, i.e., anthracite and bituminous, decides the chemical content of the flyash. When older and harder anthracite burns, they produce Class F flyash, which is pozzolanic in nature with 10% lime (CaO). In order to possess cementitious property, Class F flyash with the glassy silica and alumina requires a cementing agent like Portland cement, quicklime, or hydrated lime. The addition of sodium silicate (water glass) as a chemical activator to a Class F flyash may lead to the geopolymer formation. The burning of younger sub-bituminous coal produces flyash with pozzolanic and self-cementing properties. The presence of water hardens Class C flyash and enhances its strength with time. More than 20% lime is possessed by Class C flyash. Like self-cementing Class C flyash does not require an activator unlike Class F. The percentage of alkali and sulfate (SO_4) are high in Class C flyash. Till a decade, flyash was considered as an industrial waste pollution and was disposed off in ash ponds. Several cumulative researches have been carried out worldwide for flyash management and disposal. The quality of flyash depends on various factors like coal, percentage of ash in coal, coal particle fineness, combustion technique used, air/fuel ratio, burners used, and type of boiler. During last few decades, innovative and alternate building materials with economic construction techniques have been developed by intensive research efforts which provide an opportunity to bring down the cost of construction. Flyash is the by-product of industry generated about 112 million tons from Thermal Power Plants (TPPs) with proven suitability for various applications such as admixture in cement/concrete/mortar, lime pozzolana mixture (bricks/blocks), etc. The cement industry is responsible for 50% utilization of flyash, the total utilization of this accounts for 30MT (28%). The other application areas include low lying area fill (17%), roads and embankments (15%), dyke raising (4%), brick manufacturing (2%) and in paint industry, agriculture, etc. [3]. According to 2001 census, India is the first largest country for the production of flyash worldwide. **Figure 1** shows the flyash production (million tons/year) in various countries.

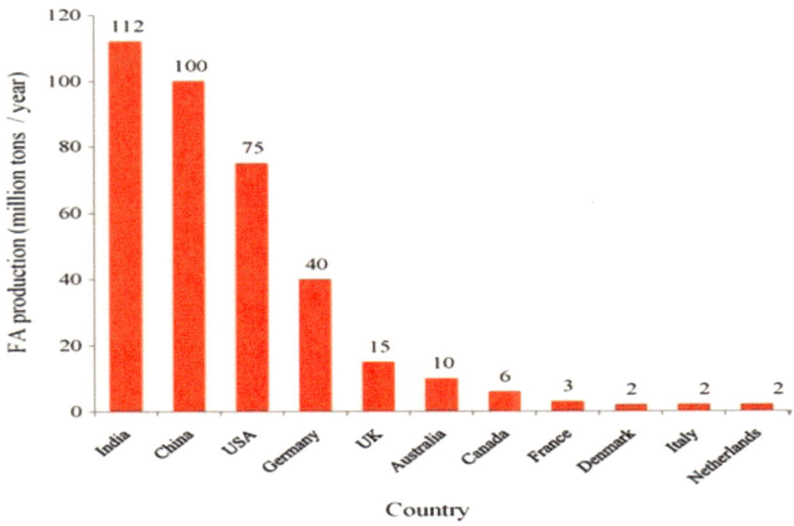

Figure 1. Representation of the production (million tons/year) of flyash in different countries.

2. Current scenario of flyash in India

Recent production of flyash, a by-product from coal-based TPPs annually, is about 160 million tons (MT) and is expected to be about 600 MT by 2030 according to the current ENVIS CSIR report. As per the reports, India is among the largest producer of coal and contributes about 70% of the total installed capacity for power generation [4]. The bituminous and sub-bituminous coal-use contains over 40% ash content. About 120–150 million tons of coal flyash are generated from 120 existing coal-based TPPs in India [5]. The emerging amount of average generation and utilization data received during the last 5 years by ENVIS Centre on flyash [6] is 166 million and 96 million tons. Flyash generation and utilization during the year 2011–2012, 2012–2013, 2013–2014, 2014–2015, and the first half year of 2015–2016 are shown in **Table 1**.

Flyash generation and utilization data for the first half of 2015–2016 (April 2015 to September 2015) is expected to be 132 coal/lignite-based thermal power stations of different power utilities in the country. As per the data received on March 15, 2016, a conclusion was derived on the current status of flyash generation and its utilization in the country [7]. Flyash utilization percentage (of 146 thermal power stations) has increased during the first half of 2015–2016 compared to utilization during first half of the previous year 2014–2015 [7]. ENVIS Centre on flyash hosted by CSIR-Central Building Research Institute, Roorkee sponsored by the Ministry of Environment, Forests and Climate Change, Government of India shows the correlation between production and utilization from 1990 to 2030. The flyash generation and utilization scenario in India is shown in **Figure 2**.

2.1. Materials and method

The flyash utilized in the study was collected through electrostatics precipitators of Harduaganj Thermal Power Station Aligarh (UP), India [8]. The chemical analysis of flyash was done to determine the oxides of silica, iron, calcium, aluminum, total sodium, magnesium, and total potassium. The procedures used to determine the abovementioned elements are as per the specification of ASTM Designation: D 4326. About 0.50 g of the ashed flyash sample and 6.5 g

Description	2011- 12	2012-13	2013-14	2014-15	1st Half Year 2015-16
Nos. of Thermal Power Stations from which data was received	124	138	143	145	132
Installed capacity (MW)	1,05,925.3	1,20312.30	1,33,381.30	1,38,915.80	1,30,428.80
Coal Consumed (million tons)	437.41	482.97	523.52	549.72	251.69
Average Ash Content (%)	33.24	33.87	33.02	33.50	33.23
Flyash Generation (million tons)	145.42	163.56	172.87	184.14	83.64
Flyash Utilisation (million tons)	85.05	100.37	99.62	102.54	46.87
Percentage Utilization	58.48	61.37	57.63	55.69	56.04

Source: ENVIS Centre on Flyash Hosted by CSIR-Central Building Research Institute, Roorkee. Sponsored by Ministry of Environment, Forests & Climate Change, Govt. of India.

Table 1. Flyash generation and utilization current scenario in India.

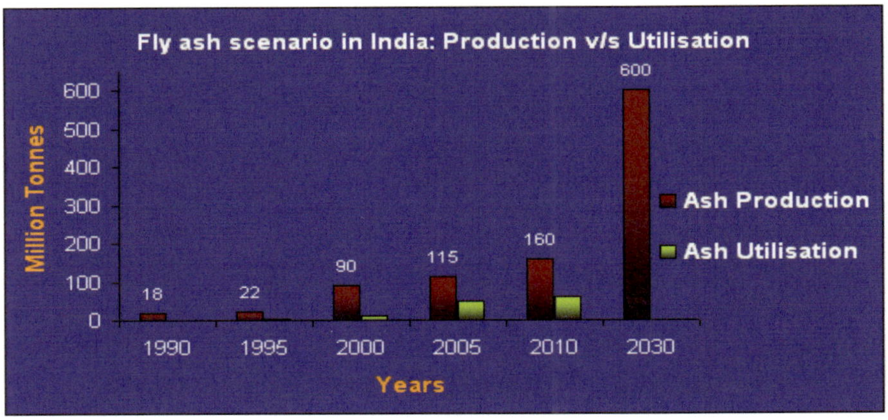

Figure 2. Current flyash scenario generation vs. utilization. Source: ENVIS Centre on Flyash Hosted by CSIR-Central Building Research Institute, Roorkee sponsored by Ministry of Environment, Forests & Climate Change, Govt. of India.

of fluxing material were taken into the crucible with 400 μL of liquid lithium bromide before fusing. In order to determine the loss on ignition, available sodium oxide, potassium oxide and alkali were used in accordance with ASTM designation C311.3. The sulfur trioxide with the carbon sulfur determinator was used to determine sulfur. XRF (X-ray fluorescence), X-ray diffraction (XRD), wavelength dispersive spectroscopy (microprobe WDS), and X-ray micro-analysis (EDS) techniques are used in the chemical analyses of flyash. The physical properties were determined by Indian standards.

2.2. Chemical analysis of flyash

Determination of chemical composition of flyash is mandatory for classification of flyash as per the standards. The chemical composition of flyash consists of silicon, aluminum, calcium, iron, sulfur, and magnesium oxides along with carbon and other trace elements. These elements found in the ash possess high melting points and the short duration of the ash particles remain in furnace during combustion process. The SiO_2 causes the initiation of the combustion process and are maintained as quartz in the coal ash. The clay mineral loses water molecule and melt forming alumino-silicate crystalline and noncrystalline (glassy) materials. Fe, Ca, and Mg along with oxygen in the air form oxides, such as magnetite (Fe_2O_3), lime (CaO), and periclase (MgO).

2.2.1. Chemical analysis of Class F flyash for present study

Locally available samples were examined in the present study. The detailed image analysis about the surface morphology and texture of individual particles were studied by Scanning Electron Microscopy (SEM) with Energy Dispersive Spectroscopy (EDS). SEM is one of the widely used techniques for the chemical characterization of ash. The exact chemical composition is discussed in **Table 2**. The elements identified in the flyash samples were found to be C, O, Al, Si, K, Ca, and Ti in various compound forms (Al_2O_3, SiO_2, K_2O, CaO, TiO_2, etc.). The quantitative result from point analysis of Class F flyash used in the study is shown in **Figure 3**.

http://dx.doi.org/10.5772/intechopen.82078

Element	Weight %	Atomic%
C	14.42	21.05
O	54.59	59.83
Al	11.96	7.77
Si	16.47	10.29
K	0.73	0.33
Ca	0.94	0.41
Ti	0.88	0.32

Table 2. List of Elements with their weight (%) and atomic (%) present in Class F flyash.

2.2.2. SEM/EDS analysis for present study

The surface morphology characterization has a vital role to play in understanding the physical and chemical behavior of the material. There detailed knowledge of the physical nature of solid surfaces in the fields of material science and surface chemistry is of utmost importance. SEM is a technique to study different modes of association of particle for surface irregularities detection. SEM is used to investigate the surface morphology of the sample. **Figure 4** shows SEM images recorded on Class F flyash sample surface at ×2000, ×5000, and ×6000 magnifications. As, it can be seen, **Figure 4a** and **b** clearly indicates the presence of irregular shaped particles of variable size, covered with relatively smooth grains of quartz. The micrographs in **Figure 4c** and **d** also designated dark areas as organic materials, light areas as mineral matter, and gray as mixture of coal and ash. The solid and porous part indicated the presence of mineral matter most likely quartz, partially burnt coal particles were shown by irregular black porous parts. Particles size 10 μm at WD 13 mm appeared to be spherical with small bulging of siliceous and aluminous glass **Figure 4e**. EDS of flyash sample suggested the presence

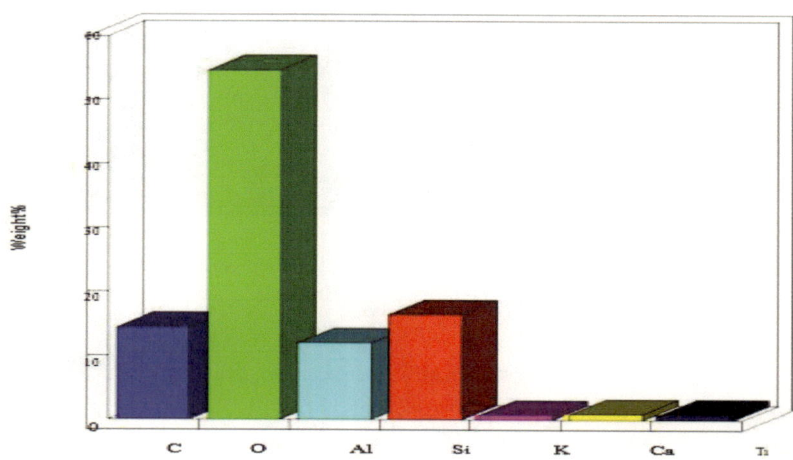

Figure 3. Quantitative results from point analysis of Class F flyash.

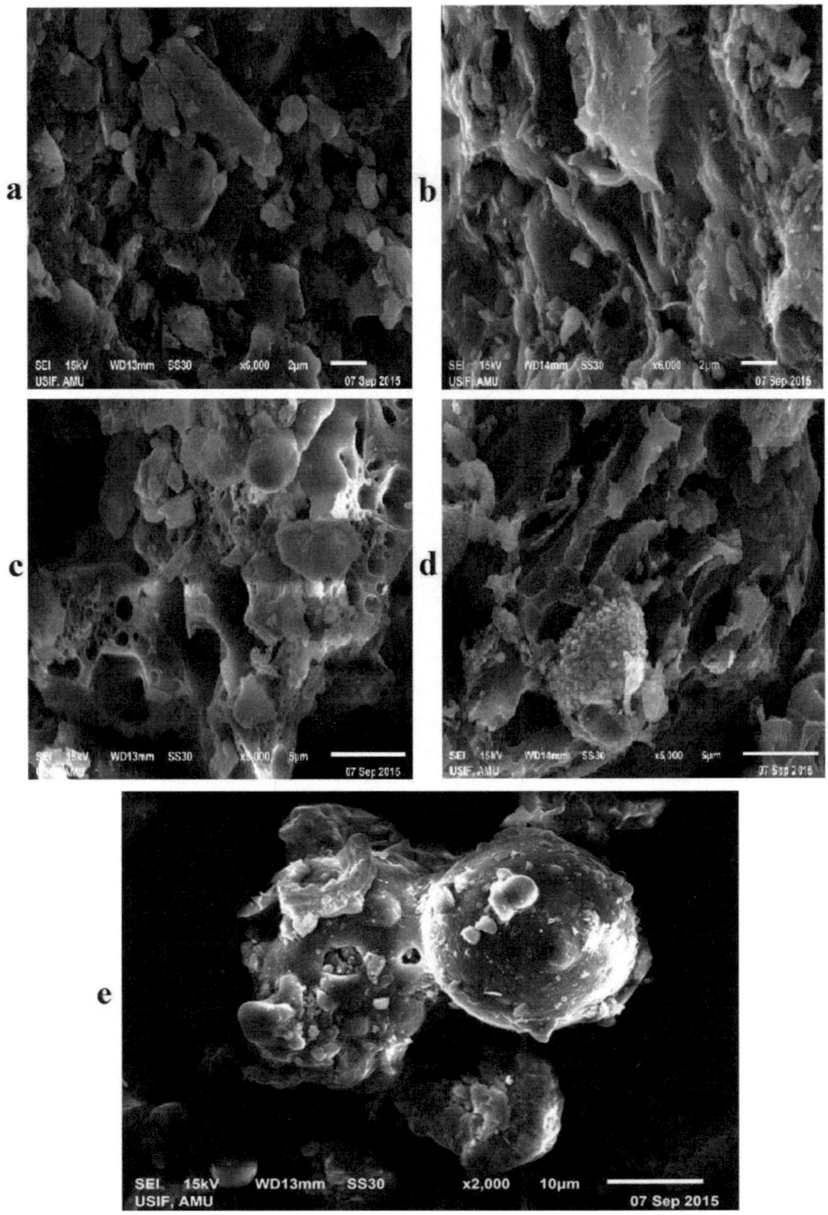

Figure 4. SEM micrograph of Class F Flyash with different particles size (a) 2 μm at WD 13 mm; (b) 2 μm at WD 14 mm (c); 5 μm at WD 13 mm; (d) 5 μm at WD 14 mm; and (e) 10 μm at WD 13 mm.

of carbon, oxygen, aluminum, silicon, potassium, calcium, and titanium as the primary elements. Hence, SEM/EDS are one of the important and widely used techniques for physicochemical analysis of flyash samples. The identified elements in the flyash samples were found to be C, O, Al, Si, K, Ca, and Ti in various compounds (Al_2O_3, SiO_2, K_2O, CaO, TiO_2, etc.) as determined by EDS, as shown in **Figure 5**.

http://dx.doi.org/10.5772/intechopen.82078

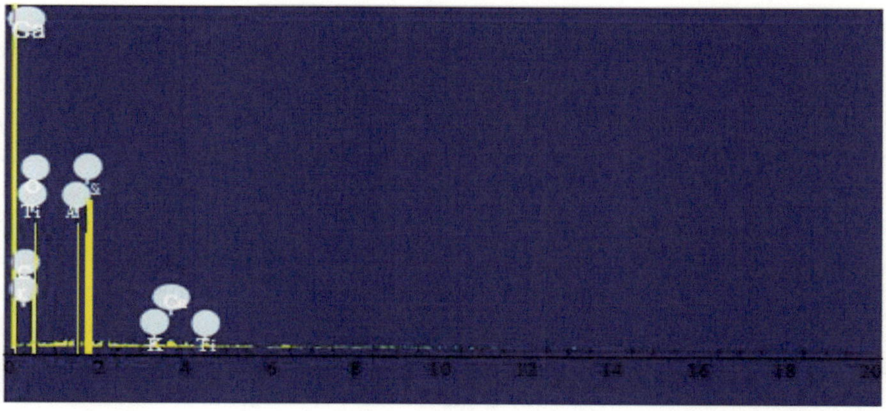

Figure 5. Identified element of Class F flyash by EDS.

2.3. Physical analysis of Class F flyash for present study

Physical analysis is one of the important parameter for selection and consideration of material in the civil engineering construction industry. Its geo-technical properties make it a good substitute of soil and the required percentage provide the general range of physical geo-technical properties available in the flyash sample. As determined in the present study, the physical properties of Class F flyash are listed in **Table 3**. The physical properties of common soils

Parameter	Range Experiment Present Study	CBRI (2016)
	Class F Flyash	Common soil
Bulk Density (gm/cc)	1.25	1.3–1.7
Specific Gravity	2.12	2.5-2.8
Plasticity	Lower or non-plastic	Lower or non-plastic
Maximum Dry Density (gm/cc)	1.2875	1.3-2.4
Optimum Moisture Content (%)	18	5.0-30.0
Angle of Internal	28	30-40
Friction(degrees)	2.1	Negligible
Cohesion (kN/m²)	1.650×10^{-5}	8×10^{-6} - 7×10^{-4}
Permeability (m/sec)	Higher	Low - high
Shrinkage Limit (Vol stability)	Major fine sand range / and	Major sand size fraction /
Grain size	very small per cent of clay size particles	silt and clay fraction and small per cent of gravel size fraction
Clay (percent)	Negligible	Low- medium
Free Swell Index	Very low	very low
Classification (Texture)	Sandy silt to silty loam	sandy clay, silty clay, clay loam and silt loam
Water Holding Capacity (WHC)	40-60	10-70
(per cent)	30-65	15-75
Porosity (per cent)		

Table 3. Summary of test results for different experiments on Class F Flyash.

mentioned in [6] for Indian soil are selected for the comparison also provided in **Table 3**, the typical variations of comparative study obtained by **Table 3**. The sample of locally available common soil also tested for comparison with Class F sample. It is found in the study that the physical properties of Class F flyash is very close to the relative values of common soils though it may differ from one country to another on their geographical conditions. There are several factors as to why soils differ with respect to region. The parent material are the source of most influential factors, others include climate and terrain of the region as well as the type of plant life and vegetation present. As per the study, the focus is on the geo-technical functions of the flyash and its comparison with the common soil which is an important criterion to replace any natural material. The strength and durability are two important factors to replace any material in construction industry. A material is considered as a building material if it requires properties suitable enough to be engineered for construction. A series of experiments conducted [9] showed that the addition of flyash improved the soil properties, viz., texture, structure, and bulk density. Permeability of clay loam soil increased from 0.54 to 2.14 cm/h by 50% addition of flyash, whereas it decreased from 23.80 to 9.67 cm/h in sandy soil by 50% flyash addition. Water retaining capacity of sandy soil increased from 0.38 to 0.53 cm/cm at 50% level. The characterization studies have been performed by researchers to evaluate the suitability of flyash for other fields of applications. The nature of Indian flyash is alkaline, hence, its use in agricultural soils increase pH of the soil thereby, neutralizing acidic soil [10].

2.4. Published studies on flyash

The production of high performance concrete (HPC) needs to incorporate the supplementary cementations materials such as flyash and kaolin in the concrete mix [10]. The fiber combination in the premix reinforced flyash lime stone dust brick (10FRFALSDB3') to have highest compressive strength (9.155 MPa) with 10% stone dust, and sand combination at 10% cement [11]. Ref. [12] carried out several tests on flyash-based brick tiles of Class F type. After experimental investigation, it was proved that the tiles failed on the lower compressive strength as compared to the conventional clay roof tiles. The highest compressive strength reached 6.896 MPa for 15TFASDBT combination. The experimental investigation [13] discloses that on increasing the content of Cement (C) at the fixed percentage of Treated Flyash (T.F.A) and Radish Stone Dust (R.S.D), the permeability values lowered down. Although the permeability (k) dropped with variation of the coarse sand (C.S) with C, k has been found to be in the range 10^{-7} much closer to the value of clay readily available in the market for making bricks and roof tiles. Through new sets of experiment conducted by [14], its compressive strength increased up to 30.65% as compared to past learning of [12]. The Class C flyash category was employed as a raw material for full replacement of clay for making flyash bricks. In that particular study, the effect of flyash with high replacement of clay mixed with different materials were studied at a constant percentage of cement, i.e., 10%. It also revealed that the highest compressive strength (79 kg/cm^2) was obtained in the case of 25TFASDB which was compared to be less than the maximum strength (105 kg/cm^2) of standard first class clay brick. Moreover, the study was important as it replaced 50% of top soil by flyash. The 25TFASDB bricks can be safely used in frame structure buildings as a non-load bearing walls and also in load bearing walls in case of single storey constructions [15]. Ref. [16] suggested that the flyash-scrap

http://dx.doi.org/10.5772/intechopen.82078

tire fiber composite offered a sustainable supplement to traditional insulation. This not only increased the competence of traditional insulation, but also helped in significantly reducing the environmental issues related to disposal of waste products.

2.5. Flyash utilization in different sectors

Since the last 30 years, flyash is not considered as hazardous waste; and researchers are extensively exploring the area of research related to utilization of flyash in various sectors. The programs concerned with flyash utilization may be broadly analyzed as extenuating the effects of environment and considering disposal issues (low value–high volume utilization) [17]. Below are addressed some of the potential areas of flyash utilization.

2.5.1. Development of flyash-based polymer composites as substitute for wood

Recently, flyash-based polymer composites have been developed with jute cloth as reinforcement and flyash as filler. This technology may be used as a substitute for wood which may have applications in door shutters, flooring tiles, partition panels, ceiling, wall paneling, etc. The technology of flyash-based polymer composite was developed by Regional Research Laboratory, Bhopal in collaboration with Building Materials and Technology Promotion Council (BMTPC) and TIFAC. A commercial plant is being based near Chennai for the above developed technology [18].

2.5.2. Flyash-based cement

According to the specifications of Bureau of Indian Standards, up to 35% flyash is permissible in PPC manufacture in India; whereas worldwide other countries permit up to 55% flyash utilization in PPC production. In such applications, 25% of cement production is meant for OPC and the rest 75% may be for PPC with 30% flyash content [18]. This would utilize around 25 MT flyash, thereby replacing cement clinker with gross net saving of Rs. 2500 crores [19].

2.5.3. Role of bio-amelioration of flyash on soil

Recent studies recommend the use of flyash with organic compounds like cow manure, paper factory sludge, press mud, sewage sludge, farmyard manure, organic compost, and crop residues for upgrading degraded/marginal soil [20]. The beneficial effects of flyash along with soil organic matter are reduction in the availability of heavy metal and sludge pathogens [21]; improvement in nutrient concentration of soil and better texture quality, lesser bulk density, better porosity and content of mass moisture, and increase in content of fine-grained minerals [22]; improvement in biological activity of the soil [23]; reduction in the leaching of major nutrients [24]; and favorable for vegetation growth [25]. The utilization of swine manure along with flyash improved the accessibility of Ca and Mg in the soil, thereby balancing the monovalent and bivalent cations ratio (Na^{++} K^{+}/Ca^{2+} Mg^{2+}) that is detrimental to the soil [26]. The utilization of "slash" in a mixture of flyash, sewage sludge, and lime (60:30:10) had valuable soil ameliorating effect. The incorporation of "slash" in soil demonstrated positive effects on soil pH and Ca, Mg, and P content and reduction in the translocation of Ni and Cd [27] and enhanced growth and yield of corn, potatoes, and beans in pot trials. Therefore, changes in flyash may enhance the quality of agricultural sector for crop production [28].

2.5.4. *Flyash bricks*

According to a study by Central Fuel Research Institute, Dhanbad, India, a technology is proposed for the use of flyash in building bricks [17]. About 40–70% flyash is used in building bricks. The current scenario suggests a market with production of 100 billion clay bricks a year. In case, on technical acceptance of flyash brick which are environment friendly and economically viable, it may target at least 2 billion flyash bricks per year. This would consume approximately 5 million ton of flyash/year, hence generating a saving of around Rs. 200 million per year. There are several advantages of flyash bricks over conventional clay bricks. Apart from this, unglazed tiles for footpaths can be made from flyash bricks. The Government has to provide special incentives for such ecofriendly approach in the near future [25].

2.5.5. *Flyash in distemper*

The use of flyash in manufacture of distemper has replaced white cement. Several buildings in Neyveli, Tamil Nadu have been reported to use flyash distemper in the interior surface with satisfactory performance. This would decrease the cost of production of distemper by 50% [17].

2.5.6. *Flyash-based ceramics*

In order to provide superior resistance to abrasion, National Metallurgical Laboratory, Jamshedpur, India has developed ceramics from flyash [17].

2.5.7. *Ready-mixed flyash concrete*

In developed countries, ready-mix concrete is pretty popular; but in India, it utilizes less than 5% of total cement consumption. Recently, the application of ready-mixed concrete has grown at a rapid rate. Several ingredients and quality parameters are strictly controlled in ready-mix concrete, which is quite not possible in concrete production at site and it may accommodate higher quantity of flyash [18].

2.5.8. *Minefills*

Almost one-third of Indian TPPs are near to pit heads. These mines heave sand for backfilling from river beds, normally 50–80 km away. The cost includes not only royalty, but also transportation cost of sand. An estimation suggests that about 15–20 million ton of ash per annum can be safely consumed in minefills generating an economy of about Rs. 1500 million per annum [19].

2.5.9. *Flyash in road construction*

Flyash is used in construction of roads as well. It saves surface soil which is usually used, avoiding formation of low lying areas [4]. Flyash is used in road construction for stabilizing and constructing sub-base or base; upper layers of pavements; filling purposes, etc. The use of

http://dx.doi.org/10.5772/intechopen.82078

concrete with flyash (10–20% by wt.) is quite cost effective and enhances performance of rigid pavement. The mixture of soil with flyash and lime enhances California Bearing Ratio (CBR) to (84.6%). As per the reports, National Highway Authority of India (NHAI) is using 60 lakh m^3 of flyash currently and proposed the consumption of another 67 lakh m^3 in future projects.

2.6. Embankment

As an engineering material, the properties of flyash are somewhat unique. For the construction of embankment, soils are used; but unlikely flyash has a large uniformity coefficient due to clay-sized particles. The properties that affect flyash use in embankments include compaction characteristics, grain size distribution, shear strength, compressibility, permeability, and frost susceptibility. Among all, flyash used in embankments are Class F flyashes [17]. In order to develop road infrastructure in the country, conservative approximates about 15–20 MT of flyash consumption in construction of road and flyover embankments per year in the environs of TPPs. It would yield a saving of around Rs. 100 crore per annum [20].

2.6.1. Roller compacted concrete

Flyash may be also consumed in compacted concrete dams. There are many dams in US which are reported to be constructed with high flyash contents. The heat of hydration is lowered by the use of flyash to allow thicker placements. Data regarding this may be found at the US Bureau of Reclamation. This may also be justified by the data reports of Ghatghar Dam Project in India [29].

2.6.2. Asphalt concrete

Asphalt concrete is basically a composite made of an asphalt binder and mineral aggregate. Both Class F and Class C flyash are used as a mineral filler to fill the voids and provide contact points between larger aggregate particles in asphalt concrete mixes. This application is used in conjunction or as a replacement for, other binders (such as Portland cement or hydrated lime) [29]. For use in asphalt pavement, the flyash must meet mineral filler specifications outlined in ASTM D242. The hydrophobic nature of flyash gives pavements better resistance to stripping. Flyash increases the stiffness of the asphalt matrix, thereby improving rutting resistance and mix durability [3].

2.6.3. Use of flyash in agriculture

In India, agriculture and waste land have been dumped with flyash. This accidental activity of waste management has improved permeability, texture, and fertility of soil; reduced bulk density of soil; improved water holding capacity/porosity; optimized pH value; improved soil aeration; reduced crust formation, provided micro nutrients like Fe, Zn, Cu, Mo, B, Mn; provided macro nutrients like K, P, Ca, Mg, S, etc.; worked as a part substitute of gypsum for reclamation of saline alkali soil and lime for reclamation of acidic soils; ash ponds provided suitable conditions and essential nutrients for plant growth; and helped to improve the economic condition of local inhabitants. Apart from this, crops yielded on flyash mixed soil are found to be safe for human consumption and to maintain groundwater quality [3]. The improvement in crop yield is recorded with flyash doses varying from 20 to 100 ton/hectare.

About 20–30% yield has increased out of 150 million hectare of land under cultivation, 10 million hectares of land can safely be taken up for application of flyash per year. In order to take a moderate flyash dose of 20 m per hectare, it would consume 200 million ton flyash per year. The expected uptake is more than the annual availability of flyash; therefore the shortfalls may be met from 1500 million ton stock of flyash available in ash ponds. The field treated with flyash is expected to give additional yield of 5 million ton foodgrains per annum valuing about Rs. 3000 crore income [19].

3. Conclusions

As per Indian scenario, there is a vast difference in the production and utilization of flyash which is still considered as waste material. In India, the majority of flyash produced fall in Class F category. This chapter shows the chemical and physical analysis of Class F flyash for construction industry. Flyash has the capability to hold free lime that leads to lesser bleed voids and reduction of permeability to water and sulfate. Furthermore, the experimental data suggests that the strength of Class F flyash-based bricks and tiles with cement are far more as compared to standard clay and bricks. It also clarifies that physical properties of flyash are similar to local soil. With the use of Class F flyash, we may reduce the amount of soil used in the production of construction materials like blocks, tiles, roof materials, etc. Apart from this, the utilization of flyash as pozzolanic material in concrete mixture may prove to be economical. The use of additional waste material may provide both durability as well as economical means for ecological balance. For Indian condition, it is recommended that the Class F flyash may be used as filling material in construction work, i.e., buildings, roads, embankments, and low lying areas. Flyash is a lightweight material when compared to local soils, which is commonly used as fill material. As flyash is a lightweight material, as compared to commonly used fill material (local soils), it causes lesser settlements. India is an agriculture-based country. The overuse and inappropriate use of soil results in nutrient depletion, erosion, and other forms of degradation, hence the soil productivity declines; it also reduces the area available for agricultural use. By utilizing flyash as a fill material, an equal volume of top soil, which will otherwise be used in filling, can be saved. Recycling of flyash will conserve the natural raw materials and reduce the disposal cost. It will also create new revenues and business opportunities in civil engineering construction industry while protecting the environment.

Acknowledgements

The authors acknowledge the help of the staff at the Soil Mechanics Laboratory at A.M.U Aligarh, India, in conducting the experiments. They are also thankful to USIF, A.M.U Aligarh, India for providing SEM/EDX data.

http://dx.doi.org/10.5772/intechopen.82078

Author details

Mohammad Nadeem Akhtar[1]* and Nazia Tarannum[2]

*Address all correspondence to: nakhtar@fbsu.edu.sa

1 Department of Civil Engineering, Fahad Bin Sultan University, Tabuk, Saudi Arabia

2 Department of Chemistry, Chaudhary Charan Singh University, Meerut, Uttar Pradesh, India

References

[1] American Standards for Testing and Materials; 2005

[2] American Standards for Testing and Materials; 2006

[3] Alam J, Akhtar MN. Fly ash utilisation in different sectors in Indian scenario. International Journal of Emerging Trends in Engineering and Development. 2011;**1**(1):1-14

[4] Sahu SK et al. Characterization and quantification of persistent organic pollutants in flyash from coal fueled thermal power stations in India. Microchemical Journal. 2009; **92**:92-96

[5] Dwived A, Kumar Jain M. Fly ash—Waste management and overview: A Review. Recent Research in Science and Technology. 2014;**6**:30-35

[6] The Built Envison. Central Building Research Institute Roorkee; 2016

[7] Report on fly ash generation at coal/lignite based thermal power stations and its utilization in the country for the year 2015-16. Central Electricity Authority, New Delhi; 2016

[8] I.S: 3812. Specification for Fly Ash as Pozollana and Admixture (First Revision); Bureau of Indian Standard Manak Bhavan, Bahadur Shah Zafar Marg, New Delhi. 1983

[9] Dhindsa HS, Sharma RD, Kumar R. Role of fly ash in improving soil physical properties and yield of wheat (*Triticum aestivum*). Agricultural Science Digest. 2016;**36**:97-101

[10] Phung HT, Lund IJ, Page AL. Potential use of flyash as a liming material. In: Environmental Chemistry and Cycling Processes; 1978. Conf-760429 504-515

[11] Akhtar JN, Alam J, Akhtar MN. An experimental study on fibre reinforced fly ash based lime bricks. International Journal of Physical Sciences. 2010;**5**:1688-1695

[12] Alam J, Khan M, Akhtar M. Fly ash based brick tiles: An experimental study. International Journal of Emerging Trends in Engineering and Development. 2013;**6**:35-44

[13] Akhtar MN, Khan MA, Akhtar J. Use of the falling-head method to assess permeability of fly ash based roof tiles with waste polythene fibre. International Journal of Scientific and Engineering Research. 2014;**5**:476-483

[14] Akhtar MN, Akhtar JN, Al Hattamleh OH, Halahla A. Sustainable fly ash based roof tiles with waste polythene fibre: An experimental study. Open Journal of Civil Engineering. 2016;**6**:314-327

[15] Akhtar JN et al. Bricks with total replacement of clay by flyash mixed with different materials. International Journal of Engineering Science and Technology (IJEST). 2011;**3**(10):7338-7346. ISSN: 0975-5462

[16] Van de Lindt JW, Carraro JAH, Heyliger PR, Choi C. Application and feasibility of coal fly ash and scrap tire fiber as wood wall insulation supplements in residential buildings. Resources, Conservation and Recycling. 2008;**52**:1235-1240

[17] Senapati MR. Fly ash from thermal power plants—Waste management and overview. Current Science. 2011;**100**(25):1792

[18] Saxena M, Prabhakar J. Emerging technologies for third millennium on wood substitute and paint from coal ash. In: 2nd International Conference on "Fly Ash Disposal & Utilization"; February; New Delhi, India; 2000

[19] Kumar V, Mathur M, Sharma P. Fly ash: A billion dollar resource–Wasted so far. Mission beyond 2000 A. D., Fly Ash disposal and deposition: beyond 2000 A.D. Narosa Publishing House

[20] Tripathi RD, Singh SN, Singh N, Vajpayee P, Kumar A. Reclamation of fly ash landfills by successive plantation, soil amendments and/or through biotechnological approach. Final Technical Report. Directorate of Environment, U.P., India; 2000 (unpublished material)

[21] Wong JWC. The production of artificial soil mix from coal fly ash and sewage sludge. Environmental Technology. 1995;**16**:741-751

[22] Shen JF, Zhou XW, Sun DS, Fang JG, Liu ZJ, Li Z. Soil improvement with coal ash and sewage sludge: A field experiment. Environmental Geology. 2008;**53**:1777-1785

[23] Kumpiene J, Lagerkvist A, Maurice C. Stabilization of Pb and Cu contaminated soil using coal fly ash and peat. Environmental Pollution. 2007;**145**:365-373

[24] Sajwan KS, Paramasivam S, Alva AK, Adriano DC, Hooda PS. Assessing the feasibility of land application of fly ash, sewage sludge and their mixtures. Advances in Environmental Research. 2003;**8**:77-91

[25] Rautaray SK, Ghosh BC, Mittra BN. Effect of fly ash, organic wastes and chemical fertilizers on yield, nutrient uptake, heavy metal content and residual fertility in a rice–mustard cropping sequence under acid lateritic soils. Bioresource Technology. 2003;**90**:275-283

[26] Giardini L. Aspetti agronomici della gestione dei reflui zootecnici. Rivista di Ingegnaria Agraria. 1991;**12**:679-689

[27] Rethman NFG, Truter WF. Plant responses on soils ameliorated with waste products. In: 16th National Meeting of ASSMR; Albuquerque, NM, USA; 2001. p. 425

[28] Pandey VC, Singh N. Impact of fly ash incorporation in soil systems. Agriculture, Ecosystems and Environment. 2010;**136**:16-27

[29] Kumar V, Mathur M, Sharma P. Fly Ash Disposal: Mission Beyond 2000 A. D., Fly Ash Disposal and Deposition: Beyond 2000 A.D. Narosa Publishing House; 1999

Chapter 11

Degradation Monitoring Systems for a BIM Maintenance Approach

Pedro Romano and Paulo Brito

Additional information is available at the end of the chapter

http://dx.doi.org/10.5772/intechopen.81433

Abstract

Digitization allows to develop unprecedented technological systems based on the use of sensors, robotics, and automation. The construction industry is involved in this process of integrating new technologies through a platform called Building Information Modeling (BIM), which simplifies the management of the increasing complexity of construction processes. This methodology aims to create a global interactive system of information sharing between the different actors in the construction process. The integration of the processes creates economic and environmental opportunities, which can translate into increased efficiency of the sector. The information collected can be used throughout the construction lifecycle, which together with the monitoring of the buildings will support maintenance decisions. The monitoring of reinforced concrete structures with sensors allows the identification and quantification of the degradation processes, through the monitoring of several characteristic parameters of the reinforced concrete over time, and the determination of significant changes that indicate the existence of a degradative process in development. Obtaining this type of information, and its integration into BIM models, will allow intervention at an early stage in order to limit damages and costs associated with the maintenance of the structure, contributing to increase in the structure's useful life.

Keywords: concrete, monitoring, BIM, maintenance, innovative construction

1. Introduction

Industrialization has been a dynamic process that began at the end of the eighteenth century through the mechanization of manufacturing processes and which has extended to the present day. The initial phase of this process is called the industrial revolution for having introduced deep changes in the society at the time. Industrialization involved continuous

IntechOpen

© 2018 The Author(s). Licensee IntechOpen. This chapter is distributed under the terms of the Creative Commons Attribution License (http://creativecommons.org/licenses/by/3.0), which permits unrestricted use, distribution, and reproduction in any medium, provided the original work is properly cited. (cc) BY

progress, and also disruptive processes such as the electrification of the factory in the early twentieth century, called the 2nd industrial revolution, or automation using electronic and computer media that began in the 1970s (3rd industrial revolution) [1]. At present, we are going through the so-called 4th industrial revolution also known as digitization, which is characterized by the use of Cyber Physical Systems (CPS), based on heterogeneous data and knowledge integration [2], as shown in **Figure 1**.

These systems aim to respond rapidly to production requirements, which are subject to frequent changes, and to the improvement of industry performance in terms of efficiency and effectiveness [3]. The digitization will allow to create unprecedented technological platforms for consumer products systems, urban infrastructures, and industrial assets, based on the use of intelligent sensors, robotics, and automation [4]. The process of evolution that the industry was developing and implementing was formalized in Germany in 2013, with the presentation of the program called "Industry 4.0," which involves digital data, digital access, automation, and connectivity. Subsequently, other countries presented their own industrial development strategies, the USA with the "Internet Industrial Project," France with the "New French Industrial" strategy, England with the "British Industrial Strategy of 2050", Japan with the "Revitalization Strategy of Manufacturing", and China with "Made in China 2025" [5].

All these strategies seek to accelerate the integration of new technologies into production chains in order to improve their overall operation. The intelligent factory model, according to the Industry 4.0 concept, is based on the widespread use of ciber physical system, which allow the vertical integration of its various components to implement a flexible yet reconfigurable, though intelligent, production system. This type of factory, like any other factory, is equipped with physical objects (machines, conveyors, and products), and also with automation and information systems to implement flexible and agile production. In the Industry 4.0 concept, a factory consists of four levels of resources, which are the physical resources, the industrial network, the cloud, and the control and supervisory terminals. The former are equipped with intelligent systems that communicate with each other through the industrial network. The cloud supports the integrated information system, where the data obtained at the physical resources level are collected and processed and interact with the operators through supervisory control terminals [6]. Computational data processing techniques, usually called Big Data, allow the processing of large amounts of information, extracting useful results according to a specific objective [7].

Like all industries, the construction industry is also involved in this process of integrating new technologies and reformulating supply networks [8]. However, the construction sector

1st industrial revolution	2nd industrial revolution	1rd industrial revolution	4th industrial revolution
End of 18th century	Beginning of 20th century	Decade of 1970s	Today
Introduction of mechanical production	Industrial electrification	Automation of production	Cyber-physical production systems

Figure 1. Graduation scheme of the industrial revolution.

has specific characteristics that, a priori, make it difficult to adapt to these new technologies. Among these peculiarities is the fact that the construction process is temporary, unrepeatable (both in terms of content and location), phased, and involves a large number of actors with different levels of expertise. These variability and fragmentation constitute a challenge for the efficient coordination of projects, limit the possibility of learning from experience, make it difficult to implement a process of continuous improvement through repetition, and impose complex coordination of communication and exchange of information between stakeholders [9]. This sector is also characterized by having a very strong and rigid culture, which presents little openness to change [10]. A construction work can be analyzed in physical terms by the set of steps or processes that take place outside the work site, before starting and during its execution, the activities that take place at the construction site during the execution of the work, and later during the use phase of the construction.

To join all phases of a construction process, which involves the activities that take place before, during, and after construction, which are naturally linked, but which need a support that shows the whole process in a clear way, in order to be able to improve effectiveness and overall efficiency of the system. Most of the studies on Industry 4.0 consider the BIM methodology as the base technology for the digitization of the construction sector, by allowing the simulation and modeling of the constructions, which facilitates the management of the increasing complexity of the construction processes. On the other hand, this technology will allow in the future the integration of all the information produced from the design phase until the end of the life of the construction process [2].

2. Building information modeling (BIM)

The BIM methodology focuses on the creation of a global interactive system of information sharing between the different actors in the construction process [11]. This methodology allows to develop synergies in the various phases of this process, especially in the construction phase, and also in the design phase when critical issues need to be decided. Integrating the whole process creates important opportunities for reducing economic and environmental impacts and may in the future produce significant increases in efficiency in the construction industry, which is considered to be one of the least efficient industries [12]. This methodology aims to obtain a global view of buildings (and infrastructures), creating an information base that can be used throughout its life cycle, but that imposes changes in the way they are designed, constructed, and managed.

The BIM methodology is based on the use of a database infrastructure to incorporate the information about a construction according to the specific interest of the actors in the process, having as support the 3D modeling of the building. It integrates digital descriptions of all objects of construction and the interconnections between the different objects, so that those interested can consult the information, simulate the behavior, and estimate the activities and the processes of construction in a logical study of the life cycle [10]. This methodology leads to the development of the projects in a more interactive way among the different specialties, allowing to identify the design incompatibilities in an initial phase. The centralization of information optimizes the preparation of the work both in terms of measurements, budgeting, and preparation of the planning of the work.

In the work phase, the use of 3D modeling allows visualizing, both 2D and 3D, the works to be developed, facilitating their understanding and simplifying their execution. A study published by Hosseini et al. shows that interest in research on off-site construction has increased in recent years, focusing primarily on operations and management [13]. In this context, it is worth highlighting the use of the LEAN methodology as a management tool, as it promotes the reduction of waste in the production chain. This methodology focuses on identifying the products and services that have value for customers, with the objective of eliminating the parts that are not valued, classified as waste. It is a process that involves defining the value chain of the product, promoting the continuous flow of the product/service through the value chain (reducing production times and removing obstacles), seeking to produce exclusively what customers want (reducing stock), and promotes a process of continuous improvement, aiming to reduce the time of production cycles and obtain the best relation between quality and quantity, always focusing on the interest of the customer [12].

After construction, the information produced is available to the developer, so the process of maintenance and monitoring of the buildings is supported on a solid basis, which allows to support the decisions of interventions. This information can range from product datasheets to equipment manuals to enable service technicians to access specific equipment data for their interventions. In order for maintenance processes to be consistent, it is also important to monitor buildings in structural and environmental terms in order to collect information over time that will support maintenance and eventually end-of-life decisions.

The implementation of information modeling of buildings will contribute to the massive production of new data related to buildings, which will necessarily have to be treated as intrinsic value for the performance of the sector, will require the use of computational techniques for data processing, such as Big Data, which has applications for almost all industrial activities, including the construction industry. Currently, large volumes of heterogeneous data are already being worked on in the industry, which tend to increase exponentially as new systems are introduced, including sensor networks for data capture, which tend to increase user convenience [7].

3. Maintenance management tools

The maintenance of a building should be done using tools that allow the definition and implementation of clear guidelines to be able to develop the necessary maintenance actions. This activity can occur in three distinct but complementary forms, called preventive, corrective, and predictive maintenance, depending on its operationalization of a set of elements, namely:

- preventive maintenance action plans;
- issuance of work orders;
- recording of occurrences and failures;

- recording of actions taken for use in fault diagnosis;
- description of how the failures were identified;
- recording of possible causes (predictive capacity);
- resources made available and used;
- collection of relevant information (support for future events).

Preventive maintenance occurs in a planned manner, systematically with defined time intervals, or conditionally, depending on certain predefined conditions. The planning of interventions is supported in the prior knowledge of the durability characteristics of the materials and the specificities of the existing equipment in the building. It is sought to avoid the appearance of problems and, in this way, to limit the consequences that result from the normal deterioration of the materials. Preventive maintenance reduces operating costs, increases the life of buildings, and ensures user safety.

Corrective maintenance is the reaction to a certain occurrence, which may be urgent or not, whose purpose is to restore normal building use. This type of maintenance may involve high costs, both in terms of the materials and labor required to be mobilized, and costs associated with the discontinuity of the current use of the building.

The third type of maintenance is the predictive maintenance and results from the analysis of the information collected through the monitoring of the building and the inspections carried out. Predictive maintenance allows us to anticipate the need for maintenance services for a particular component of the building, contributing to:

- reduce corrective maintenance work;
- prevent the multiplication of damages;
- limit the impact of interventions on the use of equipment;
- extend the use of the different building components;
- mitigate the impact of inspections;
- increase the time of use of the building without constraints;
- raise the degree of confidence in the performance of the building.

For the development of models of predictive maintenance in buildings, it is necessary to use systems of measurement of relevant indicators, in a more or less continuous way in time. Through the use of the BIM methodology, it is possible to integrate these indicators in a single information system on the state of construction over time and, simultaneously, to have access to the basic information, as well as the characteristics of the materials used in the construction. This way, we can articulate the collected information with the pre-existing information in the 3D model of the building. In this way, it is possible to make a quick and efficient analysis of the state of the building [14]. Tools that link maintenance management to BIM models are in development. With this link, it is intended to facilitate the integration of data collected for maintenance management in BIM models, in order to automize, through algorithms, the production of maintenance service orders [15].

In the specific case of the maintenance of reinforced concrete structures, several indicators are used to measure their performance and can be used to measure the conservation status and determine the need to carry out predictive maintenance interventions.

These indicators aim to identify and quantify the degradation processes to which the reinforced concrete is subjected, measuring its variation over time, in order to determine at an early stage significant changes that indicate the existence of a degradative process in development. The obtaining of this type of information allows to intervene at an early stage in order to limit damages and costs associated with the maintenance of the structure, contributing to increase the useful life of the structure and give relevant information to increase the durability of new structures.

In general, the test methods are classified as destructive and non-destructive. The first type of methods can be used in laboratory tests, but they can hardly be used systematically in operating structures. The non-destructive, or less intrusive, test methods are more interesting because they allow to follow the evolution of the phenomena of degradation of the structures.

The first sensors for the monitoring of reinforced concrete appeared in the 1990s, among which can be distinguished the sensors consisting of two electrodes, one of black steel (anode) and the other of a noble metal (cathode), embedded in the concrete that confers protection against corrosion of the anode and through which the passage of electric current could be measured, as shown in **Figure 2**. These sensors are based on the principle that at an early stage, while the concrete presents a high pH, the passage of electric current is negligible or zero and, at a later stage, either by the effect of the carbonation front or the penetration of chlorides, the pH of the low concrete creates conditions for the corrosion of the anode and, simultaneously, for the passage of electric current between the electrodes, which will increase with the unwinding of the process of corrosion of the steel [16].

In the last decades, several techniques have been proposed and applied in the monitoring of reinforced concrete, and their integration in maintenance management systems has high potential. These monitoring techniques can be classified into five main categories, namely electrochemical methods, elastic wave methods, electromagnetic methods, optical detection, and infrared thermography.

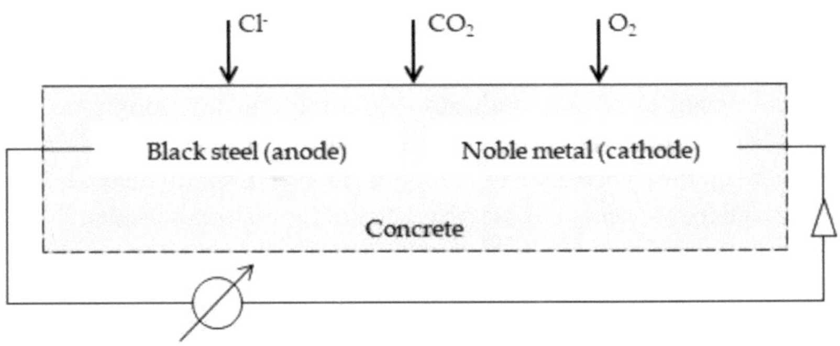

Figure 2. Sensor consisting of an anode in black steel and a cathode in noble metal.

Electrochemical monitoring techniques include open circuit potential (OCP), concrete resistivity [17], linear polarization resistance (LPR) [18], galvanostatic pulse method (GPM), galvanic cell, electrochemical impedance spectroscopy (EIS) [19], localized electrochemical impedance spectroscopy (LEIS), electrochemical noise (EN), bar electrical resistance (BER), and surface potential survey [20].

The elastic wave methods include ultrasonic pulse velocity (UPV), acoustic emission (AE), and echo ultrasonic pulse (UPE). Among the electromagnetic (EM) methods, the most important are the ground penetration radar (GPR), surface penetration radar, X-ray radiography, computed tomography, and magnetic field disturbance.

In the field of optical detection are framed Bragg fiber optic networks (FBG). The last of the categories among the monitoring methods previously listed is infrared thermograph (IRT), which allows to identify defects in concrete structures, such as cracking or delamination.

4. Monitoring techniques

The monitoring of a reinforced concrete structure can be carried out by means of the acquisition of data with different periodicities. Embedded systems allow regular readings to be collected continuously over time. Non-embedded systems, which involve the intervention of an operator, tend to be used to collect punctual readings over the life of the structure. Embedded systems have the added advantage of being able to monitor both the visible areas and the difficult-to-access areas of the structures.

4.1. Electrochemical techniques

In the following sections, the main electrochemical techniques used in the monitoring of reinforced concrete are analyzed. The techniques presented allow more or less continuous reading of the respective variables.

4.1.1. Measurement of electrochemical potential

All materials tend to interact with the surrounding environment. In the case of steel embedded in concrete, this interaction occurs between the steel and the concrete surrounding it and

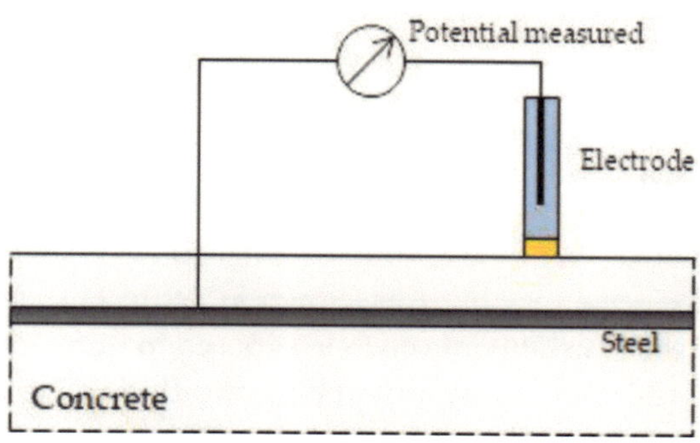

Figure 3. Measurement of electrochemical potential.

can be measured through the electrical potential of the reinforcement. The potential measurement allows estimating the corrosion state of the steel.

Potential measurement can be done through voltmeter, with high impedance, connecting the positive pole to the armature and the negative pole to a reference electrode. In embedded systems, the reference electrode is inserted in the concrete and is on the concrete surface in the case of portable systems, see **Figure 3**. In the first system, the connection to the reinforcement is made before concrete insertion, it being necessary that the steel is exposed to make the connection in the portable systems [21].

Measured potential less than −500 mV indicates that there is a high risk of the steel being corroded, the risk of corrosion being low when the potential is higher than −200 mV, according to **Table 1** [22].

4.1.2. Measurement of ionic resistivity of concrete

The ionic resistivity of the concrete has been used to determine the level of ionic contamination of the concrete, namely chloride ions, and the advance of the carbonation front. The ionic resistivity of a porous concrete depends, first, on the mobility and concentration of "free" ion, current carriers, in the solution that fills the porosity. Ionic mobility is an intrinsic characteristic of a chemical species varying only with temperature. That is, small ions, such as hydrogen, exhibit high mobility that increases with temperature. However, the "free" ionic concentration depends on the moisture content within the pores, the degree of contamination by external agents, and the solubility equilibrium of the different salts which may in the aqueous liquid fill the pores.

The resistivity of the concrete can be measured by using electrodes inserted into the concrete or placed on its surface. There are two main techniques, the two-electrode technique and the four-electrode technique. Both techniques have as presupposition the creation of an alternating or continuous electric field between electrodes. In the technique of the two electrodes embedded in concrete, the resistivity can be measured at various depths by creating an alternating electric field between the electrodes, usually in stainless steel, at the same depth, creating a current and measuring the potential difference. In the technique of the four electrodes placed on the surface of the concrete, the most common measurement process involves the creation of an electric field of alternating current between external electrodes and measurement of the potential between the inner electrodes [23], as shown in **Figure 4**. The following formula shows the relationship between the resistance measured by the electrodes and the resistivity of the concrete:

$$\rho = 2aR \tag{1}$$

with "a" being the distance between electrodes.

It is generally accepted that a concrete with a resistivity higher than 20 Ω cm is associated with a very low risk of corrosion of the reinforcement, and for that with a value smaller than 5 Ω cm, the risk of corrosion is very high, as shown in **Table 2**. There are, however, significant variations between the resistivity of concrete exposed to different environments, but the significant reduction of resistivity is an important indication of the risk of corrosion of the reinforcement [24].

Potential of corrosion (mV)	Corrosion risk
$E_{corr} > -200$	Low (10 % risk of corrosion)
$-200 > E_{corr} > -350$	Intermediate corrosion risk
$E_{corr} < -350$	High (90 % risk of corrosion)
$E_{corr} < -500$	Severe corrosion

Table 1. Corrosion potential [22].

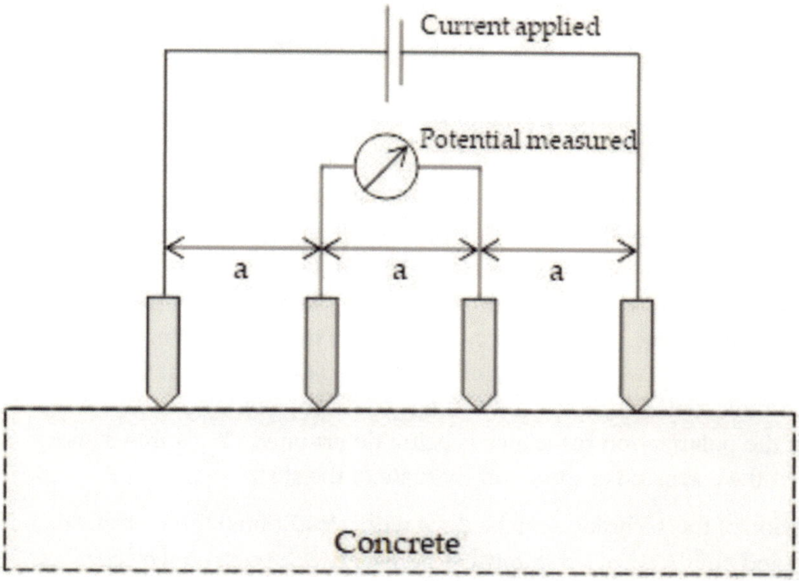

Figure 4. Technique of the four electrodes.

4.1.3. Linear polarization resistance

The polarization resistance (R_{pol}) can be defined as the slope of the polarization curve that results from the ratio of difference between the electric potential and the corrosion potential of the rebar, $E-E_{corr}$, and the intensity of the polarization current, I_E, so:

$$R_{pol} = \lim_{E \to E_{corr}} \frac{E - E_{corr}}{I_E} \tag{2}$$

The polarization resistance is related to the corrosion rate, i_{corr} (the density of the corrosion current that circulates between the anodic and cathodic zones during the corrosion process) through [25]:

$$R_{pol} = \frac{\beta_a \beta_c}{2,3\ A\, i_{corr}(\beta_a + \beta_c)} \tag{3}$$

Resistivity (kOhm.cm)	Corrosion risk
R > 20	Negligible
20 > R > 10	Low
10 > R > 5	High
R < 5	Very high

Table 2. Ionic resistivity of concrete.

where A is the area of the rebar being inspected, and β_a and β_c are the Tafel parameters for the anode and cathodic processes, respectively. The coefficient $\beta_a \beta_c/2, 3/(\beta_a + \beta_c)$ ranges from 26 mV for active-phase corrosion reinforcement to 52 mV for passive reinforcement for corrosion [26, 27]. **Figure 5** shows a linear polarization resistance measurement sensor.

R_{pol} determination is made by varying the armature potential between −10 and +10 mV of the corrosion potential and recording the intensity of the electric current flowing between the reinforcement and the auxiliary electrode placed on the surface of the concrete. In addition to the auxiliary electrode, which imposes the desired potential on the rebar, it is necessary to use a potentiostat, which allows a constant and precise potential to be imposed on a robust and stable reference electrode, also placed on the surface of the concrete, in relation to which the armature potentials, the working electrode, will be measured. As can be seen from Eq. (3), the application of this method implies a thorough knowledge of the area of the armature section from which the polarization resistance is being determined [28]. **Table 3** shows the relationship between the corrosion current and the state of the steel.

The application of this technique can be done with a traditional three-electrode scheme (reference, work, and auxiliary), used in most systems, or with two equal electrodes.

4.1.4. Galvanic pulse

The measurement of the galvanic pulse is a technique that goes back to the beginning of the studies of corrosion of the metals and was also used in the first studies related to the

Figure 5. MoniCorr system for linear polarization resistance measurement.

http://dx.doi.org/10.5772/intechopen.81433

Corrosion current (Icorr)	Condition of the rebar
Icorr < 0.1 μA/cm2	Passive condition
Icorr 0.1 - 0.5 μA/cm2	Low to moderate corrosion
Icorr 0.5 - 1.0 μA/cm2	Moderate to high corrosion
Icorr > 1.0 μA/cm2	High corrosion rate

Table 3. Corrosion rate.

concrete-steel interaction. The galvanic pulse method is based on the application of a transient bias current in a given period of time to the reinforcement inserted in the concrete. In general, a current between 10 and 100 μA with a duration of 10 s is applied [29]. In this way, the armature is polarized in the anodic direction and the potential of corrosion is compared, with the alteration of the electrochemical potential being compared with a reference electrode, which can be in stainless steel or titanium [30]. An example of a sensor for measuring the galvanic pulse is shown in **Figure 6**.

4.1.5. Electrochemical impedance spectroscopy (EIS)

The electrochemical impedance spectroscopy (EIS) is a non-destructive technique that allows to quantify the corrosion of the reinforcement inserted in the concrete. The impedance, Z, results from the relation between the voltage and the current intensity, for alternating current [31]. This technique is based on the assumption that an electric circuit can represent the behavior of the steel inserted in the concrete. Its application is based on the application to the rebar of a set of small alternating sinusoidal potential signals between 5 and 10 mV, the system response being measured in a current for a frequency range normally between 0.1 and 20,000 Hz. Impedance has one actual (Z′) and one imaginary (Z″) component; the actual impedance component represents the resistive part of the system, while the imaginary component represents the capacitive part [24]. The representation of the direction and magnitude of the vector impedance for each frequency results in the so-called Nyquist diagram, **Figure 7**.

In the steel-concrete system, the most commonly used equivalent circuit is that which is also shown in **Figure 7**. The first RC network is intended to simulate the film or layer of iron oxide which forms on the surface of the steel when it is in contact with the concrete, and the second network, the behavior of the double layer. The same figure also represents the response of

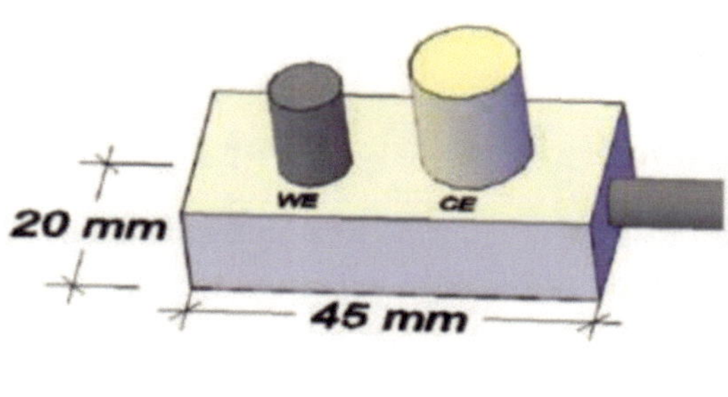

Figure 6. Galvanic sensor composed of steel and stainless steel [30].

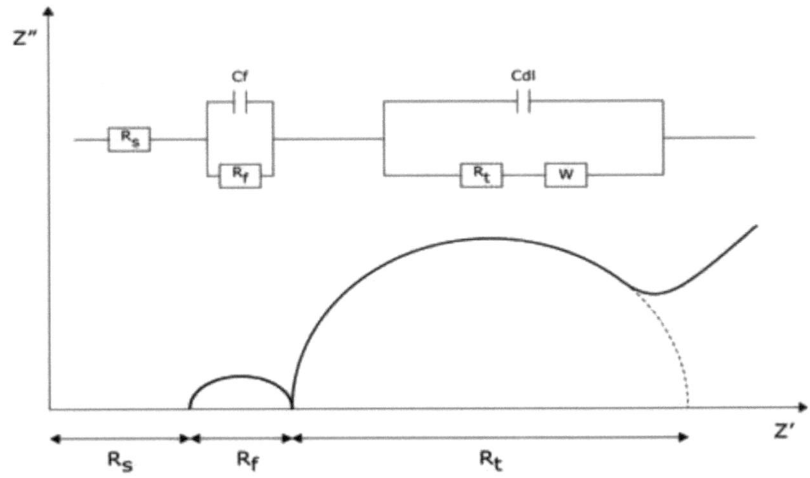

Figure 7. Equivalent circuit for the concrete-steel system and representation in the Nyquist diagram.

this circuit in the Nyquist diagram. In the Nyquist diagram obtained, the extrapolation of the diameter of the semicircle from the right to the Z′ axis allows the transfer value to be obtained at the load resistance which is equivalent to the polarization resistance used in the Stern-Geary equation and used in monitoring [32].

This technique allows to provide complementary information about the corrosive process of the reinforcement, as well as the dielectric characteristics of the concrete and the layer of oxide that forms on the reinforcements inserted therein. This technique also allows the measurement of the instantaneous corrosion rate (Rp) and the type of mechanism of the phenomenon—activation, concentration, or diffusion.

4.1.6. Electrochemical noise

Electrochemical noise measurement is a monitoring technique that can provide information on the mechanisms and corrosion rates of the reinforcement inserted in the concrete. This technique is based on the analysis of the variation of potential or current of low intensity, of the order of the microvolts, and allows to detect small variations of corrosion of the reinforcements [24].

The equipment used in this process is a micro voltmeter or a micro ammeter and a frequency analyzer that transforms the electrochemical noise as a function of the time in frequency through the Fourier transform, with the results presented graphically in the form of amplitude versus frequency [27].

4.1.7. Ag/AgCl electrode produced by immersion in sodium hypochlorite acid

For the detection of the presence of chloride ions in the concrete, Ming et al. [33] propose the use of a selective electrode of chloride ions produced by immersion of silver wire in sodium hypochlorite solution. This Ag/AgCl electrode is produced from silver wires with a diameter of 0.5 mm, connected with copper wires, the joint being sealed with epoxy resin, leaving an exposed area of 7.85 mm^2, which corresponds to a length of 5 mm. The silver wire is pre-treated, which included immersion in nitric acid (HNO3) for about 10 min, followed by immersion in anhydrous ethanol with vibration for 5 min. The finishing is obtained by immersing the yarn in an acid solution of sodium hypochlorite, resulting in a very dense coating of AgCl on its surface. The Ag/AgCl electrode is embedded in the concrete, its potential being measured in relation to a reference electrode, which in this case is SCE (saturated calomel electrode).

According to the authors, the Ag/AgCl electrode presents good reversibility, especially in solutions containing chloride ion. It should be noted that the temperature at which the electrode works is important because it influences the potential results. This electrode presents as advantages a low manufacturing cost, reduced dimensions, good performance, and robustness. **Figure 8** shows the schematic used to test this sensor.

4.2. Elastic wave methods

Measurement of waves in concrete structures can be done with external systems, such as ultrasonic pulse velocity or pulse tomography.

4.2.1. Ultrasonic pulse velocity test

The ultrasonic test is one of the oldest non-destructive methods to determine the state of the concrete, allowing information on its mechanical characteristics, homogeneity, and the existence of voids or cracks. This test is based on the measurement of the time elapsed between the emission of an ultrasonic pulse and its reception, known as the distance between the transmitter and the receiver, that is, it determines the propagation velocity of the ultrasonic pulse between two points [34, 35]. To perform this test, an ultrasonic device consisting of a transmission transducer and a receiving transducer is used. The transmission transducer produces a voltage pulse that propagates through the concrete, the signal being received by the receiving transducer. The device measures the time between the sending of the signal and its

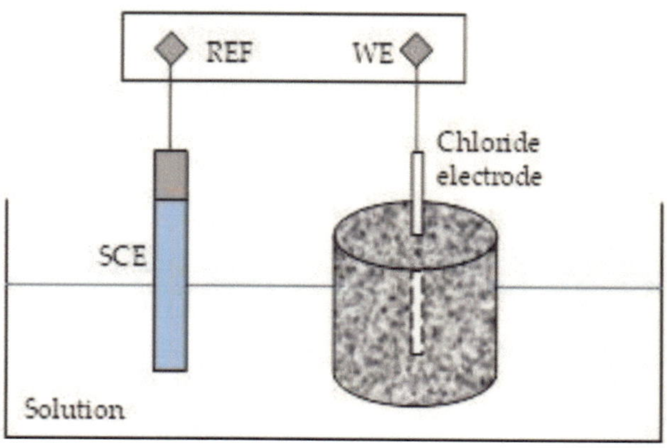

Figure 8. Ag/AgCl electrode embedded in the center of a specimen.

reception, determining the propagation speed of the wave. A small part of the emitted energy is reflected back to the surface. Wave velocity is reduced in areas where there are discontinuities. **Table 4** shows the classification of concrete as a function of ultrasonic propagation according to Whitehurst [36].

4.2.2. Ultrasonic pulse tomography

Ultrasonic tomography is a non-destructive inspection technique that allows mapping the interior of the structures and can provide reliable information about discontinuities or damage within the structures. This technique is based on the principle of reconstruction of the image with the evaluation of a series of projections of measurements made from different angles, that is, the real image of the object under study is estimated. The fidelity of this method depends on the process of acquisition and pre-processing of the data through mathematical methods of reconstruction. The generation of tomographic images can be categorized in techniques of filtered rear projection and iterative reconstruction [38].

The techniques of iterative reconstruction are based on the resolution of systems of algebraic equations generated from measurements, whose solution is a velocity map. By relating these data to the elastic characteristics of the materials, it is possible to detect discontinuities and damages in the elements being studied.

The authors [38] report that changes in compressive strength of concrete lead to different propagation velocities, with the velocity reducing with increasing resistance. In parallel, they concluded that the relatively uniform distribution of ultrasound velocities observed indicates that the tomographic technique for assessing the uniformity of concrete structures is adequate, and the analysis of the homogeneity of the concrete by ultrasonic tomography is more efficient than the simple analysis of velocity values of the waves. **Figure 9** shows a thermogram obtained by this technique.

4.3. Optical detection

Optical detection involves the installation of systems based on optical fiber, whose installation is generally carried out inside the structures, but can also be installed outdoors. In terms of reading, these systems allow continuous data collection.

Pulse velocity (m/s)	Quality of concrete
>4500	Excellent
3500 a 4500	Good
3000 a 3500	Doubtful
2000 a 3000	Poor
<2000	Very poor

Table 4. Classification of concrete quality based on the pulse velocity [36, 37].

http://dx.doi.org/10.5772/intechopen.81433

Figure 9. Thermogram obtained on the transducer at 250 kHz.

4.3.1. *Fiber Bragg grating*

Fiber Bragg grating (FBG) grid sensors can be used to monitor temperature, voltage, displacement, pressure, and, indirectly, corrosion. This type of sensor reflects at a given wavelength and transmits into another, producing a periodic variation of the refractive index of the fiber core by ultraviolet laser irradiation. The sensor itself, the Bragg grid, is a small segment of the core of the optical fiber exposed to ultraviolet light, which when actuated produces changes in the reflected wavelength. The reflected light corresponds to a specific wavelength, the light of the remaining wavelengths being transmitted through the Bragg grid [39].

Any deformation, whether by change in temperature, pressure, or vibration, applied in this grid alters the wavelength of the laser and changes in the magnitude of the reflections. These changes allow accurate measurements to be made, which can occur for long or extremely short periods of time, so that the quantification of the action on the sensor can be determined by the relation between the physical properties and the reflected wavelength of the FBG sensor [40]. **Figure 10** shows the operating scheme of an FBG sensor according to Kim et al. [41].

The unprotected FBG sensors are very fragile, so they need to be wrapped in a protective cap. The design of the capsule must be designed to allow the sensor to function and at the same time be protected. The shape and material of the capsules of the FBG sensors vary depending on the type of measurement desired and the installation site [42].

The FBG sensors have a set of characteristics that makes their use very appealing, among them being the insensitivity to the electromagnetic fields, the small size, and the weight, which make them suitable to be incorporated or fixed to any structure. Its connection to the exterior is effected by the optical fiber itself, which serves as a signal propagation channel for the control system. In addition, these sensors have excellent resolution and range, are immune to adverse weather conditions, and are resistant to water and corrosion [43].

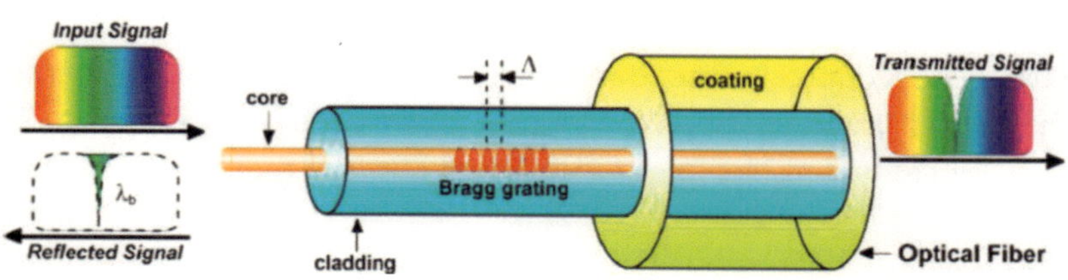

Figure 10. Operating principle of an FBG sensor [41].

FBG temperature sensors work on the basis of changing wavelengths in response to temperature changes. These sensors are protected by a small steel housing. The configuration of this transducer allows to keep one of its ends free, allowing the stainless steel rod to expand as a function of temperature variation, allowing the operation of the FBG sensor [44].

The voltage sensors work on the basis of the change in wavelength that results from the applied voltage, which produces elongation of the grid. These sensors are connected directly to the surface of the structure to be monitored, so as to have a deformation coupled with the deformation of the structure.

The displacement sensors estimate the deformation of the structure as a function of its deformation. These sensors allow you to measure the elongation, or shortening between the two points where it is anchored [44].

During the process of corrosion of the reinforcement, the volume of the rust produced is about four to six times greater than the volume of the original steel. With the reinforcement confined by concrete, this increase of volume introduces internal tensile stresses in the surrounding concrete, leading, when this tension exceeds the tensile strength of the concrete, to its cracking [45, 46]. Almubaied et al. studied the behavior of the FBG sensors installed on the face of the armature and concluded that the sensors have a good sensitivity to the progression of the corrosion process [47].

4.4. Infrared thermography

Infrared thermography is a monitoring and diagnostic technique that allows detecting superficial and internal anomalies such as voids, delamination, and cracks. This technique is based on the measurement of the surface heat transfer of a body, which occurs by radiation and conduction, by detecting the infrared radiation located in the region of the electromagnetic spectrum between the visible light and the microwaves.

Radiation is the mechanism by which a surface emits energy per unit area, which can be determined by Stefan-Boltzmann's law. Any object with an absolute nonzero temperature (−273.15°C or 0 K) radiates energy in the infrared region [48]. The factors that determine the level of the infrared radiation emitted by a material are the temperature of this material and its emissivity. The emissivity, which varies between 0 and 1, is a property of the material defined by the relation between the capacity of its surface to emit energy by radiation and the energy radiated by a black body whose emissivity is unitary ($e = 1$) due to the fact that possess null transmissivity and reflectivity. The emissivity of the concrete varies between 0.88 and 0.94 depending on surface roughness and moisture content.

The infrared camera produces a visual image from the conversion of the thermal radiation pattern of the surface under study [49]. This technology does not measure the temperature of the object under study, but identifies the different levels of radiation emission. The image obtained is influenced by the existence of other materials on the concrete surface, such as stains, water, or paints, which have different emissivities. The state of weather influences the results—if on the one hand the sunlight can increase the temperature of the surface, on the other the wind can decrease its temperature, such as rain. All these factors can contribute to highlight the temperature differences in the surface under study, allowing a better identification of the anomalies [50].

Thermography can be performed using the external heat source or natural heat source. In the first case, it is necessary to create an artificial thermal stimulus. In the second case, which is the approach suggested by the standard ASTM D 4788 [51], natural heat sources are used, with measurements being taken during the day and at night, to obtain images of thermal flows into the interior of the concrete and reverse heat flows, respectively, as shown in **Figure 11**.

Standard ASTM D 4788 [51] defines the criteria to be applied in determining the delamination of trays of reinforced concrete bridges through the use of infrared thermography. A number of authors have explored the use of this technology, namely Clark et al., who were able to identify delaminations in bridge trays with diameters less than 20 cm [50], and Cannard et al., who have been able to identify defects with areas of about 4 cm^2 [52]. More recently, Caldeira et al. used this technology to identify adhesion damages between concrete and reinforcing strips in glass fiber-reinforced polymers (GFRPs) [53]. In this study, several external heat sources were used, namely incandescent lamps, high-pressure sodium lamp, and convection heater, and concluded that the type of heating influences the quality of the thermographic evaluation and the uniformity of the heat flow in the samples.

With the proliferation of drone technology, the use of unmanned aerial vehicles to obtain thermographic images has arisen. Omar et al. have studied this technology and have identified as an immediate advantage its ability to allow the circulation of vehicles during the tests, which is extremely interesting in areas of heavy traffic and for allowing inspections to be carried out more frequently, according to the needs. This technology will also allow the expansion of the number of inspected structures, due to the time saved and the associated cost reduction. This type of technology is an important tool for the management of maintenance processes, since it allows assessing the conditions of the bridges at various stages of their useful life, contributing to the collection of supplementary information, which will allow to base the decision making regarding operation maintenance. Although most of the studies focus on the evaluation of the reinforced concrete bridges, this technology can also be used in the other components of the bridge [48].

4.5. Radio frequency identification systems

Radio frequency identification systems, other than sensors, can be very interesting support elements for a multifunctional maintenance system. These systems, usually referred to as RFID (radio frequency identification), consist of a reader connected to an antenna, which

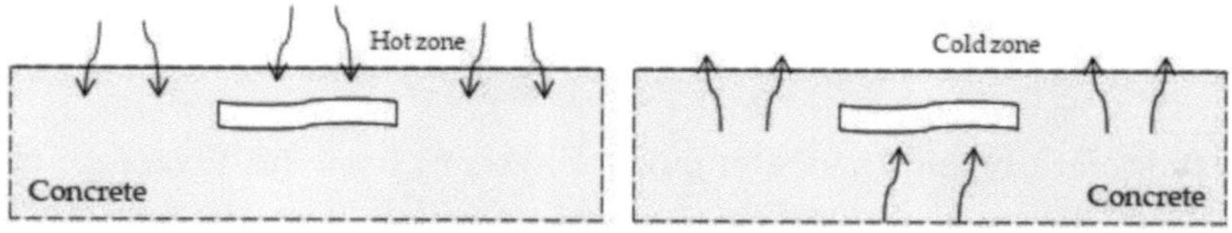

Figure 11. Effects of an interior anomaly on surface heat flux.

allows detection and reading of labels with stored information. These tags are composed of an integrated circuit and a small encapsulated antenna [54]. A computer, connected to the reader, manages the system through a suitable application [55].

The antennas can identify the labels at different distances according to the intended purpose, with systems that allow reading at small distances, a few centimeters, to systems that allow reading at greater distances, up to about 100 m. In order to choose the type of labels to adopt, its capacity of storing data and the position in which it will be placed on site is relevant. The frequency of operation of the devices depends on several parameters, namely the need to modify the contents of the labels, protection of the privacy of the data, the materials used, or the portability of the reader [54].

After construction, RFID systems can also be used to make the work of maintenance teams more efficient, through their installation in the various equipment that will be the object of maintenance throughout the life of the construction, allowing to monitor their condition or performance. According to Cheng et al., these devices may also be incorporated during construction into materials or structural elements, storing information. This would allow transferring data from the design phase to the maintenance phase, leading to the building itself providing data, increasing the efficiency of the maintenance system [56].

4.6. Analysis of monitoring techniques

For the implementation of the monitoring of a structure, it is possible to use sensors embedded in the concrete, with the readings being taken with certain periodicity, or by the use of equipment that allows obtaining punctual readings through the contact with the surface of the concrete, of singular use. For some of the techniques there are two possibilities, such as the measurement of resistivity either by sensors embedded in the concrete or by the method of the four external electrodes. **Table 5** presents the type of use for each of the techniques.

The monitoring techniques presented have been tested, in recent years, in the laboratory, and also in real structures. Most of the studies focus on the importance of using two or more techniques simultaneously, in order to guarantee a high level of reliability in the analysis of the results. This type of system has been installed in numerous real structures, some of which we mention:

- Madeira Airport and the D. Henrique bridge, whose installed systems allow to measure the galvanic current intensity, resistivity, temperature, and corrosion potential [57];
- Alto Ceira Dam, where the deformation of the concrete, the movement of the joints, and the temperature are monitored using fiber optic sensors [44];
- Bridge, infrared camera monitoring installed on drones [48].

From the standpoint of maintenance, embedded systems have the advantage of providing information on the behavior of a particular variable over time. These systems can be installed in areas difficult to access after completion of construction, such as foundations or buried walls. Monitoring with embedded systems should not inhibit the use of external equipment as the sensors do not allow coverage of the entire surface of the structure. Thus, along with

Monitoring Techniques	Sensors embedded	External sistem
Electrochemical techniques		
Measurement of electrochemical potential	X	X
Measurement of ionic resistivity of concrete	X	X
Linear polarization resistance	X	X
Galvanic pulse		X
Electrochemical impedance spectroscopy	X	
Electrochemical noise		X
Ag/AgCl electrode (sodium hypochlorite acid)	X	
Elastic wave methods		
Ultrasonic pulse velocity test		X
Ultrasonic pulse tomography		X
Optical detection		
Fiber Bragg Grating	X	
Infrared Thermography		
Infrared Thermograph		X

Table 5. Monitoring techniques.

the use of sensors, regular inspections should be carried out to detect any undetectable degradations with the installed systems, and spot inspections resulting from the information collected during monitoring. The integration of information into a 3D model, which incorporates construction information, facilitates the detailed analysis of detected degradations. The use of information identification and storage tags (RFID) embedded in structural elements facilitates access to data during inspections and increases their effectiveness.

5. Conclusions

Digitization and automation will drive the construction industry to a new level of efficiency, eliminating non-fundamental factors. In the area of building maintenance, the BIM methodology will be the hinged element of this development process, through which it will be possible to progressively integrate new work tools. This integration will have a very large impact on plant management costs, which correspond to more than 65% of the building's operating costs.

The implementation of maintenance management systems, using reinforced concrete monitoring sensors, integrated in the BIM process, involves the use of 3D models, monitoring systems, and their interconnection interface. Of these three components, the development of the interface involves relatively low costs when compared to the costs of the BIM model developments and the installation of the monitoring system. This process will have a strong impact in reducing the time spent searching for information for the implementation of maintenance tasks.

From the point of view of the suitability of the systems for their integration into BIM, the monitoring of corrosion potential, ionic resistivity of the concrete, and polarization resistance can be distinguished by the experience of use and reliability of use. These electrochemical techniques along with the fiber optic solutions give assurances of operation to the systems.

For maintenance, the existence of these new information management models will facilitate the introduction of automated monitoring systems, encouraging the use of sensors, as a result of increasing the capacity to accommodate and organize the collected information. As a result, it will be possible to acquire and manage a greater volume of information, in a continuous way in time and without human intervention. This automation process is of interest to the entire building management system, since it allows managing the environmental variables (temperature, humidity, air renewal, etc.) of the different spaces, controlling the operation of the various installed equipment or measuring and analyzing relative indicators on the conservation status of structural or non-structural elements.

The structural monitoring systems presented can be integrated as maintenance management tools into 3D models. Its incorporation in the BIM methodology has the additional advantage of allowing, in a more intuitive way, the interpretation of results using the spatial location of each of the sensors. The results of the monitoring of construction in operation demonstrate the interest of the data obtained for the maintenance processes and emphasize the necessity of the conjugation of the reading of two or more parameters, in order to guarantee a high index of reliability of the system.

Future owners will want integrated, structured, and easily accessible information for their buildings. This intent will only be gain with the use of information and systems platforms.

Acknowledgements

This study was financed by INNOACE, "Innovación abierta e inteligente en la EUROACE", (Interreg V-A Espanha-Portugal - POCTEP - 0049_INNOACE_4_E)

Conflicts of interest

The authors declare no conflict of interest.

Author details

Pedro Romano* and Paulo Brito

*Address all correspondence to: promano@ipportalegre.pt

VALORIZA - Research Centre for Endogenous Products Valorization, Instituto Politécnico de Portalegre, Portalegre, Portugal

References

[1] Qin J, Liu Y, Grosvenor R. A categorical framework of manufacturing for industry 4.0 and beyond. Procedia CIRP. 2016;**52**:173-178. DOI: 10.1016/j.procir.2016.08.005

[2] Oesterreich TD, Teuteberg F. Understanding the implications of digitisation and automation in the context of Industry 4.0: A triangulation approach and elements of a research agenda for the construction industry. Computers in Industry. 2016;**83**:121-139. DOI: 10.1016/j.compind.2016.09.006

[3] Lu Y. Industry 4.0: A survey on technologies, applications and open research issues. Journal of Industrial Information Integration. 2017;**6**:1-10. DOI: 10.1016/j.jii.2017.04.005

[4] Leviäkangas P, Mok Paik S, Moon S. Keeping up with the pace of digitization: The case of the Australian construction industry. Technology in Society. 2017;**50**:33-43. DOI: 10.1016/j.techsoc.2017.04.003

[5] Li J, Yang H. A research on development of construction industrialization based on BIM technology under the background of industry 4.0. MATEC Web of Conferences. 2017;**100**:02046. DOI: 10.1051/matecconf/201710002046

[6] Wang S, Wan J, Zhang D, Li D, Zhang C. Towards smart factory for industry 4.0: A self-organized multi-agent system with big data based feedback and coordination. Computer Networks. 2016;**101**:158-168. DOI: 10.1016/j.comnet.2015.12.017

[7] Bilal M, Oyedele LO, Qadir J, Munir K, Ajayi SO, Akinade OO, et al. Big data in the construction industry: A review of present status, opportunities, and future trends. Advanced Engineering Informatics. 2016;**30**(3):500-521. DOI: 10.1016/j.aei.2016.07.001

[8] Dallasega P, Rauch E, Linder C. Industry 4.0 as an enabler of proximity for construction supply chains: A systematic literature review. Computers in Industry. 2018;**99**:205-225. DOI: 10.1016/j.compind.2018.03.039

[9] Dubois A, Gadde LE. The construction industry as a loosely coupled system: Implications for productivity and innovation. Construction Management and Economics. 2002;**20**(7): 621-631. DOI: 10.1080/01446190210163543

[10] Arayici Y, Coates P. A system engineering perspective to knowledge transfer: A case study approach of BIM adoption. Virtual Reality–Human Computer Interaction. 2012; **2006**:179-206. DOI: 10.5772/3333

[11] Li X, Wu P, Shen GQ, Wang X, Teng Y. Mapping the knowledge domains of building information modeling (BIM): A bibliometric approach. Automation in Construction. 2017;**84**:195-206. DOI: 10.1016/j.autcon.2017.09.011

[12] Saieg P, Sotelino ED, Nascimento D, Caiado RGG. Interactions of building information modeling, lean and sustainability on the architectural, engineering and construction industry: A systematic review. Journal of Cleaner Production. 2018;**174**:788-806. DOI: 10.1016/j.jclepro.2017.11.030

[13] Hosseini MR, Martek I, Zavadskas EK, Aibinu AA, Arashpour M, Chileshe N. Critical evaluation of off-site construction research: A Scientometric analysis. Automation in Construction. 2018;**87**:235-247. DOI: 10.1016/j.autcon.2017.12.002

[14] Chen W, Chen K, Cheng JCP, Wang Q, Gan VJL. BIM-based framework for automatic scheduling of facility maintenance work orders. Automation in Construction, Elsevier. 2018;**91**:15-30. DOI: 10.1016/j.autcon.2018.03.007

[15] Pishdad-Bozorgi P, Gao X, Eastman C, Self AP. Planning and developing facility management-enabled building information model (FM-enabled BIM). Automation in Construction. 2018;**87**:22-38. DOI: 10.1016/j.autcon.2017.12.004

[16] Schiessl P, Raupach M. Monitoring of the corrosion risk for the reinforcement of bridges. In: Bridge Rehabilitation. Proceedings of the 3rd International Workshop On Bridge Rehabilitation, June 14-17; Technical University Darmstadt and The University of Michigan; 1992. pp. 741-752

[17] Morris W, Vico A, Vázquez M. Chloride induced corrosion of reinforcing steel evaluated by concrete resistivity measurements. Electrochimica Acta. 2004;**49**:4447-4453. DOI: 10.1016/j.electacta.2004.05.001

[18] Birbilis N, Cherry BW. Alternative methodology for on site monitoring of corrosion and remediation of reinforced concrete. Corrosion Engineering, Science and Technology. 2004;**39**(4):321-326. DOI: 10.1179/174327804X13154

[19] Angst U, Elsener B, Larsen CK, Vennesland Ø. Critical chloride content in reinforced concrete - A review. Cement and Concrete Research. 2009;**39**:1122-1138. DOI: 10.1016/j.cemconres.2009.08.006

[20] Qiao G, Liu T. Characterization of the reinforcing steel corrosion by potentialdynamic scan approach. Journal Wuhan University of Technology, Materials Science Edition. 2012;**27**(3):418-421. DOI: 10.1007/s11595-012-0477-4

[21] Brenna A, Lazzari L, Ormellese M. Monitoring chloride-induced corrosion of carbon steel tendons in concrete using a multi-electrode system. Construction and Building Materials. 2015;**96**:434-441. DOI: 10.1016/j.conbuildmat.2015.08.037

[22] ASTM International. 1999. ASTM C 876-1991 - Standard Test Method for Corrosion Potentials of Uncoated Reinforcing Steel in Concrete. West Conshohocken, PA: ASTM International; 1999. DOI: 10.1520/C0876-09.2

[23] Sengul O, Gjørv OE. Electrical resistivity measurements for quality control during concrete construction. ACI Materials Journal. 2009;**105**:541-547

[24] Song H, Saraswathy V. Corrosion monitoring of reinforced concrete structures - A review. International Journal of Electrochemical Science. 2007;**2**:1-28

[25] Andrade CU, Alonso C. On-site measurements of corrosion rate of reinforcements. Construction and Building Materials. 2001;**15**:141-145

[26] Soleymani HR, Ismail ME. Comparing corrosion measurement methods to assess the corrosion activity of laboratory OPC and HPC concrete specimens. Cement and Concrete Research. 2004;**34**(11):2037-2044. DOI: 10.1016/j.cemconres.2004.03.008

[27] Andrade C, Alonso C. Corrosion rate monitoring in the laboratory and on-site. Construction and Building Materials. 1996;**10**(5):315-328

[28] Brito P, Rodrigues L, Cunha P. Apresentação de um Sistema de Monitorização da Degradação de Estruturas de Betão Armado. Revista Ingenium. 2007;**98**:132-136

[29] Elsener B. Corrosion rate of steel in concrete-measurements beyond the Tafel law. Corrosion Science. 2005;**47**(12):3019-3033

[30] Pereira EV, Figueira RB, Salta MML, da Fonseca ITE. A galvanic sensor for monitoring the corrosion condition of the concrete reinforcing steel: Relationship between the galvanic and the corrosion currents. Sensors. 2009;**9**(11):8391-8398. DOI: 10.3390/s91108391

[31] Merioua A, Bezzar A, Ghomari F. Non-destructive electrical methods for measuring the physical characteristics of porous materials. Journal of Nondestructive Evaluation. 2015;**34**(2):13. DOI: 10.1007/s10921-015-0287-7

[32] Brito PSD, Romano P. Monitoring electrochemical degradation of concrete structures. Recent Patents on Corrosion Science. 2013. pp. 156-163. DOI: 10.2174/2210683903666131227193149

[33] Jin M, Jiang L, Tao D, Bai S. Characterization of Ag/AgCl electrode manufactured by immersion in sodium hypochloride acid for monitoring chloride content in concrete. Construction and Building Materials. 2016;**122**:310-319. DOI: 10.1016/j.conbuildmat.2016.05.163

[34] NP EN 12504-4. Testing concrete in structures. Part 4: Determination of ultrasonic pulse velocity. Caparica: IPQ; 2007

[35] ASTM International. ASTM C597-16, Standard Test Method for Pulse Velocity Through Concrete. West Conshohocken, PA: ASTM International; 2016

[36] Whitehurst EA. The Soniscope—A Device for Field Testing of Concrete. Indiana, United States: Purdue University; 1951. https://docs.lib.purdue.edu/cgi/viewcontent.cgi?article=3009&context=roadschool

[37] IAEA IAEA. Guidebook on non-destructive testing of concrete structures. Training Course Series. 2002;**17**(17):231. Available at: http://200.10.161.33/cirsoc/pdf/ensayos/tcs-17_web.pdf

[38] Haach VG, Ramirez FC. Qualitative assessment of concrete by ultrasound tomography. Construction and Building Materials, Elsevier Ltd. 2016;**119**:61-70. DOI: 10.1016/j.conbuildmat.2016.05.056

[39] Torres B, Payá-Zaforteza Ignacio I, Calderón PA, Adam JM. Analysis of the strain transfer in a new FBG sensor for structural health monitoring. Engineering Structures. 2011;**33**(2):539-548. DOI: 10.1016/j.engstruct.2010.11.012

[40] Kim JM, Kim CM, Choi SY, Lee BY. Enhanced strain measurement range of an FBG sensor embedded in seven-wire steel strands. Sensors (Switzerland). 2017;**17**:1654. DOI: 10.3390/s17071654

[41] Kim JM, Kim HW, Choi SY, Park SY. Measurement of prestressing force in pretensioned UHPC deck using a fiber optic FBG sensor embedded in a 7-wire strand. Journal of Sensors. 2016. p. 9. DOI: 10.1155/2016/8634080

[42] Hong CY, Zhang YF, Zhang MX, Leung LMG, Liu LQ. Application of FBG sensors for geotechnical health monitoring, a review of sensor design, implementation methods and packaging techniques. Sensors and Actuators, A: Physical. 2016;**244**:184-197. DOI: 10.1016/j.sna.2016.04.033

[43] Majumder M, Gangopadhyay TK, Chakraborty AK, Dasgupta K, Bhattacharya DK. Fibre Bragg gratings in structural health monitoring—Present status and applications. Sensors and Actuators, A: Physical. 2008;**147**(1):150-164. DOI: 10.1016/j.sna.2008.04.008

[44] Rodrigues C, Faria R, Figueiras J. Monitorização Experimental da Nova Barragem do Alto Ceira Usando a Tecnologia da Fibra Ótica', (May 2014); 2012. pp. 24-26

[45] Mao J, Chen J, Cui L, Jin W, Xu C, He Y. Monitoring the corrosion process of reinforced concrete using BOTDA and FBG sensors. Sensors (Switzerland). 2015;**15**(4):8866-8883. DOI: 10.3390/s150408866

[46] Li W, Xu C, Ho SCM, Wang B, Song G. Monitoring concrete deterioration due to reinforcement corrosion by integrating acoustic emission and FBG strain measurements. Sensors. 2017;**17**(3):657. DOI: 10.3390/s17030657

[47] Almubaied O, Chai HK, Islam MR, Lim K-S, Tan CG. Monitoring corrosion process of reinforced concrete structure using FBG strain sensor. IEEE Transactions on Instrumentation and Measurement. 2017;**66**(8, 8):1. DOI: 10.1109/TIM.2017.2676218

[48] Omar T, Nehdi ML. Remote sensing of concrete bridge decks using unmanned aerial vehicle infrared thermography. Automation in Construction, Elsevier. 2017;**83**:360-371. DOI: 10.1016/j.autcon.2017.06.024

[49] Abu Dabous S, Yaghi S, Alkass S, Moselhi O. Concrete bridge deck condition assessment using IR thermography and ground penetrating Radar technologies. Automation in Construction, Elsevier. 2017;**81**:340-354. DOI: 10.1016/j.autcon.2017.04.006

[50] Clark M, McCann D, Forde M. Application of infrared thermography to the non-destructive testing of concrete and masonry bridges. NDT and E International. 2003;**36**(4):265-275. DOI: 10.1016/S0963-8695(02)00060-9

[51] ASTM International (2007) ASTM D 4788-2007 - Standard Test Method for Detecting Delaminations in Bridge Decks Using Infrared. West Conshohocken, PA: ASTM International; 2007. DOI: 10.1520/D4788-03R07.2

[52] Cannard H, Mahrez M, Perrin T, Muzet V, Prybyla D, Brachelet F. The use of infrared thermography for defects detection on reinforced concrete bridges. In: Proceedings of 12th Quantitative InfraRed Thermography. France, Bordeaux: 12th International Conference on Quantitative Infrared Thermography; 7-11 July 2014

[53] Caldeira MM, Padaratz IJ, Caldeira MM, Padaratz IJ. Potentialities of infrared thermography to assess damage in bonding between concrete and GFRP. Revista IBRACON de Estruturas e Materiais. 2015;**8**(3):296-322. DOI: 10.1590/S1983-41952015000300004

[54] Jaselskis EJ, El-Misalami T. Implementing radio frequency identification in the construction process. Journal of Construction Engineering and Management. 2003;**129**(6):680-688. DOI: 10.1061/(ASCE)0733-9364(2003)129:6(680)

[55] Valero E, Adán A, Cerrada C. Evolution of RFID applications in construction: A literature review. Sensors (Switzerland). 2015;**15**(7):15988-16008. DOI: 10.3390/s150715988

[56] Cheng M, Chen W. Open-Building Maintenance Management Using RFID Technology. Li-Chuan Lien Ming-Hsiu Tsai', (January); 2007

[57] Pereira E, Salta M. Corrosion Monitoring Systems Installed in Portuguese Structures. Duratinet Project - Smart & green structural and repair materials, Technical Report. Lisbon, Portugal: LNEC; 2012

Chapter 12

Durability of Composite-Modified Asphalt Mixture Based on Inherent and Improved Performance

Haitao Zhang and Mingyang Gong

Additional information is available at the end of the chapter

http://dx.doi.org/10.5772/intechopen.81824

Abstract

The composite performance of modified asphalt and composite-modified asphalt mixture (CMAM) is divided into inherent performance and improved performance. The inherent performance refers to the original performance of the virgin asphalt and asphalt mixture. The improved performance refers to the performance obtained by modifying the virgin asphalt. In the study, the modified asphalt and asphalt mixture were tested through a series of experiments. The antiaging durability of the modified asphalt is explored based on the inherent performance and improved performance at high and low temperatures. The antiaging durability of the modified asphalt mixture is explored based on the inherent performance and improved performance at the mechanical performance. Meanwhile, based on inherent and improved performances, this chapter uses three kinds of CMAM (4% SBS/3% SBR, 4% SBS/15% rubber, and 4% SBR/15% rubber) as research objects to test the change rule of mechanical properties. This chapter outlines CMAM design, mechanical property tests, and comparative durability analysis.

Keywords: composite-modified asphalt mixture, inherent and improved performances, durability, test, comparative analysis

1. Introduction

With the extensive use of asphalt pavement, improving the performance of asphalt is very important for driving safety and driving comfort. The properties of asphalt are divided into many types. The service life of asphalt pavement is decided on the durability of asphalt and asphalt mixture. Study on durability of asphalt and asphalt mixture is very important in the day [1–4].

IntechOpen

© 2018 The Author(s). Licensee IntechOpen. This chapter is distributed under the terms of the Creative Commons Attribution License (http://creativecommons.org/licenses/by/3.0), which permits unrestricted use, distribution, and reproduction in any medium, provided the original work is properly cited. (cc) BY

Zhang et al. [5] investigated the change in molecular weight during the aging process of asphalt and found some new functional groups that were produced inside the asphalt; they also found that the change in molecular structure will lead to changes in the morphology and performance of the asphalt. Chen's paper concluded that temperature has a great influence on the aging of asphalt. When the temperature was above 100°C, a dehydrogenated chemical reaction of the asphalt was produced. When the temperature was below 100°C, an oxidation reaction of the asphalt was produced, as well as some oxygenated compounds. The chemical reaction produced by asphalt aging has an important influence on the performance of the asphalt [6]. According to Zhu et al., as the aging time is increased, the residual needle penetration and residual ductility are decreased and the softening point is increased. This means that the high-temperature performance of the asphalt is increased and the low-temperature performance is decreased after aging [7]. Zhang et al. [8] indicated that the variation of the performance of the asphalt during the aging progress is mainly due to the change of its interior components. When asphalt is aging, the content of its lighter components is decreased, such as oil and gelatin, and the content of the heavier components is increased, such as asphaltene. This results in a harder asphalt, which is easily broken at low temperatures. Meanwhile, the aging of the asphalt threatens the safety of the road surface. In the aging process, the low-temperature performance and water stability of the asphalt mixture are decreased. The asphalt pavement creates the risk of road fractures and the formation of a loose mixture of asphalt in cold and wet areas [9–11]. Therefore, it is of great significance to improve the service life and performance of asphalt pavement via improving the aging durability of asphalt.

This chapter will discuss the formation mechanism and durability of two properties of modified asphalt, and further analyze the technical characteristics of different modifiers. On the basis of this research, a new modified asphalt material based on composite modifiers is proposed to optimize the road performance of modified asphalt. And in this work, SBS, SBR, and rubber powder modifiers were selected as composite modifiers. 4% SBS/3% SBR, 4% SBS/15% rubber, and 4% SBR/15% rubber were used for testing. Based on the inherent and improved performance of each mixture, a comparative analysis was carried out to study the changing mechanical performance of CMAM. Findings offer theoretical and practical value for research on asphalt mixture durability [12–20].

2. Test materials and methodology

2.1. Test materials

2.1.1. Asphalt

Virgin asphalt 90# was selected in this test; asphalt indices are displayed in **Table 1**.

2.1.2. Modifiers

Overall, SBS, SBR, TPS, Sasobit and rubber modifiers were tested (**Figure 1**). Corresponding indices are shown in **Tables 2–6**.

http://dx.doi.org/10.5772/intechopen.81824

Asphalt	Penetration (25°C/0.1 mm)	PI	Ductility (cm)		Softening point (°C)	Viscosity (135°C/Pa s)
			5°C	15°C		
90#	93	−0.77	9	165	44.5	0.328

Table 1. Technical indexes of virgin asphalt.

Figure 1. Modifier types. (a) SBS, (b) SBR, (c) rubber (80 mesh); (d) TPS; (e) Sasobit.

Tensile strength (MPa)	Elongation at break (%)	Hardness (A)	Styrene/butadiene	Density (kg/cm^3)
33	800	76	31/69	0.94

Table 2. Technical indices of SBS modifier (LG501).

Moisture quality fraction (%)	Molecular weight ($\times10^4$)	Tensile strength (MPa)	Elongation at break (%)	Ethylene propylene mass fraction (%)	Dry rubber (%)	Grains (mesh)
≤2	20–30	31	320	22.5–24.5	≥75	<40

Table 3. Technical indices of SBR modifier (1502).

Physical index				Chemical index			
Relative density (kg/cm^3)	Moisture (%)	Metal (%)	Fiber (%)	Ash (%)	Acetone extract (%)	Carbon black (%)	Rubber hydrocarbon (%)
289	0.27	0.02	0	10	10.02	32.86	51

Table 4. Technical indices of rubber.

2.1.3. *Aggregates*

The CMAM (AC-13) included the above aggregate gradations. Optimum asphalt content was determined to be 5.6% using the Marshall design method. Asphalt mixture AC-13 was selected, with the aggregate gradation shown in **Table 7**.

	Particle size (mm)	Density (g/cm³)	Water absorption rate (%)
Technical indicators	≤5	0.7~1.0	<1

Table 5. Technical indexes of modifier TPS.

	Freezing point (°C)	Penetration (0.1 mm)		Viscosity (135°C/cp)	Density (25°C/(g/cm³))	Melting point (°C)
		25°C	65°C			
Prescribed value	≤99	≤1	≤10	10~14	—	98~110
Measured value	98	<0.7	7	12	0.94	100

Table 6. Technical indexes of modifier Sasobit.

Mixture type	Percentage of passing through sieve (mm)/%											
	26.5	19	16	13.2	9.5	4.75	2.36	1.18	0.6	0.3	0.15	0.075
AC-13			100	95	76.5	53	37	26.5	19	13.5	10	6

Table 7. Aggregate gradations.

2.2. Test methodology

2.2.1. Modified asphalt test methodology

The 4% SBS asphalt, 4% TPS asphalt, 16% rubber asphalt, and 2% Sasobit asphalt were selected as the research objects to carry out aging tests (aging times were 0, 5, 20 h). Then, the durability comparison of the asphalt was carried out based on inherent durability and inherent performance.

The antiaged durability of the modified asphalt based on inherent and improved performances is compared and analyzed. The penetration index was used to evaluate the performance at high temperature of the asphalt and the ductility index was used to evaluate the performance at low temperature. The change curves of modified asphalt with different aging times are linear regressions, and it is also used to evaluate the inherent and improved performances of the modified asphalt.

2.2.2. CMAM test methodology

The performance of composite-modified asphalt mixture was tested and analyzed based on inherent and improved performances. A comparative study was tested through the aging durability, plastic deformation durability, and fatigue durability.

1. In terms of aging durability, taking dynamic stability, bending strain, TSR as indicators to explore the change rule of inherent, improved, and composite performances of

different composite-modified asphalt mixtures in aging process by controlling the aging time (0, 5, 20 hours).

2. In terms of plastic deformation durability, taking plastic deformation of asphalt mixture as an indicator to do a comparative analysis on the change rule of inherent, improved, and composite performances of different composite-modified asphalt mixtures by controlling repeated load times.

3. In terms of fatigue durability, taking fatigue life of different asphalt mixtures as indicators to do a comparative analysis on the change rule of inherent, improved, and composite performances of different composite-modified asphalt mixtures.

3. Comparative analysis on inherent and improved performances of asphalt

3.1. Comparative analysis of the penetration and PI base on inherent and improved performances

The penetration and PI of the SBS asphalt (90#), TPS asphalt (90#), rubber asphalt (90#), and Sasobit asphalt (90#) based on different performance and aging times are shown in **Figures 2–5**.

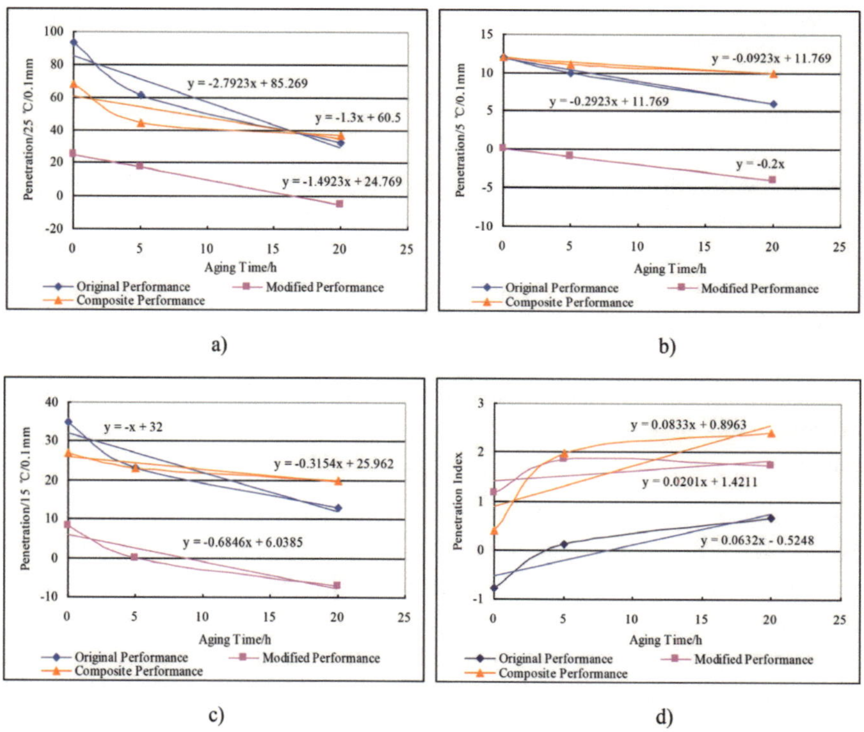

Figure 2. Relationship between penetration, PI, and aging times of 4% SBS asphalt (90#) with different performances: (a) penetration/25°C; (b) penetration/15°C; (c) penetration/5°C; (d) penetration index.

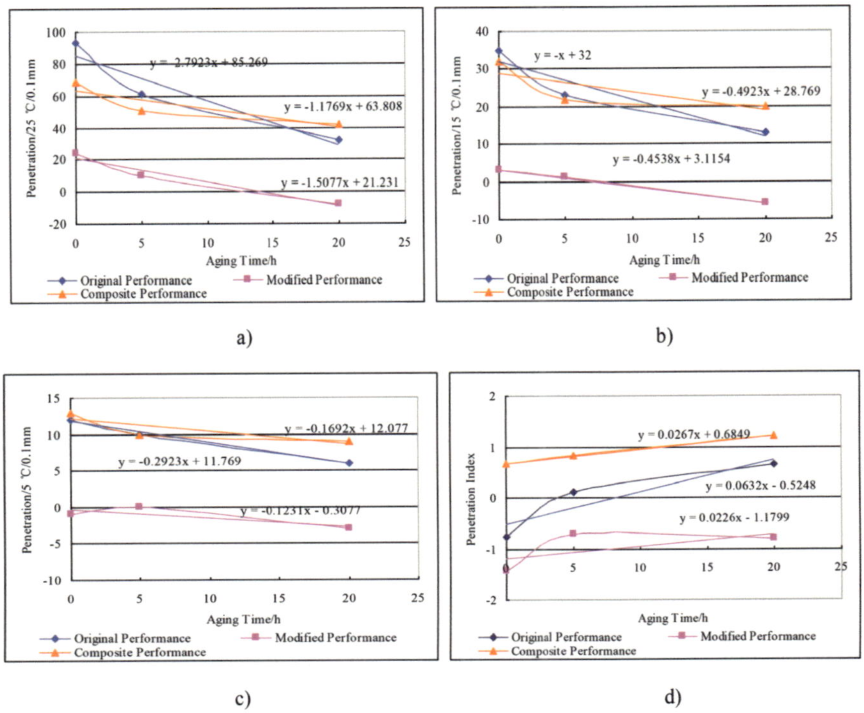

Figure 3. Penetration and PI of TPS asphalt (90#) based on different performances and aging times: (a) penetration/25°C; (b) penetration/15°C; (c) penetration/5°C; (d) penetration index.

The following conclusions can be drawn by comparing the data in **Figures 2–5**.

1. In regard to the penetration data (25, 15, and 5°C), the results show that the penetration change rate of the SBS asphalt (90#), TPS asphalt (90#), rubber asphalt (90#), and Sasobit asphalt (90#) based on improved performance (performance at high temperatures) at different temperatures is smaller than that of inherent performance. But the smaller the penetration change rate is, the better the performance at high temperatures. If penetration drops too rapidly during the aging process, this indicates a poor antiaging durability of penetration. So the inherent performance of the SBS asphalt (90#), TPS asphalt (90#), rubber asphalt (90#), and Sasobit asphalt (90#) was better than that of improved performance.

2. In regard to the change rate of the PI regression equation, the results show that the penetration change rate of the SBS asphalt (90#), TPS asphalt (90#), and Sasobit asphalt (90#) based on improved performance at the value of PI is smaller than that of inherent performance. The rubber asphalt (90#) based on improved performance at the value of PI is bigger than that of inherent performance. The smaller the value of PI change rate is, the better the temperature sensitivity of asphalt. So the SBS asphalt (90#), TPS asphalt (90#), and Sasobit asphalt (90#) based on improved performance at the temperature sensitivity is smaller than that of inherent performance. The rubber asphalt (90#) based on improved performance at

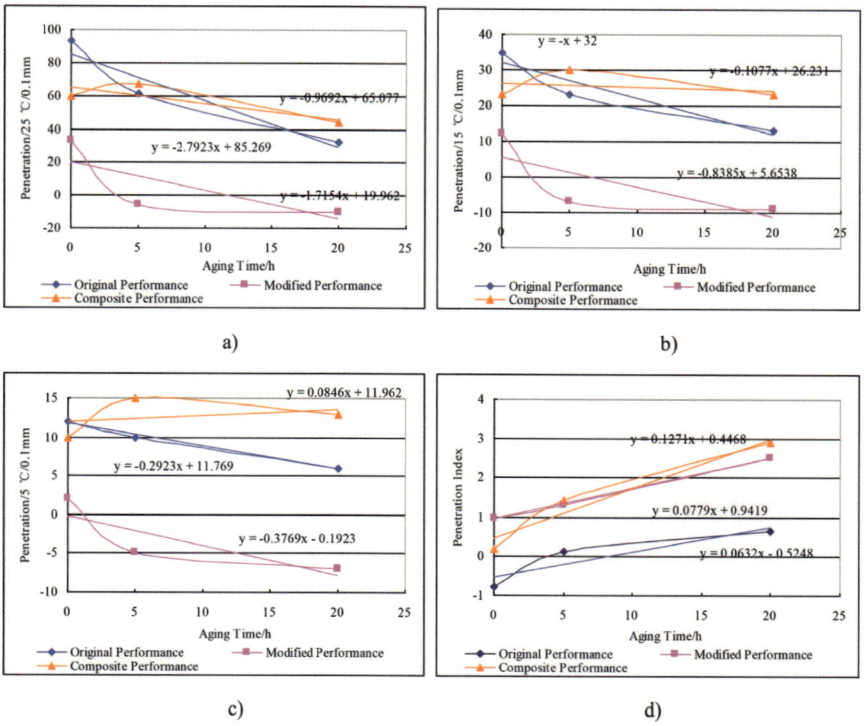

Figure 4. Penetration and PI of rubber asphalt (90#) based on different performance and aging times: (a) penetration/25°C; (b) penetration/15°C; (c) penetration/5°C; (d) penetration index.

the temperature sensitivity is bigger than that of inherent performance. The results show that the SBS asphalt (90#), TPS asphalt (90#), and Sasobit asphalt (90#) at the temperature sensitivity of improved performance is weak and the antiaging durability of performance at high temperature is better.

3.2. Comparative analysis of the ductility based on inherent and improved performances

The ductility of SBS asphalt (90#) based on different performance and aging times is shown in **Figure 6**.

The following conclusions can be obtained by comparing the data in **Figures 6–9**.

1. In terms of the ductility (5°C, 15°C) data, the results show that the ductility change rate of the SBS asphalt (90#) and TPS asphalt (90#) at 5°C ductility based on inherent performance is smaller than that of the improved performance. The ductility change rate of the rubber asphalt (90#) and Sasobit asphalt (90#) at 5°C ductility based on improved performance is bigger than that of the inherent performance. The larger the ductility at 5°C, the better the performance at low temperatures. So the results show that the antiaging durability of the

SBS asphalt (90#) and TPS asphalt (90#) at 5°C ductility based on inherent performance (performance at low temperatures) is better than that of the improved performance. The antiaging durability of the rubber asphalt (90#) and Sasobit asphalt (90#) at 5°C ductility based on improved performance (performance at low temperatures) is better than that of the inherent performance.

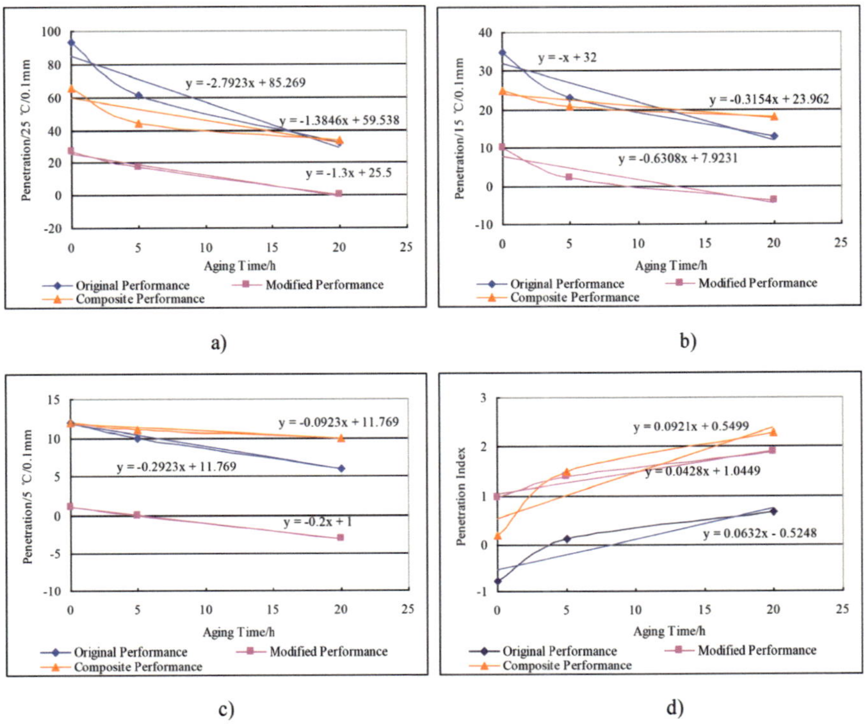

Figure 5. Penetration and PI of Sasobit asphalt (90#) based on different performances and aging times: (a) penetration/25°C; (b) penetration/15°C; (c) penetration/5°C; (d) penetration index.

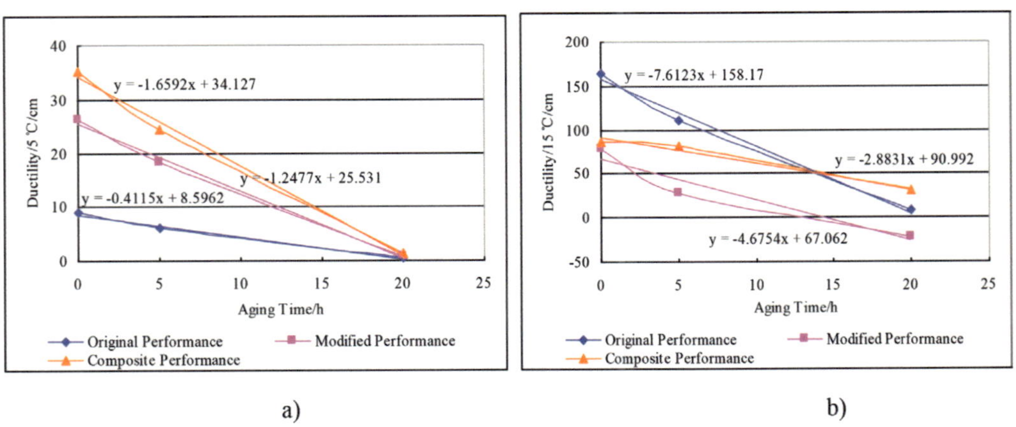

Figure 6. Relationship between ductility and aging times of 4% SBS asphalt (90#) with different performances: (a) ductility 5°C and (b) ductility 15°C.

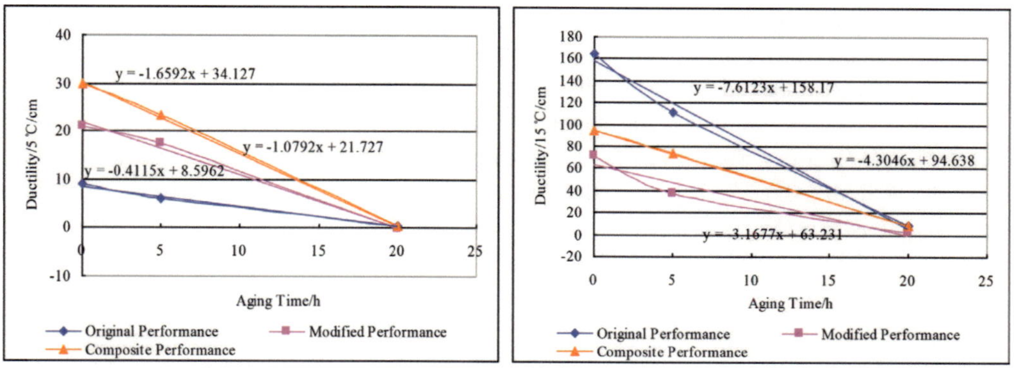

a) b)

Figure 7. Ductility of TPS asphalt (90#) based on different performance and aging times: (a) ductility 5°C and (b) ductility 15°C.

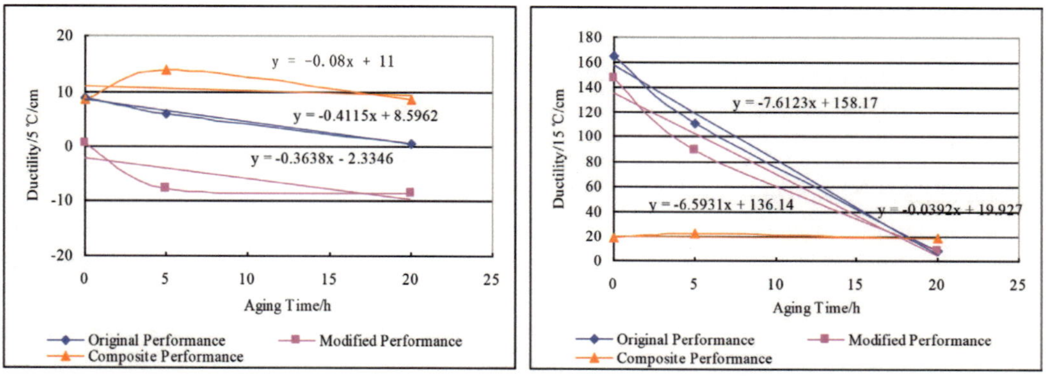

a) b)

Figure 8. Ductility of rubber asphalt (90#) based on different performance and aging times: (a) ductility 5°C and (b) ductility 15°C.

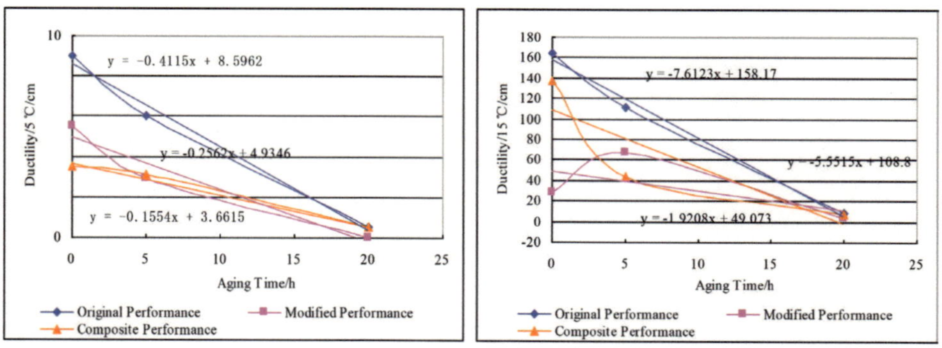

a) b)

Figure 9. Ductility of Sasobit asphalt (90#) based on different performances and aging times: (a) ductility 5°C and (b) ductility 15°C.

2. According to the ductility of 15°C, the results show that the ductility change rate of the SBS asphalt (90#), TPS asphalt (90#), rubber asphalt (90#), and Sasobit asphalt (90#) at 15°C ductility based on improved performance is smaller than that of the inherent performance. The ductility of 15°C is replaced as the elasticity of asphalt. The larger the ductility at 15°C is, the better the performance at the elasticity of asphalt is. So according to the elasticity of asphalt, the SBS asphalt (90#), TPS asphalt (90#), rubber asphalt (90#), and Sasobit asphalt (90#) at the elasticity based on improved performance is smaller than that of the inherent performance.

3.3. Comparative analysis of the softening point and viscosity based on inherent and improved performances

The test results are shown in **Figures 10–13**.

1. In regard to the comparison of the data, the results show that the softening point change rate of the SBS asphalt (90#), TPS asphalt (90#), rubber asphalt (90#), and Sasobit asphalt (90#) at 5°C ductility based on improved performance is smaller than that of the inherent performance. The larger the softening point, the better the performance at high temperatures. So the high temperatures of the SBS asphalt (90#), TPS asphalt (90#), rubber asphalt (90#), and Sasobit asphalt (90#) based on improved performance are better than that of the inherent performance.

2. According to the 135°C viscosity of asphalt, the results show that the viscosity (135°C) change rate of the TPS asphalt (90#) and Sasobit asphalt (90#) at 135°C viscosity based on improved performance is smaller than that of the inherent performance. The SBS asphalt (90#) and rubber asphalt (90#) based on inherent performance is bigger than that of the improved performance. The results of the 135°C viscosity replace that the hardness of asphalt. The larger the 135°C viscosity is, the better the performance of resistance to load is. So the performance of resistance to load of the TPS asphalt (90#) and Sasobit asphalt (90#) based on improved performance is better than that of the inherent performance. The SBS asphalt (90#) and rubber asphalt (90#) based on inherent performance is better than that of the improved performance.

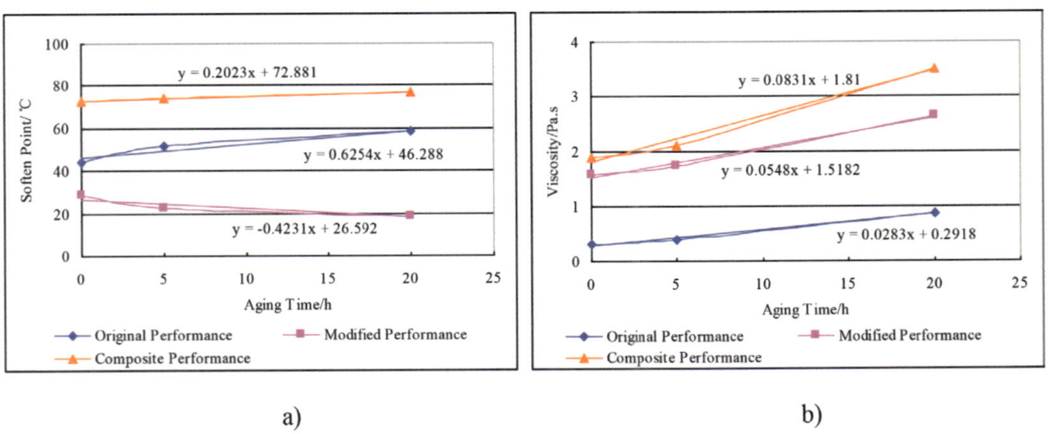

Figure 10. Softening point and viscosity of SBS asphalt (90#) based on different performances and aging times: (a) softening point and (b) viscosity.

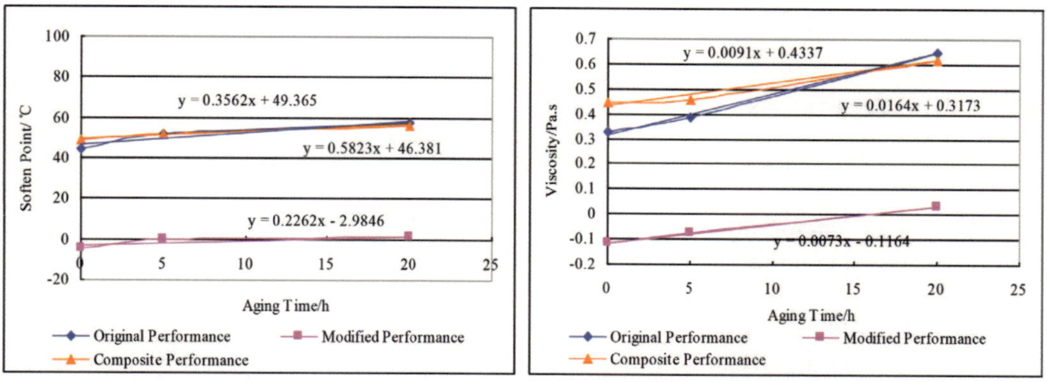

a) b)

Figure 11. Softening point and viscosity of TPS asphalt (90#) based on different performances and aging times: (a) softening point and (b) viscosity.

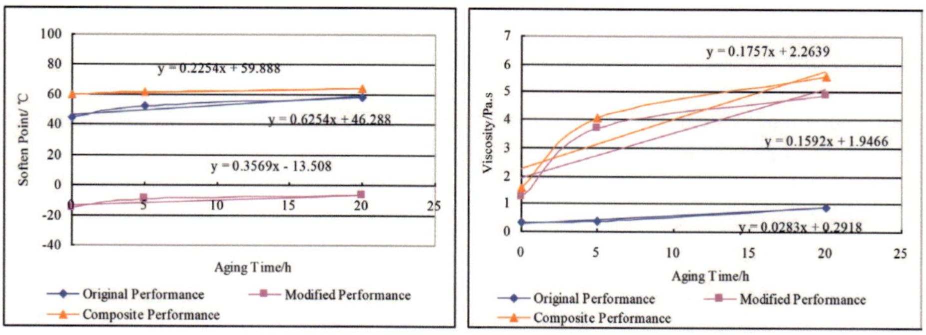

a) b)

Figure 12. Softening point and viscosity of rubber asphalt (90#) based on different performance and aging times: (a) softening point and (b) viscosity.

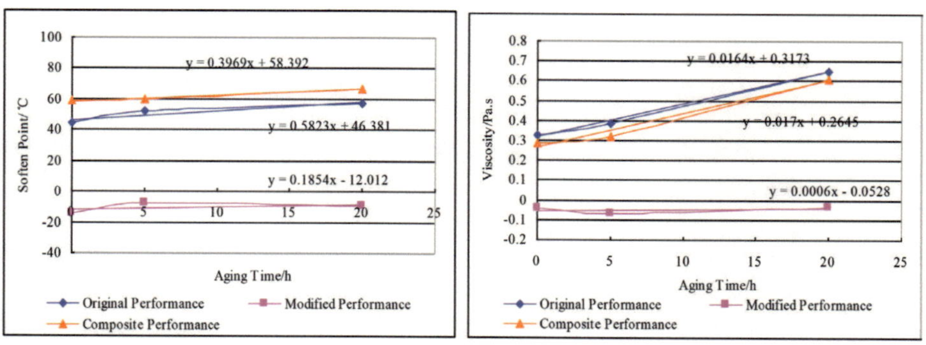

a) b)

Figure 13. Softening point and viscosity of Sasobit asphalt (90#) based on different performance and aging times: (a) softening point and (b) viscosity.

4. Comparative analysis on inherent and improved performances of HMA

4.1. Comparative analysis of aging durability

4.1.1. Comparative analysis of durability of 4% SBS/3% SBR asphalt mixture

Test results of the aging durability of the 4% SBS/15% SBR asphalt under different loading times are shown in **Figure 14**.

1. In regard to high-temperature stability and low-temperature stability, the change rate of improved performance under different aging times was better than that of inherent performance. Thus, the aging durability of the 4% SBS/3% SBR asphalt mixture on inherent performance was greater than that of improved performance.
2. In regard to water stability, the change rate of the inherent performance with different aging times was better than the improved performance. As such, the aging durability of the mixture based on improved performance was better than the inherent performance.

4.1.2. Comparative analysis of durability of 4% SBS/15% rubber asphalt mixture

Test results of the aging durability of the 4% SBS/15% rubber asphalt under different loading times are shown in **Figure 15**.

1. Regarding to high-temperature stability, in terms of change rate of the 4% SBS/15% rubber asphalt mixture, the 4%SBS/15% rubber asphalt mixture of inherent performance was less than that of improved performance. Thus, the aging durability of the inherent performance of the mixture was better than the improved performance.
2. Regarding to low temperature and the water stability change rate, the change rate of inherent performance became better than that of improved performance with different aging times. Therefore, in terms of the aging durability of the 4% SBS/15% rubber asphalt mixture, the improved performance of the mixture was better than that of inherent performance.

4.1.3. Comparative analysis of durability of 4% SBR/15% rubber asphalt mixture

Test results of the aging durability of the 4% SBR/15% rubber asphalt mixture at different aging times are given in **Figure 16**.

1. Regarding to high- and low-temperature stability, in terms of change rate of the 4% SBR/15% rubber asphalt mixture, the change rate of inherent performance was less than improved performance as aging time prolonged. The inherent performance of the mixture was better than that of improved performance.
2. Regarding to water stability, in terms of change rate of the 4% SBR/15% rubber asphalt mixture, the rate change of inherent performance was greater than that of improved performance with different aging times. The aging durability of the improved performance of the 4% SBR/15% rubber asphalt mixture was better than the inherent performance.

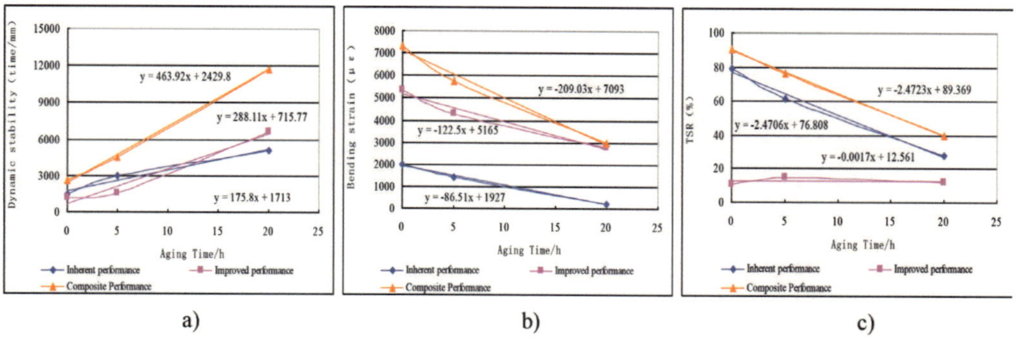

Figure 14. Relationship between mechanical properties of 4% SBS/3% SBR asphalt mixture and aging time: (a) dynamic stability; (b) bending strain; (c) TSR.

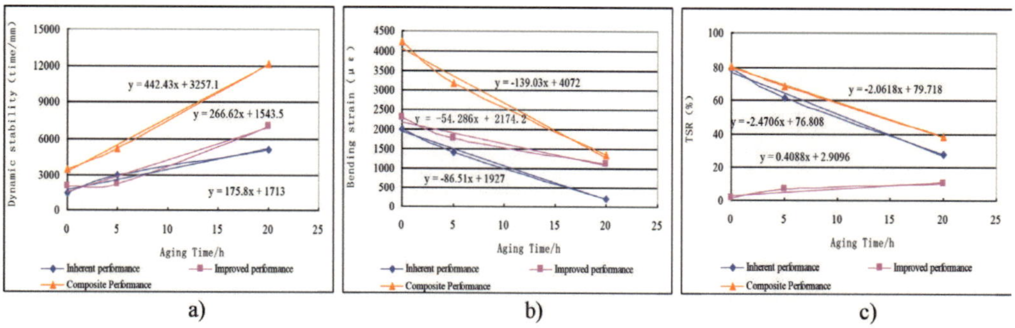

Figure 15. Relationship between mechanical properties of 4% SBS/15% rubber asphalt mixture and aging time: (a) dynamic stability; (b) bending strain; (c) TSR.

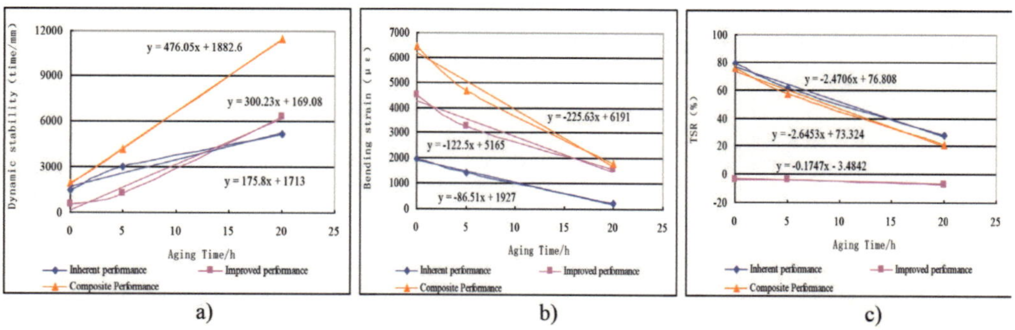

Figure 16. Relationship between mechanical properties of 4% SBR/15% rubber asphalt mixture and aging time: (a) dynamic stability; (b) bending strain; (c) TSR.

4.2. Comparative analysis of plastic deformation durability

The test results are shown in **Figures 17–19.** With the different time, in terms of the plastic deformation durability, the change rate of 4% SBS/3% SBR, 4% SBS/15% rubber, and 4% SBR/15% rubber of the inherent performance was greater than that of improved performance, implying that the improved performance of the mixture was better than the inherent performance in terms of plastic deformation durability **Tables 8–12**.

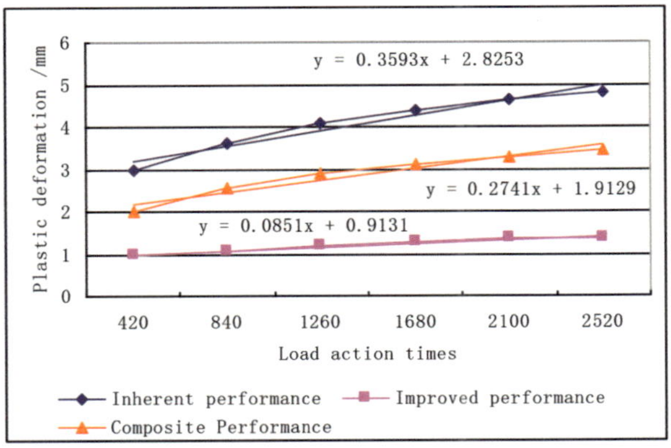

Figure 17. Test results of plastic deformation durability of 4% SBS/3% SBR asphalt mixture.

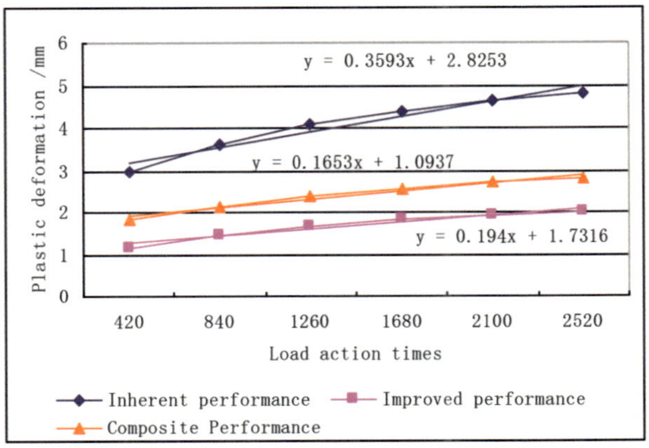

Figure 18. Test results of plastic deformation durability of 4% SBS/15% rubber asphalt mixture.

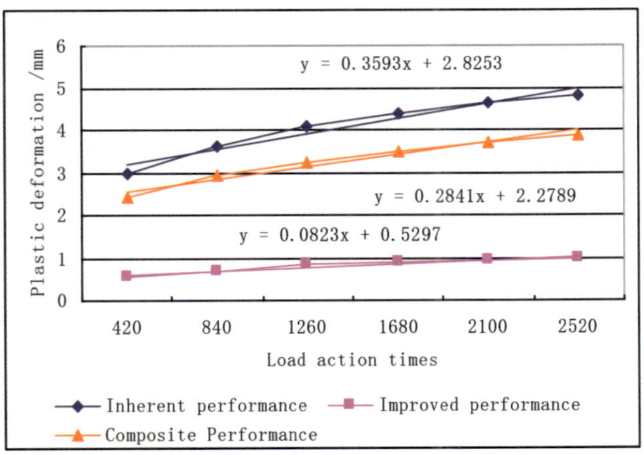

Figure 19. Test results of plastic deformation durability of 4% SBR/15% rubber asphalt mixture.

Asphalt types	Penetration at 5°C	Penetration at 15°C	Penetration at 25°C	PI
SBS asphalt (90#)	Improving performance is better than inherent performance	Improving performance is better than inherent performance	Improving performance is better than inherent performance	Improving performance is better than inherent performance
TPS asphalt (90#)	Improving performance is better than inherent performance	Improving performance is better than inherent performance	Improving performance is better than inherent performance	Improving performance is better than inherent performance
Rubber asphalt (90#)	Inherent performance is better than improving performance	Improving performance is better than inherent performance	Improving performance is better than inherent performance	Inherent performance is better than improving performance
Sasobit asphalt (90#)	Improving performance is better than inherent performance	Improving performance is better than inherent performance	Improving performance is better than inherent performance	Improving performance is better than inherent performance

Table 8. Comparison of penetration and PI based on inherent and improved performances.

Asphalt types	Ductility 5°C	Ductility 15°C
SBS asphalt (90#)	Inherent performance is better than improving performance	Improving performance is better than inherent performance
TPS asphalt (90#)	Inherent performance is better than improving performance	Improving performance is better than inherent performance
Rubber asphalt (90#)	Improving performance is better than inherent performance	Improving performance is better than inherent performance
Sasobit asphalt (90#)	Improving performance is better than inherent performance	Improving performance is better than inherent performance

Table 9. Comparison of ductility (5°C, 15°C) based on inherent and improved performances.

Asphalt types	Softening point	Viscosity
SBS asphalt (90#)	Improving performance is better than inherent performance	Inherent performance is better than improving performance
TPS asphalt (90#)	Improving performance is better than inherent performance	Improving performance is better than inherent performance
Rubber asphalt (90#)	Improving performance is better than inherent performance	Inherent performance is better than improving performance
Sasobit asphalt (90#)	Improving performance is better than inherent performance	Improving performance is better than inherent performance

Table 10. Comparison of softening point and viscosity based on inherent and improved performances.

Asphalt types	Dynamic stability	Bending strain	TSR
4% SBS/3% SBR asphalt mixture	Improving performance is better than inherent performance	Improving performance is better than inherent performance	Improving performance is better than inherent performance
4% SBS/15% rubber asphalt mixture	Improving performance is better than inherent performance	Improving performance is better than inherent performance	Improving performance is better than inherent performance
4% SBR/15% rubber asphalt mixture	Improving performance is better than inherent performance	Improving performance is better than inherent performance	Improving performance is better than inherent performance

Table 11. Comparison of penetration rate based on inherent and improved performance.

Asphalt types	Dynamic stability	Bending strain
4% SBS/3% SBR asphalt mixture	Plastic deformation durability	Fatigue durability
4% SBS/15% rubber asphalt mixture	Improving performance is better than inherent performance	Improving performance is better than inherent performance
4% SBR/15% rubber asphalt mixture	Improving performance is better than inherent performance	Improving performance is better than inherent performance

Table 12. Comparison of dynamic stability and bending strain based on inherent and improved performance.

4.3. Comparative analysis of fatigue durability

The test results are shown in **Figures 20–22**. The smaller the *k*-value is, the better the performance of fatigue durability is. The test results show that the *k*-value of 4% SBS/3% SBR, 4% SBS/15% rubber, and 4% SBR/15% rubber based on the inherent performance of the mixture exceeded that of improved performance. Therefore, in regard to fatigue life, the inherent performance of

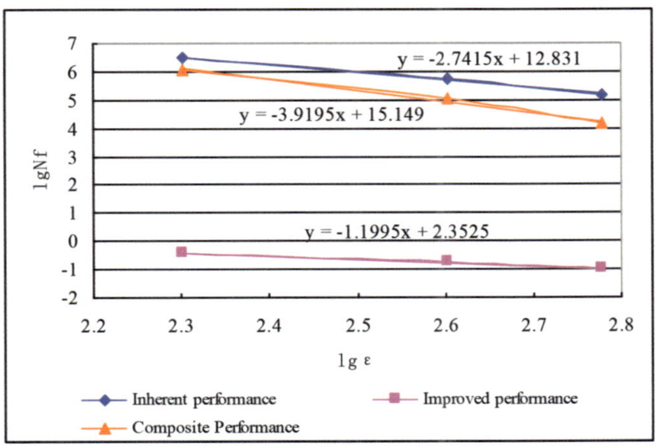

Figure 20. Test results of fatigue durability of 4% SBS/3% SBR asphalt mixture.

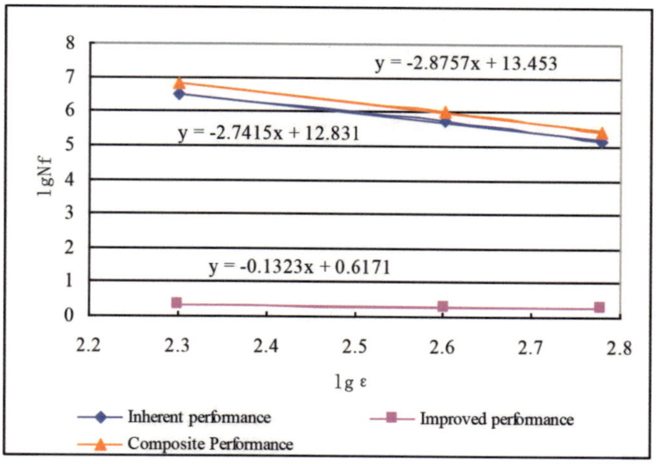

Figure 21. Test results of fatigue durability of 4% SBS/15% rubber asphalt mixture.

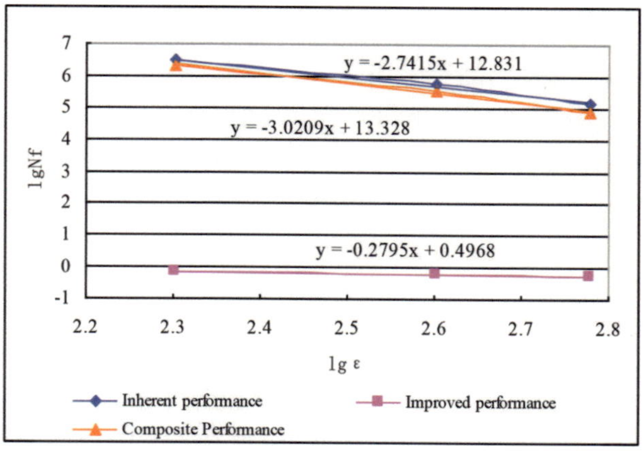

Figure 22. Test results of fatigue durability of 4% SBS/15% rubber asphalt mixture.

4% SBS/3% SBR, 4% SBS/15% rubber, and 4% SBR/15% rubber was better than the improved performance. According to the *n*-value, the inherent performance of the mixture exceeded the improved performance. The fatigue sensitivity based on the inherent performance of the asphalt mixture was larger than that of improved performance. In regard to the greater the fatigue sensitivity, the worse the fatigue durability. The fatigue durability of the 4% SBS/3% SBR asphalt mixture on improved performance was hence better than the inherent performance.

5. Conclusions

1. Through comparative analysis of the antiaged durability of different modified asphalts based on inherent and improved performances, the conclusion can be obtained: for SBS- and TPS-modified asphalt, taking penetration (performance at high temperature) and

5°C ductility (performance at low temperature) as indexes, the antiaged durability of the penetration based on inherent performance is less than that of the improved performance. The antiaged durability of 5°C ductility based on inherent performance is better than that of improved performance. The analysis result of rubber asphalt is contrary to this, which shows the particularity of rubber-modified asphalt.

2. The comparative test results for the durability of composite-modified asphalt based on inherent and improved performances reveal the following: In regard to aging progress, the high and low stabilities of the 4% SBS/3% SBR asphalt mixture and 4% SBR/15% rubber mixture based on inherent performance were better than that of improved performance. Water stability based on improved performance was better than inherent performance. The high-temperature stability of the 4% SBS/15% rubber asphalt mixture based on improved performance was better than that of inherent performance. Low-temperature and water stability based on improved performance were also better than that of inherent performance. Plastic deformation and fatigue durability based on improved performance of these three asphalt mixtures were better than that of inherent performance.

3. The test results of the improved performance durability of CMAMs reveal that the 4% SBS/15% rubber composite modifier was more stable in improving aging durability at high and low temperatures than the other two modifiers. The 4% SBS/3% SBR modifier was more stable in improving water stability. The 4% SBS/15% rubber composite modifier was more stable in improving plastic deformation durability and fatigue durability and should be adopted when selecting composite modifiers with better stability.

Acknowledgements

The authors gratefully appreciate the support of the Province Key Laboratory of Road at Northeast Forestry University and the foundations of the project from the National Natural Science Foundation of China (E080703) and the project of the Heilongjiang Traffic and Transportation Department.

Conflict of interest

The authors claim no conflicts of interest.

Author details

Haitao Zhang* and Mingyang Gong

*Address all correspondence to: zht6781@163.com

College of Civil Engineering, Northeast Forestry University, Harbin, China

References

[1] Yildirim Y. Polymer modified asphalt binders. Construction and Building Materials. 2007; **21**(1):66-72

[2] Feng Q. Superfical discussion on how to increase durability of asphalt pavement of high-type highway. Northern Communications; 2008

[3] Zhao F, Zhao Y, Min Q. Pavement performance of asphalt mixture modified by rock asphalt. Journal of Central South University. 2010;**41**(6):2407-2411

[4] Mirzaei E, Motlagh A. Effect of using fibre on the durability of asphalt pavement. Civil Engineering Journal. 2016;**2**(2):63-72

[5] Zhang H, Yu J, Feng Z, Xue L, WU S. Effect of aging on the morphology of bitumen by atomic force microscopy. Journal of Microscopy. 2012;**246**(1):11-19

[6] Chen P, Jiang S, Huang X, Li H. The influence of aging factors on the performance of asphalt. Traffic Technology. 2005;(2):104-107

[7] Zhu J, Yang J. Analysis of performance and index of asphalt aging. Traffic Science and Technology. 2006;(5):86-88

[8] Shi H, Xu T, Zhou P, Jiang R. Combustion properties of saturates, aromatics, resins, and asphaltenes in asphalt binder. Construction and Building Materials. 2017;**136**:515-523. DOI: 10.1016/j.conbuildmat.2017.01.064

[9] Lee S, Amirkhanian S, Kim K. Laboratory evaluation of the effects of short-term oven aging on asphalt binders in asphalt mixtures using HP-GPC. Construction and Building Materials. 2009;**23**(9):3087-3093. DOI: 10.1016/j.conbuildmat.2009.03.012

[10] Zeng W, Wu S, Pang L, Chen H, Hu J. Research on ultraviolet (UV) aging depth of asphalts. Construction and Building Materials. 2018;**160**:620-627

[11] Xu X, Yu J, Zhang C, Cao Z, Gu Y, Xue L. Effect of reactive rejuvenators on structure and properties of UV-aged SBS modified bitumen. Construction and Building Materials. 2017;**155**:780-788. DOI: 10.1016/j.conbuildmat.2017.08.068

[12] Zhang F, Yu J. The research for high-performance SBR compound modified asphalt. Construction and Building Materials. 2010;**24**(3):410-418. DOI: 10.1016/j.conbuildmat.2009.10.003

[13] Jiang K. Experimental Study on Water Stability and High Temperature Stability of SBS Modified Asphalt Mixture. Lanzhou University of Technology; 2012

[14] Wang H, Dang Z, Li L. Analysis on fatigue crack growth laws for crumb rubber modified (CRM) asphalt mixture. Construction and Building Materials. 2013;**47**(5):1342-1349

[15] Wang X, Qiu Y, Xue S. Study on durability of high-modulus asphalt mixture based on TLA and fibre composite modification technology. International Journal of Pavement Engineering. 2016:1-7. DOI: 10.1080/10298436.2016.1224411

[16] Zhang F, Hu C, Zhang Y. Research for SEBS/PPA compound-modified asphalt. Journal of Applied Polymer Science. 2018;**135**. DOI: 10.1002/app.46085

[17] Cheng Y, Zhu C, Tan G. Laboratory study on properties of diatomite and basalt fiber compound modified asphalt mastic. Advances in Materials Science & Engineering. 2017; **2017**(3):1-10. DOI: 10.1155/2017/4175167

[18] Guoc Q, Li L, Cheng Y. Laboratory evaluation on performance of diatomite and glass fiber compound modified asphalt mixture. Materials & Design. 2015;**66**:51-59. DOI: 10.1016/j.matdes.2014.10.033

[19] Huang W, Gao C, Li K. Study on influence factors of fatigue properties of rubber asphalt mixture. Journal of Tongji University (Natural Science Edition). 2009;**37**(12):1608-1614

[20] Zhang D, Zhang H, Shi C. Investigation of aging performance of SBS modified asphalt with various aging methods. Construction and Building Materials. 2017;**145**:445-451. DOI: 10.1016/j.conbuildmat.2017.04.055

Chapter 13

Bitumen and Its Modifier for Use in Pavement Engineering

Mehrdad Honarmand, Javad Tanzadeh and Mohamad Beiranvand

Additional information is available at the end of the chapter

http://dx.doi.org/10.5772/intechopen.82489

Abstract

This chapter focuses on bitumen specifically. This chapter consists of several parts that can be mentioned, including the history of the appearance of bitumen and the types of constituent elements, as well as its mechanical properties and chemical structure and its thermal sensitivity. In all parts, the effects of bitumen on asphalt are discussed. In the following sections, the bitumen modification mechanism, polymer modifiers, and their behavior on the bitumen resistance to asphalt failures are also discussed. This chapter is very suitable for students and researchers interested in improving polymerization asphalt and bitumen and will help them to carry out research and concepts.

Keywords: bitumen, modify, asphalt, polymer

1. Introduction

Traffic loading and environmental factors are one of the most important destructive causes in asphalt pavement. Traffic loading leads to breakdowns such as rutting and fatigue cracks, and environmental factors such as temperature are the main cause of the refrigerant cracks. Traffic loading can cause tensile, compression, shear stresses, or a combination of them in different pavement points, depending on factors such as load size, contact surface, temperature, hardness, and pavement thickness.

Typically, the repetition of these stresses and strains leads to damage the pavement. Fatigue cracks are, in fact, fine cracks, which increase with the continuity of loading in the pavement system and ultimately expand into fatigue cracks. The accumulation of these cracks

IntechOpen

© 2019 The Author(s). Licensee IntechOpen. This chapter is distributed under the terms of the Creative Commons Attribution License (http://creativecommons.org/licenses/by/3.0), which permits unrestricted use, distribution, and reproduction in any medium, provided the original work is properly cited. (cc) BY

eventually disrupts the pavement. Therefore, the ability to predict the behavior of pavement against the phenomenon of fatigue is important. Since the fatigue phenomenon occurs more in the bitumen phase of the asphalt mixture [1], identifying the structure of the bitumen in the asphalt is very important to develop durability and life span.

1.1. History of bitumen and its reformation

The Sumerians, the Assyrians, and many earlier civilizations have used bitumen widely. The beginning of the modern bitumen industry can be attributed to 1712 when bituminous stones were discovered in France. At that time, the bitumen was simply distributed as the clod on the surface of the local roads, and under load traffic, they were rub and consolidated. This technique was quite successful, and shortly afterward, improvements were made in the form of powdering and warming the material before using. Then they, slamming and flattening the asphalt, tightened and consolidated the asphalt, known as compressed rock asphalt, and used on the streets of Europe. Such streets than earthy roads were more durable, healthier, and interesting. Their only drawback was that they were loose, volatile, and slippery during heavy traffic [2].

The first bitumen reformation dates back to 1843. In the 1930s, experimental bitumen refinement project experiments were carried out in Europe, and in the 1950s, the use of Neoprene Latex, as bitumen modifier in North America, was begun [3]. In 1963, the first experimental modified bitumen roads in France were made in order to understand the behavior of modified bitumen with various types of natural and synthetic rubbers [4].

In the late 1970s, Europe was better than the United States in the process of refining bitumen. One of its reasons was the requirement for European contractors to provide a guarantee for the durability and shelf life of the pavement, which would have reduced the costs of the lifespan of the road, even at the expense of increasing initial costs. The relatively high initial cost in performing refurbished asphalt had limited its consumption in the United States. In the mid-1980s, European technology introduced new polymers that increased the consumption of polymeric bitumen in the United States [3].

1.2. Bitumen sources

The bitumen used in the road construction industry is divided into two general categories:

1. Bitumen (mineral pitch)
2. Bitumen produced from distillation of crude oil (petroleum pitch)

1.2.1. Bitumen (natural pitch)

When volatile crude oil components is vaporized in the depths of the earth, over time and under factors such as high temperature and pressure, the black substance remains in place, which is called natural bitumen. This type of pitches is not usually pure and is a mixture of bitumen and mineral materials.

1.2.2. Petroleum pitch

Oil pitch is obtained from the refinery of crude oil in distillation towers. In fact, what remains at the bottom of the distillation tower above 380°C is pure oil pitch. Bitumen with different stiffness degrees for different road applications can be obtained by setting the temperature and pressure inside the distillation towers [5].

1.3. Chemical bitumen building

Bitumen is a very complex chemical compound, composed of various hydrocarbons types, which are in colloidal form.

The chemical structure of bitumen is not fully known, and the chemical composition of the bitumen and its structure is heavily dependent on the initial crude oil from which it is derived and the processes applied for its production. Chemical compositions and crude oil structure are also very diverse. In this regard, considering the type of predominant hydrocarbons in crude oil, it can be divided into three groups of paraffins, asphaltenes, and naphthenes (**Table 1**).

As mentioned, the chemical structure of the pitches differs according to the location and source of the crude oil. The operation and production process in the refinery (such as aeration) also influence the chemical structure of bitumen.

Generally, the components constituting pitch include carbon, hydrogen, sulfur, oxygen, nitrogen, and some metallic elements found in very small percentage. The percentage of each element is given in **Table 2**.

Hydrocarbons	Paraffinic crude oil	Naftens crude oil	Asphaltenes crude oil
Paraffins	40	12	5
Naftens	48	75	15
Aromatics	10	10	20
Asphaltenes	2	3	60

Table 1. Chemical structure of different types of crude oil.

Elements	Percent
Carbon	82–88
Hydrogen	18–11
Sulfur	0–6
Oxygen	0–5/1
Nitrogen	0–1

Table 2. Bituminous ingredients.

1.3.1. Components of bitumen

Bitumen consists of various hydrocarbon compounds, which according to the method of separation its compounds can be divided into different chemical groups. Bituminous components can be divided into two distinct chemical groups of asphaltenes and maltenes (petrolenes). Maltenes are also classified into groups of aromatic compounds, polar aromatics (resins), and saturated aromatics (paraffins). These four groups are not always consistent, and there is some interference between them.

To investigate the chemical properties of bitumen, it must first be separated from each other. There are several methods for separating and identifying the quantitative and qualitative chemical bitumen composition. For this purpose, separation using the chromatographic column is the most common method. Since bitumen components have different degrees of solubility; this separation can be done using appropriate solvents with different abilities associated with chromatographic column. Each group has different physical and chemical properties and has independent and separate effects in relation to the physical and chemical properties of bitumen, which resulted from the total influences of these components, according to their quantitative ratios.

1.3.1.1. Asphaltenes

Asphaltenes are a brittle solid substance in black to brown; in addition to carbon and hydrogen, they contain nitrogen, sulfur, and oxygen and usually have very polar compounds that contain highly aromatic complex components with high molecular weight. Asphaltenes comprise 5–25% the composition of bitumen, and in terms of building, a combination of compressed aromatic rings with aliphatic lateral chains and the number of aromatic rings is 6–20 and even sometimes more of this.

Asphaltene rate in bitumen has many effects on rheological properties of bitumen. The higher the bitumen asphalt, the more the stiffness of the bitumen, and subsequently the degree of penetration would be less. In addition, it will result in a greater softening point and more penetration of bitumen.

1.3.1.2. Polar aromatics (resins)

These compounds are usually composed of carbon and hydrogen, which also contain a small amount of oxygen, nitrogen, and sulfur. The compounds are solid and semisolid and in dark brown color. They are anticoagulants for asphaltene, and the ratio of the resin to the asphaltenes may be as high as that from the sol[1] or gel[2] mode. The resins in bitumen have a molecular weight of 500–50,000. The adhesive properties of bitumen result from its resins.

[1] Mode (sol): in this case, the particles are dispersed uniformly and homogeneously in solution. The pitches obtained from the refinery are usually of this category.

[2] Mode (gel): in this case, colloidal pitch particles create a regular sponge-like network that propagates throughout the system. In this case, the system is semisolid, and this is where the material constituting the maltene in terms of chemical compositions has the most properties of oil or paraffin hydrocarbons.

1.3.1.3. Oil aromatics

The aromatic oils are the least molecular weight ring compounds that contain aromatic or petroleum lateral chains and are an excellent anticoagulant for asphaltenes. The group, which consists of 40–60% bitumen, is a viscous liquid in dark brown color with an average molecular weight between 300 and 2000. The aromatics have polar carbon chains.

1.3.1.4. Saturated materials (paraffins)

Saturated compounds are composed of straight and branched chain aliphatic hydrocarbons with alkyl naphthenes and some aromatic alkyls. These compounds are often nonpolar and colorless viscose oils. The average molecular weight of these is similar to that of aromatics. The components of the saturated compositions include saturated waxy and non-wax[3] materials. Saturated compositions comprise 5–20% bitumen.

1.4. Mechanical and physical bitumen properties

These two properties of bitumen are not only important during the fabrication, distribution, and density of asphalt mixtures, but these properties are also very important during servicing. Considering the mechanical and physical properties of bitumen along with its chemical properties can be one of the factors for achieving durable asphalt pavement.

1.4.1. Bitumen behavior

Bitumen from the behavior viewpoint is a viscoelastic substance that in the environment heat and under the specified load neither the behavior of an elastic material nor the behavior of a viscous material but is a combination of these two states, which is viscoelastic. The pure bitumen behavior depends on, due to its viscoelastic nature, the temperature and loading time. In other words, the temperature and time of loading have a combined effect on bitumen behavior.

In high-temperature conditions or long loading times, bitumen behaves as a viscous material (such as moving trucks on the asphalt at low speed or at a stop). Under these conditions, asphalt mix materials tolerate the load. At low temperatures or in fast loading mode (such as fast traffic motion), pure bitumen behaves like an elastic solid. In the elastic behavior resulted from the loading, the deformation is achieved, and the made change is brought back to its original state by loading. Of course, if it is more than a power or compressive resistance, it is possible to break the elastic material. Even the pure bitumen has a low elastic behavior at a low temperature, but when it is overloaded, it would be brittle and fragile, leading to the cracking. Therefore, in cold weather at low temperature, a cracking in the in asphalt pavements may occur.

In ambient conditions, between very low cold temperatures and very high warm temperatures, i.e., the average temperature of bitumen, it has the behavior of viscous fluids and the

[3] Wax: part of the paraffins is crystalline. The method for determining the amount of bitumen wax is given in DIN 52015.

behavior of plastic solids simultaneously. Bitumen, in this case, has either viscous behavior or elastic behavior, depending on the temperature and loading time.

In environmental conditions, between high cold temperatures and high heat temperatures, i.e., the average temperature of bitumen, the behavior of viscous fluids, and the behavior of plastic solids are simultaneously obtained. In this case, bitumen has both viscous and elastic behaviors depending on the temperature and loading time.

1.4.1.1. Newtonian behavior of bitumen

When there is a linear relationship between shear stress applied to fluids and velocity gradient or shearing strain rate, they are called Newtonian fluids. Road bitumen often shows Newtonian behavior at high temperatures. **Figure 1** shows stress ratio with strain rat that is a feature of Newtonian materials.

Another condition required for a Newtonian material is that it does not show any elastic behavior. For that bitumen show elastic behavior at low temperatures; they cannot be classified as Newtonian materials, although they display Newtonian flow behavior (n = 1) [6].

1.4.1.2. Pseudo-plastic behavior and Bingham plastic

A type of behavior which results in shear thinning of fluid (decrease of viscosity with increase of stress or reduce of strength with increase of shear stress) is called pseudo-plastic. Fluid which shows solid behavior under stresses less than yield stress and shows fluid behavior under stresses more than yield stress; Bingham fluid and this behavior are said Bingham plastic that in this case (1 = n). Bitumen, which in the low temperatures and the low shear stresses (low shear velocities) have been tested, generally displays pseudo-plastic behavior or Bingham plastic behavior.

Figure 2 shows that the shear velocity in high stresses is increased faster that leads to reduce of viscosity. When (n < 1), the bitumen is sensitive to stress and acts as a pseudo-plastic material. Blown bitumen as roof coverings has a pseudo-plastic behavior. At low temperatures, often the material displays one special yield point, which is seen in Bingham plastic behavior

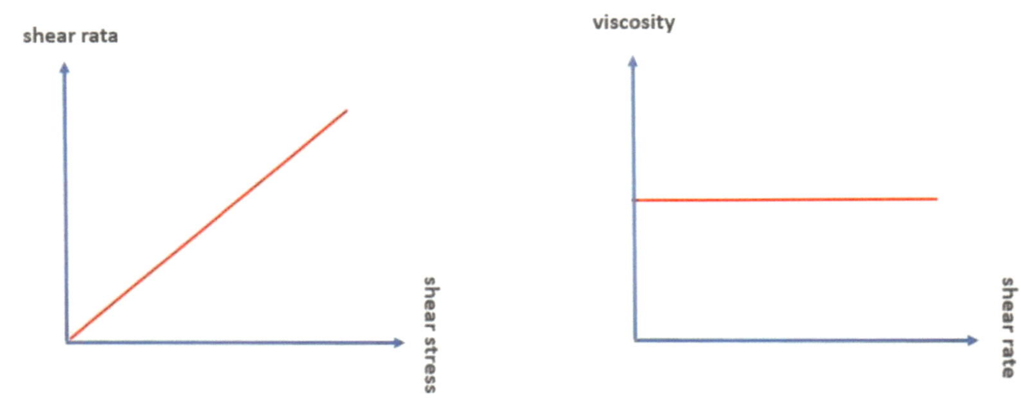

Figure 1. Newtonian behavior [6].

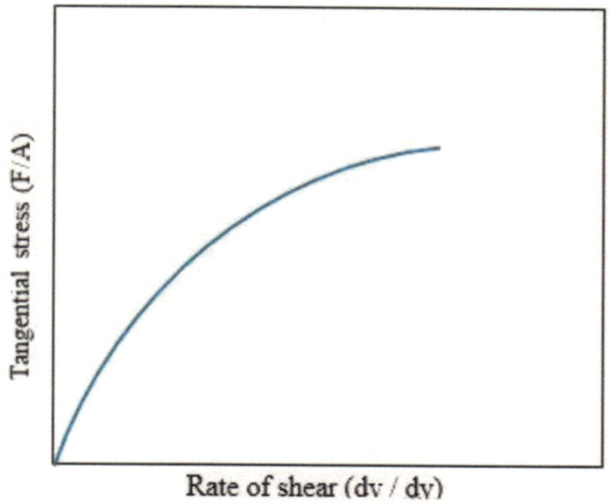

Figure 2. Pseudo-plastic behavior [7].

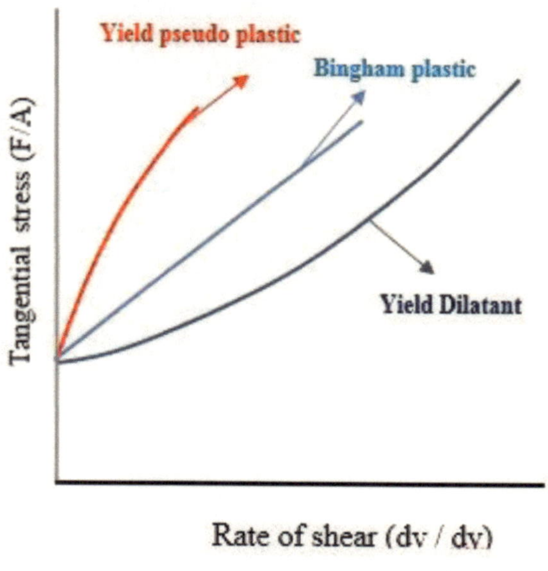

Figure 3. Bingham plastic behavior [7].

in **Figure 3**. All bitumen at very low temperatures display the behavior of a friable elastic solid with flow potential or very low creep. Usually bitumen is considered as glass solids at temperatures below the glass transition temperature (Tg)[27] [7].

1.4.2. Bitumen temperature susceptibility

Temperature susceptibility which may be defined as how changes made in consistency or stiffness of bitumen (viscosity or degree of penetration) at different temperatures is one of the effective parameters in the behavior of bitumen that for a variety of bitumen is classified in one group (classification based on viscosity or classification based on degree of penetration) may vary. For example, the classification based on the degree of penetration, some bitumen that have the same degree of penetration at temperature of 25°C may not be the same degree of penetration at the other temperature. Also in terms of classification based on viscosity, several asphalts may have the same viscosity at the temperature of 60°C, but their viscosity or stiffness is not the same at higher or lower temperatures; therefore, their behavior at times of diffusion, density, and service will be different. This can affect the durability and stability of asphalt mixtures. The greater the changes of stiffness or consistency with changes of temperature, the higher temperature susceptibility in bitumen, and the lower the change, lower temperature susceptibility in bitumen. Bitumen with temperature susceptibility will have better performance in asphalt mixtures.

1.4.3. Bitumen hardening

Bitumen hardening is a phenomenon that may occur under various conditions and factors in the short or long terms. The reason for this is the rapid or gradual changes in the shape and chemical compounds of bitumen due to various factors. Different processes can interfere with the chemical changes of bitumen, among which can be indicated to sublimation of light materials and escape of bitumen, oxidation, polymerization, carbonization, the absorption of bitumen oils by aggregates, the chemical reaction between bitumen and mineral components of aggregates, and so on.

All abovementioned processes are dependent on environmental conditions, temperature, and thickness of bitumen lining on aggregates in asphalt mixtures. Bitumen hardening in asphalt occurs in two stages: one is short term and fast, and the other is long term and gradual. The hardening stage occurs in the short term with high speed and intensity over a short period during heating bitumen and mixing it with rocky materials at high temperatures. The second stage of hardening of the bitumen is gradually and slowly made during service. The results of the physical hardening of the bitumen reduce the degree of penetration, increase the softening point, reduce the plasticity, increase viscosity, decrease adhesion property, and increase the bitumen fragility property. The main factor in the hardening of bitumen is the changes, which occur in the chemical components of bitumen during heating the bitumen, mixing it with stone materials and at the time of service due to the abovementioned factors. In **Figures 2–4**, the rate of change in each of the chemical components of bitumen has been shown in each of the different stages of the pavement life. As it is seen in the figure, the amount of asphalt gradually increases, while the amount of resin and aromatic compounds decreases; as a result, at these stages, the aging index (the ratio of the viscosity of bitumen available to the initial bitumen) increases. The aging index in these shapes is the ratio of viscosity of bitumen extracted from samples to the initial bitumen at 25°C. Investigations show that the most changes in the bitumen viscosity occur during mixing and diffusing the mixture, and after that, the bitumen viscosity changes are insignificant in the period of the service. Regarding the chemical compositions of bitumen, asphalt is increased while mixing and density as shown in **Figure 4**, and this process continues in the period of the service. Resin and aromatics are decreased in the course of time. The tested samples have been gotten with a thickness of 3 mm more than the samples taken from the pavement.

http://dx.doi.org/10.5772/intechopen.82489

1.4.4. Bitumen aging

The gradual changes in the physical and chemical properties of oil bitumen have been accepted as a principle. Researchers believe in this field that the most important factor of these changes is the phenomenon of aging hardening that causes the appearance of various types of damages in the asphalt. Various and different processes interfere with the gradual changes or the bitumen hardening such as oxidation, photochemical reactions, sublimation of bitumen light weight and volatile materials, polymerization, carbonization, absorption of oils in bitumen by aggregates, chemical reactions between bituminous components and mineral compounds of rocky materials, and microbiological transformations. All of the processes mentioned are dependent in time, ambient temperature, heat, thickness, and membrane around aggregates. For example, the rate of oxidation, which is the most important factor of the hardening of the materials, which it doubles with an increase 10 at 100°C. Physical hardening is another factor of aging or usual hardening in bitumen. Physical hardening occurs when bitumen is at ambient temperature and is usually attributed to the rearrangement of bitumen molecules and slow crystallinity of waxes. Of course, physical hardening is reversible, and the initial viscosity of bitumen is obtained by heating. **Figure 5** shows the dependence of viscosity in temperature

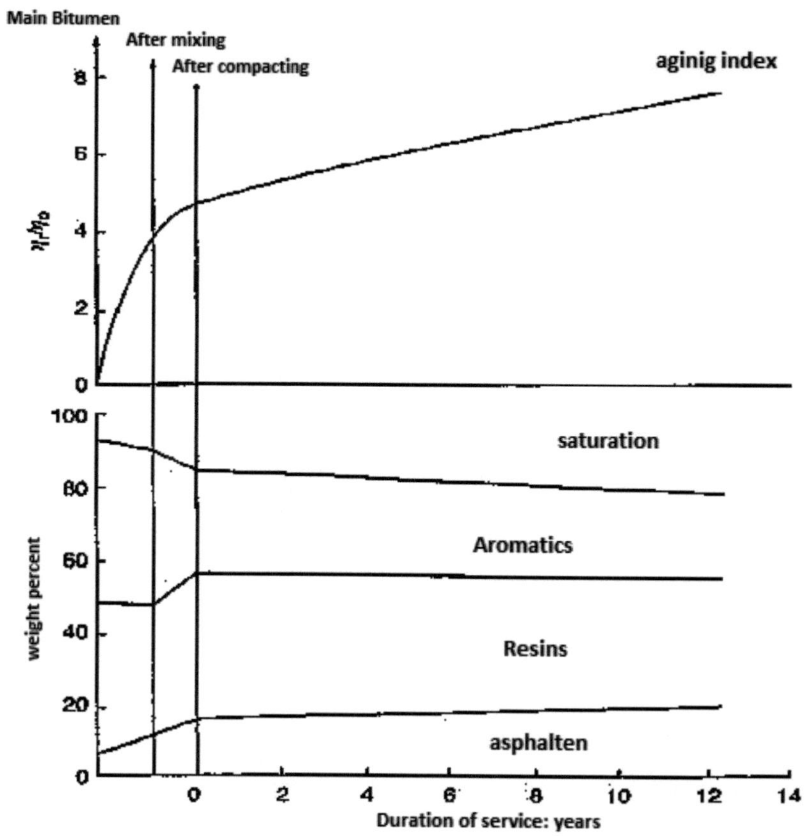

Figure 4. The changes of bitumen components during mixing, diffusing, compacting, and serving.

for two samples of bitumen. These two figures also show that all bitumen are not hardened equally or their physical and chemical properties do not change with the same rate.

High and long-term studies have been done on bitumen used in making the road in order to determine the chemical compositions of bitumen changed over time; the results of this research were definition of the aging index of bitumen:

$$\text{Aging index} = \eta_r/\eta_o \quad (1)$$

The aging index is calculated according to the ratio of the viscosity of bitumen recycled (η_r) to the viscosity of main bitumen (η_o) at a temperature of 25°C.

1.5. Bitumen modification

As mentioned before, there is a complex relation between bitumen chemical structure, bituminous colloid structure, and its physical and rheological properties. Any material which changes the chemical structure of bitumen consequently changes the properties of bitumen, and as a result, it can be a modifier [4].

Ideally, modified bitumen has more adhesion than pure bitumen and lower thermal sensitivity in the range of service temperatures and sufficient viscosity at execution temperature. In addition, its sensitivity to the time of loading is low, and its strength to plastic deformation, fatigue, and cryogenic cracks is high. Eventually, its properties after aging are good for execution and service [4].

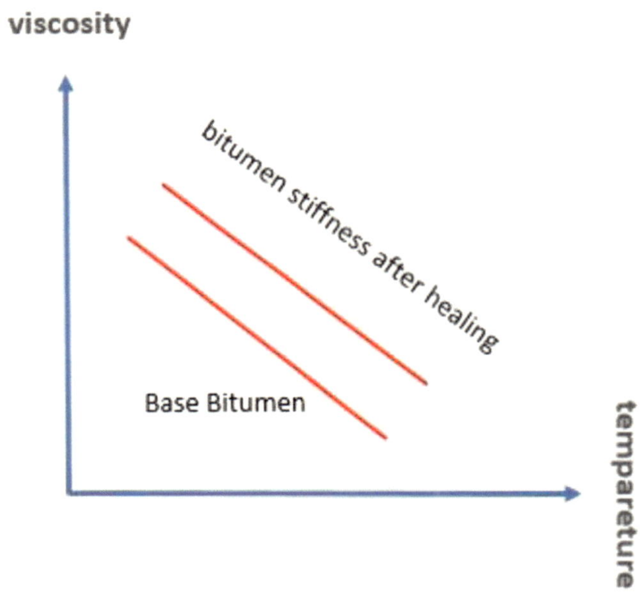

Figure 5. Dependence of viscosity in temperature for two bitumen samples.

Thermoplastic elastomers	Styrene butadiene diblock (SB)
	Styrene butadiene triblock/radial (SBS)
	Styrene isoprene (SIS)
	Styrene ethylbutylene styrene (SEBS)
	Styrene butadiene rubber latex (SBR)
	Polychloroprene latex
Plastomers	Ethylene vinyl acetate (EVA)
	Ethylene propylene diene monomer (EPDM)
	Polyisobutylene
	Polyethylene (low density and high density)

Table 3. Types of thermoplastic elastomers and common plastomers in rendering bitumen [12].

1.5.1. Bitumen modification by polymer

By adding polymers, the chain of small molecules is repeated, and as a result, the pavement performance improves. Polymer-modified bitumen increases strength against rutting and fatigue and cryogenic cracks and reduces damask and thermal sensitivity. So polymer-modified bitumen are used successfully in places with high stress such as intersections, airports, truck weighing areas, and race routes. The positive properties of polymeric bitumen include increase of elastic recovery, viscosity, softening point, adhesion, and flexibility [3].

Polymers used to accommodate bitumen are usually in three forms: thermoplastic elastomers, plastomers, and reactive polymers [8, 9]. Thermoplastic elastomers usually have more elasticity against pavement at low temperature, while reactive polymers and plastomers cause the hardening and strength to be increased against permanent deformations. It should be considered that reactive polymers react with bitumen due to the presence of operating groups and show greater compatibility than the other two polymers [8–11].

Experience shows that only a few polymers have desirable compatibility with bitumen. Industrial polymers used to modify bitumen are usually divided into two general categories:

- One of the most common types of plastomers is copolymer ethylene vinyl acetate (EVA) in which its category is determined according to the percentage of vinyl acetate and its molecular weight (determined in terms of the melting temperature) [4].
- One of the most common types of thermoplastic elastomers is copolymer styrene-block styrene-butadiene (SBS). These polymers are classified according to percentage of styrene, molecular weight, and structure (linear and radial) [4].

Types of elastomers and plastomers common in modifying bitumen are shown in **Table 3** [12].

The result of the mixing of bitumen with a polymer of thermoplastic elastomers at high temperature can be in three types:

- Heterogeneous mixture: in this type of mixture which is most likely, polymers and bitumen are incompatible and cause the separation of mixed components. This bitumen mixture does not have any proper properties for pavement application [4].
- Homogeneous mixture: this mixture is completely homogeneous to the molecular level in which bitumen and polymer are completely compatible. The oil in the bitumen completely resolves the polymer and eliminates any interaction at inter-micromolecular level. In such a situation, the bitumen is completely stable, and its properties change than the base bitumen quite low, in which the viscosity is increased. As a result, such a situation is not desirable [4].
- Micro-heterogeneous mixture: this mixture contains two distinct parts, which are simply locked and clamp together. Such a level of compatibility is desirable and improves the properties of bitumen. In this status, the polymer particles that are compatible absorb the oil-based phase of the bitumen, and they swell to create a polymer phase distinct from residual bitumen components (consisting of heavier components of bitumen such as oil, resin, and asphaltene) [4].

1.5.2. The mechanism of bituminous modification with a polymer

As described below, three general statuses can be defined for the mixing of bitumen and polymer:

- The first status—low polymer content (less than 4%). In this situation, the bitumen is the base phase, and the polymer phase is dispersed in it. In this status, in order to maintain the stability during the storage, the polymer should absorb a significant amount of light materials and volatile asphalt; thus in comparison to its initial volume, it swells five to ten times. When the polymer absorbs the bitumen oil, the asphaltene percentage in bitumen increases, so the adhesion and the elasticity of bitumen increase. At a high service temperature (about 60°C), the polymer phase hardness modulus is higher than the bitumen matrix modulus, which enhances the mechanical performance of bitumen at that temperature. At low temperatures, the polymer hardness modulus is less than bitumen, which results in a reduction in the brittleness of modified bitumen. Therefore, it can be concluded that the polymer dispersed in bitumen (discontinuous phase) improves the properties of bitumen at high and low temperatures. In this status, the properties of the base bitumen significantly affect the modified bitumen properties [4].
- The second status—the polymer content is about 5%: in this status, two-phase, continuous, and interconnected microstructures are created. Controlling such a system is often difficult, and there is a problem of instability in it. In this method, the micromorphology and the properties of bitumen are usually dependent on the bitumen temperature history. In this status, as the percentage of polymer increases, the softening point increases significantly [4].
- The third status—high polymer content (more than 7% if suitable bitumen and polymer are selected): in this status, the polymer is the dominant phase and forms a continuous matrix

in the system. In fact, this system is no longer bitumen, but it is a polymer plasticized with bitumen oil in which a heavy phase of bitumen is dispersed. The properties of such a thermoplastic adhesive material are essentially different from those of bitumen, and in fact, it has the polymer properties. In this status, the softening point, which in the previous states increased with the increase of the polymer content, is fixed and may experience a minor change with the increase of polymer content [4].

The microstructure of the polymeric bitumen is very important, and as described above, there is a close relationship between polymer bitumen microstructure and its physical properties. Studies have shown that in the same percentage of bitumen and base bitumen with the same grade but different sources, polymer bitumen can have a very different microstructure and properties, especially at low temperatures [4].

1.5.3. Polymer bitumen stability

Sustainability is one of the most important issues in polymer-modified bitumen. As this type of bitumen has two separate phases, it will be studied under Stokes' law. In other words, the speed of dispersion of particles (polymer particles in a bitumen matrix or bitumen in a polymer matrix) increases with the increase of the particle size, with an increase of the difference between the density of the two phases, and with the decrease of the viscosity of the continuous phase [4]. In order to increase the stability of polymer-modified bitumen, it is necessary to control the particle size by controlling the production process. Furthermore, by controlling the chemical structure of the base bitumen and the polymer-modified bitumen, it is possible to move toward the equilibrium of the density of the two phases. For example, the stability of EVA-modified bitumen is highly dependent on the percentage of asphaltene in the base bitumen. This process is shown in **Figure 6**. In this figure, by "sustainability" parameter, the amount of delamination is meant. When this parameter is zero, it means no delamination occurred [4].

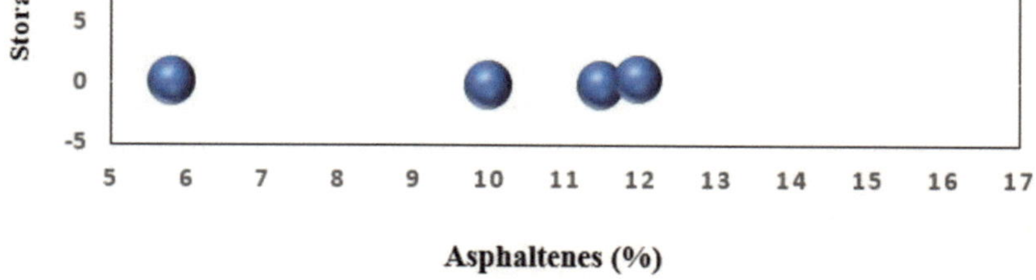

Figure 6. The effect of asphaltene content in the base bitumen on the stability of EVA polymer-modified bitumen [4].

1.5.4. Temperature history

A study in the LCPC laboratories in France showed that the formation of the continuous polymer phase in polymer-modified bitumen improves the properties of polymer bitumen, such as plasticity at low temperature. On the other hand, in polymer-modified bitumen, the plasticity of the continuous polymer phase is increased with the increase of the cooling rate in the laboratory. The polymer-modified bitumen, which was cooled at the same rate as the real conditions of the runtime (about 30°C/h), had a different structure. In this status, the bitumen had the continuous phase, and the polymer had a discontinuous and dispersed phase. This experiment showed that in the case of rapid cooling of the polymer-modified bitumen, the physical properties of polymer-modified bitumen decrease significantly so that the softening point dropped to 30°C [4].

1.6. Common polymers in bitumen modification

As shown in **Tables 2** and **3**, the different types of polymers are used in bitumen modification, each of which is different in terms of modification mechanism and properties. In this section, some of the most common polymers that are used in the modification of bitumen properties have been reviewed.

1.6.1. Styrene butadiene rubber (SBR)

SBR is a type of copolymer, in which styrene and butadiene monomers are irregularly bonded in a polymeric chain, based on their initial percentages. This copolymer has good strength due to the presence of thermoplastic styrene monomers between the layers of butadiene rubber monomers, which is why it is used extensively in the tire industry. Unfortunately, this rubber inflated in bitumen to a limited extent is not compatible with bitumen, so it cannot significantly improve the properties of bitumen [13–15].

1.6.2. Styrene butadiene styrene (SBS)

SBS is a block copolymer that increases the bitumen elasticity. This polymer is one of the most suitable bituminous modifying polymers in terms of improving the properties of bitumen, but it has some economic and technical limitations. However, after the rubber powder, SBS is the most widely used modifier in pavement [3]. The morphology of the SBS and bitumen mix may vary depending on the source of the base bitumen and the polymer. In one of the bitumen statuses, the phase is continuous, and the SBS particles are dispersed in bitumen. In the other statuses, the bitumen cells exist in the polymer as a continuous phase. In the third status (high polymer content), bitumen and polymer continuous phases interlock. In this status, a critical network is created between bitumen and polymer that increases the complex shear modulus of the mixture (G*) and thus increases the resistance to the rutting [3].

In super PIO experiments on bitumen containing 1–3% SBS, this modifier improved the high-temperature performance grade, but the intermediate-temperature performance grade remained relatively constant relative to the base bitumen. Performance tests at low temperatures by Bending Beam Rheometer (BBR) showed that this amount of SBS reduced the hardness range temperature, but due to a decrease in the creep rate (m-value), resulting in an increase in the creep rate temperature, the low temperature performance grade in these types of bitumen got worse [16]. Other research have shown that using a higher percentage SBS, it is possible to improve the low- and intermediate-temperature performance grades [17]. According to the definition of the BBR test, the hardness range temperature is a temperature at which the stiffness (bitumen hardness modulus) reaches 300 MPa and the temperature is the creep rate temperature range is a temperature at which the value of M-Value is less than 0.3.

The studies showed that in bitumen containing 5% of SBS, the property-controlling factor is the molecular weight of the polystyrene and polybutadiene blocks. On the other hand, for viscosity at 180°C, the molecular weight of polybutadiene is more effective than polystyrene. It was also observed in high-SBS-containing bitumen that the oil content of the bitumen has a determining effect on the degree of penetration and plasticity point of polymer-modified bitumen [4].

Mohammad et al. when studying samples taken from SBS-modified 8-year-old bitumen pavements in Louisiana observed hardening resulted from oxidation. The hardness of the samples at low and high temperatures was more than it was expected. By performing the test, they found that the effect of extraction and reduction of bitumen on the quality of bitumen containing SBS was very insignificant. The experiments showed that in asphalt recycling operation, increasing the percentage of reduced bitumen containing SBS increased rutting resistance, while it reduces fatigue-cracking resistance [18]. This behavior can be due to the increased hardness and brittleness of the aging bitumen containing SBS. Other findings from this study were the differences between the effect of natural aging and the effect of Pressure Aging Vessel (PAV) test on SBS-containing bitumen. Differential scanning calorimetry (DSC) analysis of bitumen showed that the aging process of bitumen tested with PAV reduces the number of paraffinic crystalline structures in bitumen, while natural aging in a sample extracted from the road increases the paraffinic crystalline structures of the bitumen, thereby in comparison with PAV tested bitumen, its hardness and brittleness increase [18].

1.6.3. Ethylene vinyl acetate (EVA)

Figure 7 shows that by increasing the percentage EVA, the degree of penetration decreases, and the softening point increases. This effect is more evident in EVA-containing less vinyl acetate (VA) [4]. In another study by Madela et al., it was found that in bitumen with a low percentage of EVA (bitumen matrix is dominant), the chemical structure of bitumen plays a determining role in the physical properties of polymer-modified bitumen. At high percentages of EVA (polymer matrix is dominant), it significantly reduces the effect of bitumen chemical structure [4].

1.6.4. Modification of bitumen with rubber powder (CRM)

The US Federal Highway Administration (FHWA) analyzed the life-cycle cost of the pavement, including rubber powder and other modifiers and showed that the use of rubber

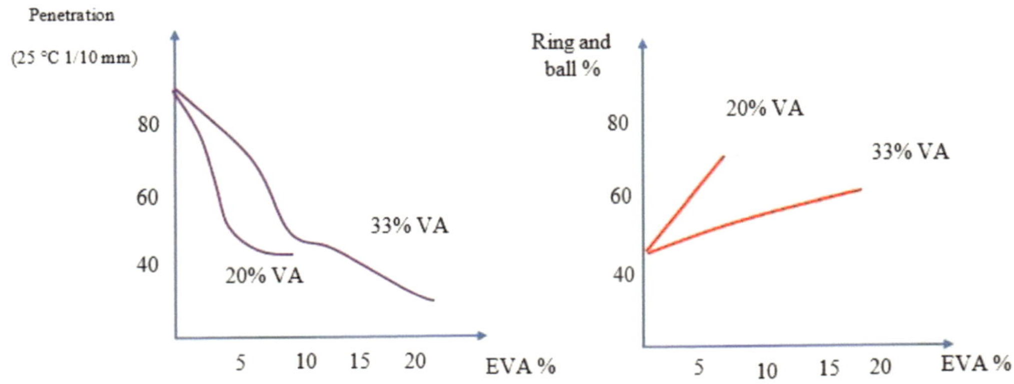

Figure 7. The effect of percentage changes of EVA in bitumen and vinyl acetate, on the degree of penetration and softening point of polymer-modified bitumen [18].

powder to modify bitumen is economical. In 1991, a US law was approved requiring the government from 1994 to use rubber powder (5%) in roads constructed with the support from the state budget. This amount should be increased to 20% by 1997. Of course, this law was later changed. Typically, rubber is made from the recycled tires. This reduces the space needed to burrow worn tires [3].

Modifying bitumen with natural rubber increases the resistance of asphalt mixture to rutting and improves its flexibility. On the other hand, it tends to be delaminated. Due to the higher molecular weight of rubber powder, this material is less compatible than bitumen. Another common problem with the use of natural rubber is the need for a higher temperature and a longer time to mix and disperse rubber in bitumen [3].

Although much research has been done about the nature of the reaction of rubber and bitumen particles, the exact mechanism of this operation is still unknown. According to the opinions of many researchers, after mixing melt pure bitumen with rubber powder, rubber powder's polymer chains absorb aromatic compounds of pure bitumen. Powdered rubber particles become soft and swollen that lead to an increase in the viscosity of the rubber-bitumen blend [19].

This swelling occurs because of the physical and chemical reaction of bitumen and particulates of rubber powder that the volume of powdered rubber particles rises two to three times the initial value. If the temperature or mixing time is high enough, the particles of the rubber are degraded and dispersed in the bitumen, and the polymeric chains are broken, which reduces the viscosity of the mixture. The swelling of powdered rubber particles and the particle breakdown of powdered rubber are two factors affecting the properties of rubber-bitumen mixtures [20].

Adding rubber powder to bitumen can increase one to three performance grades (PGs) above the operating temperature and reduce one to two performance grades to lower temperatures which is the lower temperature for the operating temperature of bitumen. Therefore, the operating temperature range of bitumen rises from both sides [21]. Other researchers also observed the same trend at high operating temperatures but did not see a change in the performance grade with the increase for rubber [22–24]. The lower limit in all cases was controlled by the value of the creep rate (m-value) which is obtained from the Bending Beam Rheometer (BBR) test. The increase in the upper limit was due to an increase in the value of the parameter G*/Sinδ from the dynamic shear rheometer test (DSR) [22, 24]. Overall, according to the results of various studies, it can be concluded that the rubber powder improves the performance grade but in the low performance, the amount of improvement is highly dependent on the bitumen, and the results are different depending on the bitumen used.

DSR testing has shown that G* is highly dependent on temperature, type of bitumen, and the percentage of rubber powdered. The effect of the rubber powder is more than other factors, while the particle size of the rubber does not affect it. In δ changes (phase angle), the aging index plays an important role than other factors. Factors such as temperature, loading frequency in the test, type of bitumen, and the percentage of plastic powder are also effective [25].

Researchers have found that adding rubber powder reduces hardness at low temperatures. Gopal showed that this event happens for the different sizes and percentages of the rubber powder, and for each compound, the size and the optimal percentage should be determined [22, 26]. The experiments, which were conducted by Bahia et al., showed that the effect of adding rubber powder is less effective in reducing the hardness at low temperatures in soft bitumen [23]. Due to the low rubber hardness at low temperatures (0–20°C), compared to bitumen hardness, its composition with bitumen reduces the hardness of the rubber bitumen [21].

1.6.5. Bitumen modification using polyphosphoric acid (PPA)

One of the modifiers of bitumen is the polyphosphoric acid, which is added to it independently or as a substitute for other modifiers in bitumen like SBS. Polyphosphoric acid acts like deflocculant of the bitumen phases, the charge of polar groups, using neutralization of the load of polar groups. This can be done by neutralizing the bases with acid or esterification [27].

One of the important issues that should be considered in the use of acids as the modifier of bitumen is reversibility of bitumen modification. The presence of lime, limestone, or anti-stripping agents can neutralize the acid added to bitumen. On the other hand, most of the acids, including PPA, are soluble in water and may be washed out by rain over time [28].

1.6.6. Bitumen modification using basalt fiber

Basalt fiber is a kind of fiber that is made of basalt stones, melting at a temperature of about 1500°C and converting into continuous fibers. Researchers have paid more attention to this fiber because of the following reasons:

- Its good performance in strength
- Its suitability for a wide range of temperature variations
- Its durability

In recent decades, basalt fibers have been used in the asphalt concrete as a highly efficient additive. Compared to two other common additions, for the arming of asphalt, namely, polyester fibers and lignin fibers, basalt fibers have a higher tensile strength and elastic modulus and a lower elongation rate [29]. Its resistance to high temperature and good chemical stability of the basalt fiber makes it an excellent additive for asphalt concrete [30].

Different researchers have carried out various experiments to evaluate the applicability of basalt fiber as a reinforcing material for structural concrete, and they have compared the efficiency of basalt fiber and glass fibers. Using the accelerated weathering test, they found that basalt fiber provides better resistance in comparison to glass fibers [31]. Basalt fibers (especially those used with Scoria) will not be dangerous to the environment, and basalt is not usable as a building material. This material is not new but can be used initially in construction with regard to mechanical and chemical properties. The base cost of basalt fibers depends on the quality and chemical compounds, and this fiber has various types and different chemical properties [33]. However basalt fiber has properties like high Marshall stability, stability against water, and stability against rutting; fewer studies have been conducted to analyze the application of basalt fibers [34–36]. Bitumen modification using basalt fiber has the potential to perform better at low and high temperatures and to increase fatigue life and reduce permanent changes [36, 37]. The use of fibers has a great influence on viscosity and mixed properties [38]. Zhao et al. added fiber basalt (with a diameter of 16 mm) to asphalt mix and using the Marshall Test and hardness with different percentages of fibers; they showed that adding 3% weight percent of basalt fibers improves it at lower temperatures [39]. The addition of basalt fiber to the asphalt mixture increases its dynamic modulus [40]. In order to increase the resistivity of the asphalt, Xiao et al. conducted different experiments, and they concluded that the basalt fibers with different lengths and values of resistance improve the crack resistance in asphalt [41].

Synthetic basalt increases the shear modulus, and the viscosity of the bitumen, as a result, increases its stiffness and decreases the difference in the phase angle. These factors have been improved in the bitumen foam synthesized with the basalt, more than bitumen modification using synthetic basalt.

In terms of molecular weight, the lower molecular weight of isocyanate used in the synthesis of basalt is the more effective synthetic basalt that is on increasing the shear modulus and reducing the phase difference angle.

Synthetic basalt is well adapted due to its chemical reaction with bitumen polar groups and the formation of new bonds in bitumen and stability in bitumen storage, which is achieved without any problem. The temperature of mixing the synthetic basalt and bitumen is 90°C, which is more favorable environmentally.

2. Conclusion

Bitumen is one of the main components of the asphalt mixture, which plays a major role in creating adhesion between the aggregates in relation to environmental factors such as temperature and traffic loading in the asphalt mixture.

Because the vast majority of asphalt failures are due to the weakness of the rheological and chemical properties and the thermal sensitivities of the bitumen in asphalt, over time, with the development in the materials science of road building. Moreover, the use of polymer

http://dx.doi.org/10.5772/intechopen.82489

modifiers could potentially improve the bitumen in the asphalt, which has led to improve the durability of asphalt against breakdowns due to rutting, permanent deformations in the high temperature, and the increasing fatigue life.

Author details

Mehrdad Honarmand[1]*, Javad Tanzadeh[2] and Mohamad Beiranvand[2]

*Address all correspondence to: mr.honarmand@hotmail.com

1 Department of Civil Engineering, North Tehran Branch, Islamic Azad University, Tehran, Iran

2 Department of Civil Engineering, Bandar Anzali Branch, Islamic Azad University, Bandar Anzali, Iran

References

[1] Bahia HU, Zhai H, Zeng M, Hu Y, Turner P. Development of binder specification parameters based on characterization of damage behavior. Journal of the Association of Asphalt Paving Technologists. 2001;**70**:442-470

[2] Bahramnejad Dariush, Riazi Mohammad Reza, Bahram Secretary, Nabibi Ali Ehsan. The possibility of mixing Iran's base beams with SBS family polymers and studying their physical-mechanical and rheological properties for road building applications [master's thesis]; 1372

[3] Yildirim Y. Polymer modified asphalt binders. Construction and Building Materials. 2007;**21**(1):66-72

[4] Brule B. Polymer-Modified Asphalt Cements Used in the Road Construction Industry: Basic Principles. Transportation Research Record 1535. Washington D.C.: TRB, National Research Council; 1997

[5] The Asphalt Handbook. 2nd ed. The Asphalt Institute; 1966. https://www.abebooks.com/servlet/BookDetailsPL?bi=147629682&searchurl=tn%3Dthe%2Basphalt%2Bhandbook%26sortby%3D17%26an%3Dthe%2Basphalt%2Binstitute&cm_sp=snippet-_-srp1-_-title9

[6] Navarro FJ, Partal P, Martínez-Boza F, Gallegos C. Thermo-rheological behaviour and storage stability of ground tire rubber-modified bitumens. Fuel. 2004;**83**(14-15):2041-2049

[7] Aklonis JJ. Introduction to Polymer Viscoelasticity. Wiley Interscience; 2005. https://www.wiley.com/en-us/Introduction+to+Polymer+Viscoelasticity%2C+3rd+Edition-p-9780471740452

[8] Giovanni P, Jiri S, Zora V, Dario B, Ludovit Z. Rheology of asphalts modified with glycidylmethacrylate functionalized polymers. Journal of Colloid and Interface Science. 2004;**280**(2):366-373

[9] Navarro FJ, Partal P, García-Morales M, Martínez-Boza FJ, Gallegos C. Bitumen modification with a low-molecular-weight reactive isocyanate-terminated polymer. Fuel. 2007;**86**(15):2291-2299

[10] Becker Y, Müller AJ, Rodríguez Y. Use of rheological compatibility criteria to study SBS modified asphalts. Journal of Applied Polymer Science. 2003;**90**(7):1772-1782

[11] Iqbal MH, Hussein IA, Al-Abdul Wahhab HI, Amin HB. Rheological Investigation of the influence of acrylate polymers on the modification of asphalt. Journal of Applied Polymer Science. 2006;**102**(4):3446-3456

[12] Bahia HU, Hanson DH, Zeng M, Zhai H, Katri MA, Anderson RM. Characterization of Modified Asphalt Binders in Superpave Mix Design. NCHRP Report 459. Washington, D.C.: National Academy Press; 2001

[13] Whiteoak D, Read JM. The Shell Bitumen Handbook. London: Thomas Telford Services Ltd; 2003

[14] Mohammad S. Asphalt. Dehkhoda Publications; 1377

[15] Sengoz B, Topal A, Isikyakar G. Morphology and image analysis of polymer modified bitumens. Construction and Building Materials. 2009;**23**(5):1986-1992

[16] Aflaki S, Tabatabaee N. Proposals for modification of Iranian bitumen to meet the climatic requirements of Iran. Construction and Building Materials. 2009;**23**(6):2141-2150

[17] Ho S, Zanzotto L. The low temperature properties of conventional and modified asphalt binders evaluated by the failure energy and secant modulus from direct tension tests. Materials and Structures. 2005;**38**(1):137-143

[18] Mohammad LN, Negulescu II, Wu Z, Daranga C, Daly WH, Abadie C. Investigation of the use of recycled polymer modified asphalt binder in asphalt concrete pavements. Journal of the Association of Asphalt Paving Technologists. 2003;**72**:551-594

[19] Michael H. Design and construction of asphalt paving materials with crumb rubber modifier. Transportation Research Record. 1992;**1339**:1-8

[20] Abdolrahman MA, Carpenter SH. Mechanism of interaction of asphalt cement with crumb rubber modifier. Transportation Research Record. 1999;**1661**:106-113

[21] Troy K, Sebaaly PE, Epps JA. Evaluation systems for crumb rubber modified binders and mixtures. Transportation Research Record. 1996;**1530**:3-10

[22] Tabatabaee N, Tabatabaee HA, Sabouri MR, Teymourpour P. Evaluation of performance grading parameters for crumb rubber modified asphalt binders and mixtures. In: Proceedings of the 7th International RILEM Symposium on Advanced Testing and Characterization of Bituminous Materials, Rhodes, Greece. Vol. 1. 2009. pp. 597-606

[23] Bahia HU, Davies R, Martinez V. Recycling of Scrap Tires and Plastic Waste in Modification of Asphalt Binders. Transportation Congress, American Society of Civil Engineers; 1995. https://books.google.com/books?id=4yg3TOWVXRAC&pg=PA606&dq=B

ahia+HU,+Davies+R,+Martinez+V.+Recycling+of+Scrap+Tires+and+Plastic+Waste+in+Modification+of+Asphalt+Binders.&hl=fa&sa=X&ved=0ahUKEwi70YySupbfAhVEgM4BHcZlBMQQ6AEIJTAA#v=onepage&q=Bahia%20HU%2C%20Davies%20R%2C%20Martinez%20V.%20Recycling%20of%20Scrap%20Tires%20and%20Plastic%20Waste%20in%20Modification%20of%20Asphalt%20Binders.&f=false

[24] Madapati RR, Lee KW, Franco CA. Evaluation of Crumb Rubber in Asphalt. Transportation Congress, American Society of Civil Engineers; 1994. https://cedb.asce.org/CEDBsearch/record.jsp?dockey=0096981

[25] Kim S, Loh S, Zhai H, Bahia HU. Advanced characterization of crumb rubber modified asphalts using protocols developed for complex binders. In: Transportation Research Board 80th Annual Meeting; Washington D.C. Vol. 1767. 2001. pp. 15-24

[26] Gopal VT, Sebaaly PE, Epps J. Effect of crumb rubber particle size and content on the low temperature rheological properties of binders. In: Transportation Research Board 81st Annual Meeting; Washington D.C. 2002. pp. 13-17

[27] Edwards Y, Tasdemir Y, Isacsson U. Influence of commercial waxes and polyphosphoric acid on bitumen and asphalt concrete performance at low and medium temperatures. Materials and Structures. 2006;**39**(7):725-737

[28] Ho S, Zanzotto L, MacLeod D. Impact of different types of modification on low temperature tensile strength and critical of asphalt binders. In: Transportation Research Board 81st Annual Meeting; Washington D.C. Vol. 1810. 2002. pp. 1-8

[29] Jahromi SG, Khodaii A. Effects of nanoclay on rheological properties of bitumen binder. Construction and Building Materials. 2009;**23**:2894-2904

[30] Sim J, Park C, Moon DY. Characteristics of basalt fiber as a strengthening material for concrete structures. Composites Part B Engineering. 2005;**36**(6-7):504-512

[31] Ludovico MD, Prota A, Manfredi G. Structural upgrade using basalt fibers for concrete confinement. Journal of Composites for Construction. 2010;**14**(5):541-552

[32] Celauro C, Praticò FG. Asphalt mixtures modified with basalt fibres for surface courses. Construction and Building Materials. 2018;**170**:245-253

[33] Fiore V, Scalici T, Di Bella G, Valenza A. A review on basalt fibre and its composites. Composites. Part B, Engineering. 2015;**74**(Supplement C):74-94

[34] Morova N. Investigation of usability of basalt fibers in hot mix asphalt concrete. Construction and Building Materials. 2013;**47**(Supplement C):175-180

[35] Cai YC, Zheng YX. Experiment study of water stability of fiber-reinforced asphalt mixture. Advanced Materials Research. 2011:710-716. https://www.scientific.net/AMR.243-249.710

[36] Zheng Y, Cai Y, Zhang G, Fang H. Fatigue property of basalt fiber-modified asphalt mixture under complicated environment. Journal of Wuhan University of Technology-Materials Science Edition. 2014;**29**(5):996-1004

[37] Zheng YX, Cai YC, Zhang YM. Laboratory study of pavement performance of basalt fiber-modified asphalt mixture. Advanced Materials Research. 2011:175-179. https://link.springer.com/article/10.1007/s11595-014-1033-1

[38] Huang H, White TD. Dynamic properties of fiber-modified overlay mixture. Transportation Research Record. 2001;**1545**:98-104

[39] Zhao L, Chen J, Wang S. Using mineral fibers to improve asphalt and asphalt mixture behavior. In: Seventh International Conference on Traffic and Transportation Studies (ICTTS); Kunming, China. 2010. DOI: 10.1061/41123(383)129

[40] Wu S, Ye Q, Li N, Yue H. Effects of fibers on the dynamic properties of asphalt mixtures. Journal of Wuhan University of Technology-Materials Science Edition. 2007;**22**:733-736

[41] Qin X, Shen A, Guo Y, Li Z, Lv Z. Characterization of asphalt mastics reinforced with basalt fibers. Construction and Building Materials. 2018;**159**:508-516